*A View of the River*

# A View of the River

Luna B. Leopold

*Harvard University Press*
*Cambridge, Massachusetts*
*London, England*
*1994*

Printed in the United States of America

This book is printed on acid-free paper, and its binding
materials have been chosen for strength and durability.

Library of Congress Cataloging-in-Publication Data

Leopold, Luna Bergere, 1915–
A view of the river / Luna B. Leopold.
p. cm.
Includes bibliographical references and index.
ISBN 0-674-93732-5 (acid-free)
1. River channels. 2. Rivers—United States. I. Title.
GB1215.L42 1994
551.48′3—dc20
93-34698
CIP

*For three geologists*

Parry Reiche
Kirk Bryan
Thomas B. Nolan

*in gratitude*

# Contents

# Preface

The literature on rivers and their channels is now so extensive that a summary of current thought is unavailable to readers other than scientific specialists. Yet knowledge of this information is vital to research managers, to those who would develop the resource, and to a public that wants protection of the rivers it owns. There is no theory of river action and behavior to guide river "improvement." We in the United States have acquiesced to the destruction and degradation of our rivers, in part because we have insufficient knowledge of the characteristics of rivers and the effects of our actions that alter their form and process.

I believe that the enunciation of a general hypothesis of river action would be useful to both specialists and the public. My field experience encompasses the eastern United States and the Rocky Mountains, including the semiarid Southwest. But I have not done fieldwork in the Pacific Northwest, where large dams made of organic debris are common and where debris flows and landslides are important processes. My experience in other continents is limited also, and I have not worked on very large rivers or in tropical climates.

Although my examples are therefore restricted, most of the features I describe occur on all continents and constitute important aspects of many rivers.

With these caveats, a hypothesis is presented: River form and action are determined by physical laws that do not dictate one and only one solution to the reaction of the channel as changes are imposed on it when seasons go from dry to wet and back to dry. Thus random chance plays a major role in local changes. As a result, the forms assumed and the adjustments made all tend toward the most probable form, expressed as the form having the least total variance. The most probable state is often expressed in statistics as that of the minimum variance.

Because rivers utilize energy, field observations led me to seek analogues in other aspects of physics that might help explain some of the features seen. One useful analogue is the concept of entropy, particularly that a steady state condition in an open system is one in which both minimum total work and uniform rate of energy expenditure are tendencies. But with constraints, these two tendencies cannot be met simultaneously; rivers tend to achieve an intermediate position. They tend toward both minimum work and uniform work rate in many aspects of channel form and action.

To present such a synthesis for a broad audience requires some compromise. To be most useful I think that the explanation should contain a minimum of mathematics. Nevertheless, hydraulics and geomorphology are intertwined in river study and must be presented without doing injustice to the rigor of science. Phenomena can best be understood if described in quantitative terms. Therefore this book has a large number of graphs showing the specific relations among channel parameters.

Another compromise is the virtual elimination of references in the body of the text. Individuals who have contributed significantly are named in the text, but specific citations are assembled in the References section. I have drawn heavily on the books and papers I have published over the past decades, so the generalizations express my own personal view; I make no attempt to include all the interpretations expressed by the many authors of this large literature. In this respect it is a view of the river, the scene experienced by one individual. My interpretation and my emphasis may be different from those of other observers.

Still another compromise is the choice of units of measurement. In the current scientific literature the metric system is nearly universal, but this book is directed primarily at a readership accustomed to English units of miles, feet, and pounds. Furthermore, a century of river measurement data in the United States—the preponderance available to English-speaking people—have been published by the U.S. Geological Survey in English units.

The concentration on physical form and process at the expense of the chemistry of the river, and thus the biology of river water and the life of the riparian zone, is another necessary compromise. So varied and complicated is the chemistry of river water and its biotic life that no one volume can treat them adequately. I have neither the expertise nor the space needed for a proper discussion of these factors.

This book can be called a primer. It might also have been called a book on potamology, the study of rivers, but that would imply a treatise—which it is not. In one definition, a primer is an elementary textbook, a simple treatment. But rivers are far from simple. In fact, the mathematics of hydraulics and sediment movement, for instance, have become so complex that even many experts find them difficult to understand. What makes this volume elementary is the attempt to describe the complexities with the minimum possible mathematics.

In referring to research carried out in collaboration with my closest associates, I often use the word "we" to indicate that the data and ideas came from a joint effort. The persons referred to at various times include Ralph A. Bagnold, William B. Bull, David R. Dawdy, Thomas Dunne, William W. Emmett, Walter B. Langbein, John P. Miller, Robert M. Myrick, David L. Rosgen, Herbert E. Skibitzke, and M. Gordon Wolman.

The extraordinary assistance of John E. Ross of the University of Wisconsin in editing the manuscript is gratefully acknowledged. Equally generous was the contribution of my compadre Thomas Dunne, who aided me in clarifying concepts and looked carefully at the facts presented. For her work in manuscript preparation, my thanks go to Susan Noble of Pinedale, Wyoming. The drawings were prepared by Megan Allen Keelaghan of Eureka Cartography.

By this book I pay tribute to my fellow rafters and scientists who work continuously for the protection and appreciation of rivers, especially Martin Litton, Cort Conley, Bill Trush, Dave Rosgen, Rod Nash, Bob Doppelt, Verne Huser, Philip Williams, Eliot Porter, Stanley Schumm, Charles Belt, Roger Corbett, Bill Emmett, Michael Church, Richard Hey, Mark DuBois, and Tim Palmer.

More important than all the others, Barbara has shared my view of the river. Without her help, none of my work on this book would have been possible.

# *A View of the River*

CHAPTER ONE

# The River Channel

## The Grand Circle

We live on the surface of a planet that is in slow but constant change. The processes accomplishing that change operate because the planet is very special—special in position in the solar system and special in size. Earth moves in an orbit nearer to the sun than Mars, but more distant than Mercury or Venus. If Earth were appreciably closer to the sun, liquid water would not exist; it would occur only as vapor. And if Earth were much farther from the sun, water would be forever frozen. Moreover, Earth is just the right size, large enough to have a semimolten mantle from which volcanoes can erupt, bringing water vapor to the surface. Through this mechanism, it is believed, the oceans of the Earth were slowly developed. The moon is too small to have such volcanic activity and cannot form or hold either oceans or atmosphere.

Thus, by coincidence of favorable size and location in the solar system, Earth alone among the planets has oceans, an atmosphere, and thus a hydrologic cycle. The grand circle of movement of water from ocean to atmosphere to continent and back to ocean is the essential mechanism that allows organisms—including humans—to emerge, to develop, and to live on Earth.

Water plays a part in all physical and biological processes. It is essential to the actions that have developed the Earth's surface as we now observe it. Mountains are forced up by the collision of the great plates that make up the Earth's crust. But mountains on the continental surfaces are gradually worn away by the ubiquitous weathering of their rocks, and the transport of weathered products downhill by the action of water, wind, and gravity. The weathering processes that change hard rocks to erodible material incorporate water at every stage. Furthermore, water is the principal agent of movement of the weathered material that makes up the soil and supports vegetation, of the sedimentary rocks formed by the accumulation of the weathering products, and of the channels along which they are carried.

All the water presently on and in the surface of the Earth was brought there by volcanic action. What we see and use is derived from precipitation. That grand pattern of circulation of water called the hydrologic cycle describes in general terms what happens through time as water evaporates from ocean, plants, and soil, moves in the atmospheric circulation, and reprecipitates locally or far from its point of evaporation.

When precipitation falls on a continent, it separates into that which infiltrates the ground, that which immediately evaporates, and that which runs off the ground surface. The runoff carves or maintains channels of rill, stream, and river. This water on the surface may infiltrate, evaporate, or somewhere else be augmented by emerging groundwater. The terms "groundwater" and "surface water" refer merely to the location of water at a given moment. Water often moves between surface and subsurface depending on local conditions.

Rivers are both the means and the routes by which the products of continental weathering are carried to the oceans of the world. More water falls as precipitation than is lost by evaporation and transpiration from the land surface to the atmosphere. Thus there is an excess of water, which must flow to the oceans. Rivers, then, are the routes by which this extra water flows to the ultimate base level. The excess of precipitation over evaporation and transpiration provides the flow of rivers and springs, recharges groundwater storage, and is the supply from which humans draw to meet their needs.

A good deal of the water that appears as river flow is not transmitted into the river channels immediately after falling as precipitation. A large percentage is infiltrated into the ground and flows underground to the river channels. This process provides, then, a form of storage and thus regulation that sustains the flow of streams during nonstorm or dry periods of bright, sunny weather. The discharge represents water that has fallen during previous storm periods and has been stored in the rocks and in the soils of the drainage basin.

The excess of precipitation over evapo-transpiration loss to the atmosphere is a surprisingly small percentage of the average precipitation. The average amount of water that falls as precipitation over the United States annually is 30 inches. Of this total, 21 inches are returned to the atmosphere in the form of water vapor through the processes of evaporation and transpiration from plants. The balance of 9 inches contributes to the maintenance of groundwater and the flow of rivers.

About 40 percent of the runoff from the continental United States is carried by the Mississippi River system alone. The amount of deep seepage from groundwater to ocean is not known but is believed to be quite small, probably much less than 0.1 inch per year.

For the land area of the continent the water cycle balances: credit, 30 inches of precipitation; debit, 9 inches of runoff plus 21 inches transferred to the atmosphere. In the atmosphere, however, the budget appears out of balance because 30 inches are delivered to the land as rain and snow, but only 21 inches are received back as vapor by evaporation and transpiration. Accordingly, 9 inches of moisture must be transported by the air from the oceans to the continent, to balance the discharge of rivers to the sea. It is estimated that each year the atmosphere brings about 150 inches from the oceans over the land area of the United States and carries back 141 inches.

The precipitation represented by surface runoff, about one-third, flows from the hillslope or valley bottom to definite channels—usually to small channels that join to form larger ones, which in turn meet to form still larger channels. By convention, the smallest of these are called rills; they meet to form creeks, runs, or streams; then, at some undefined size, they are termed rivers. Each is fed water from two sources, overland flow to a channel and groundwater emerging at the channel boundary. In nonstorm periods, all the flow in channels derives from emerging groundwater.

## Hillslope to Rill Head

Only recently has the change from overland flow on unrilled hillslope to definite rill or channel been studied. The distance from headwater divide to the upper end of the first rill may be great or small. The hydraulic conditions that lead to rill formation involve raindrop impact, erosion by raindrop splash, and depth of the overland flow. Rainfall impact on a film of water flowing overland splashes up sediment, which tends to fill and obliterate incipient rills or channels, a concept developed and measured by Thomas Dunne. Downslope, where the depth of overland flow is sufficient to shield the soil surface from the direct impact of falling rain, and where the intensity of sediment transport in the flow may be high, rills or small channels begin. This subject is elaborated by Dietrich and Dunne.

A few data are available from some areas of mixed grass and trees, including oak-grassland associations in the San Francisco Bay region,

*Table 1.1* Distance from watershed divide to upstream tip of identifiable channel

| Location | Distance to closest divide (ft) | Drainage area (sq mi) | Mean slope from rill head to divide |
|---|---|---|---|
| Contra Costa and Marin counties, California | | | |
| Briones No. 1 | 400 | 0.00074 | 0.20 |
| Olema No. 1 | 65 | .00003 | 1.00 |
| Olema No. 2 | 50 | .00033 | 0.27 |
| Olema No. 3 | 90 | .00033 | .27 |
| Sublette County, Wyoming | | | |
| Cora Hill A1 | 70 | .00017 | .17 |
| Cora Hill A2 | 115 | .00030 | .16 |
| Cora Hill A3 | 200 | .00045 | .12 |
| Cora Hill A4 | 275 | .00056 | .11 |
| Cora Hill A5 | 365 | .00064 | .08 |
| Cora Hill B1 | 200 | .00012 | .09 |
| Cora Hill C1 | 200 | .00018 | .11 |
| Arroyo del los Frijoles, Santa Fe, New Mexico | | | |
| Caliente Arroyo | N.A. | .00006 | .14 (approx) |
| Big Sweat | 30 | .0057 | .045 |
| Big Sweat | 200 | .0057 | .045 |
| Big Sweat | 190 | .0057 | .045 |
| Big Sweat | 60 | .0057 | .045 |
| Little Sweat | 220 | .0013 | .045 |

piñon-juniper woodland in New Mexico, and grass areas in west central Wyoming. They are shown in Table 1.1. In central Kenya, however, unrilled hillslopes in grasslands can be as long as 1,500 feet. In such cases drainage area is difficult to define.

## Shape of the Channel

The shape of the cross section of any river channel is a function of the flow, the quantity and character of the sediment in motion through the section, and the character or composition of the materials (including the vegetation) that make up the bed and banks of the channel. Because the flow exerts an eroding force per unit area, or shear stress, on the bed and banks, the stable form the channel can assume is one in which the shear stress at every point on the perimeter of the channel is approximately balanced by the resisting stress of the bed or bank.

A natural channel migrates laterally by erosion of one bank, maintaining on the average a constant channel cross section by deposition on the opposite bank. In other words, there is an equilibrium between erosion and deposition. The form of the cross section is stable, meaning more or less constant, but the position of the channel is not.

The effect of changes in bank material on channel form depends on the relative resistance of bed and bank material. As the threshold of erosion of the bank material increases, whether by the addition of coarse or cohesive sediments or by the presence of vegetation or bedrock, with no change in the bed material or discharge, the channel will be narrower. Thus channels with cohesive silty banks and beds will be narrower than comparable ones with sandy banks and beds.

Most rivers in cross section are not parabolic and they certainly are not semicircular. They tend more to be generally trapezoidal in straight reaches, but asymmetric at curves or bends. The appearance of rectangularity increases somewhat as the river gets larger downstream, since width increases downstream faster than does depth. Some typical cross sections are shown in Figure 1.1, where they have been drawn to different scales so that their widths on the page are the same. When cross sections are drawn without vertical exaggeration, the shapes tend to resemble channels in cohesionless materials. The relatively large width-to-depth ratio for the biggest river in Figure 1.1 is apparent. The asymmetrical cross sections at curves and bends are described in a later section.

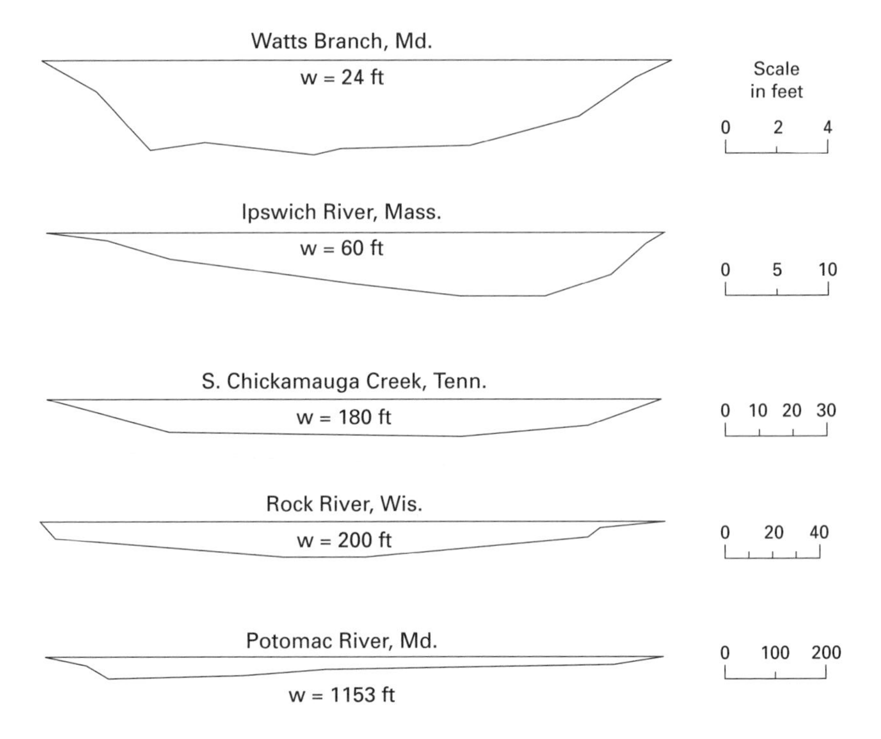


*Figure 1.1* Cross sections of some natural rivers scaled so that the width appears to be the same.

## The Floodplain

River channels are seldom straight except over short distances. A straight reach as long as 20 times channel width is a rarity. Curves, however slight, promote the tendency for erosion of the concave bank balanced by deposition near the convex bank. This tendency for erosion and deposition increases with the tightness of the curve; that is, with total angular deflection.

As the concave bank recedes due to erosion and the point bar builds outward from the convex bank, the channel width remains the same. The progressive growth of a point bar forms a flat surface or floodplain, the top of which is the level of the bankfull stage as indicated in Figures 1.2 and 1.3. The mean depth of a channel is computed as the cross-sectional area at bankfull divided by the water-surface width. The mean depth, then, is the height of a rectangle having the same area and the same width as the channel section.

A floodplain is built primarily by point-bar extension, as shown by measurements over several years at Watts Branch in Maryland (Figure

*Figure 1.2* A typical floodplain, built by extension of the point bar at the right, as the concave bank at the left is eroded. Seneca Creek, Maryland.

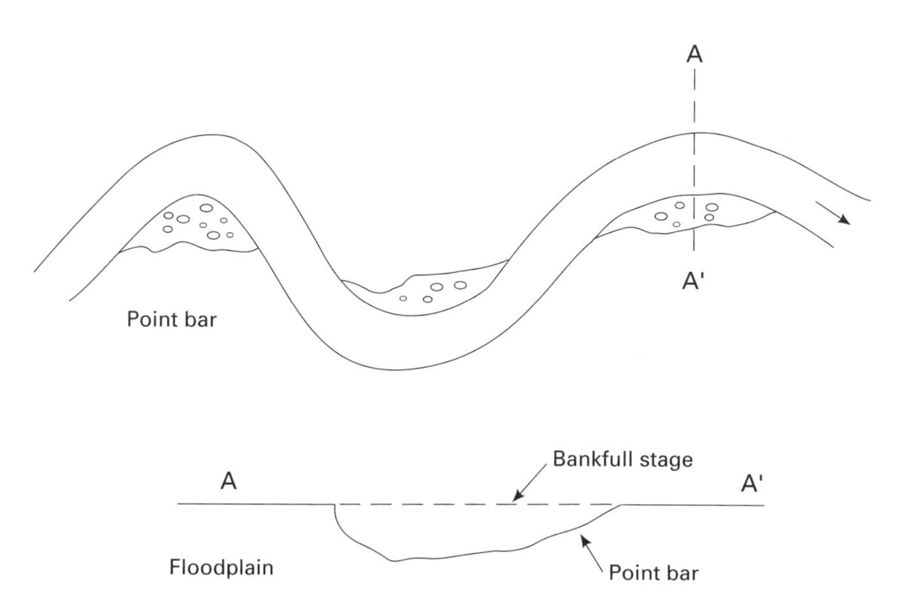


*Figure 1.3* Diagrammatic plan view and cross section indicating that the retreat of a concave bank permits the extension of a building point bar. The bankfull condition shows that the level of the floodplain is the same as the top of the point bar.

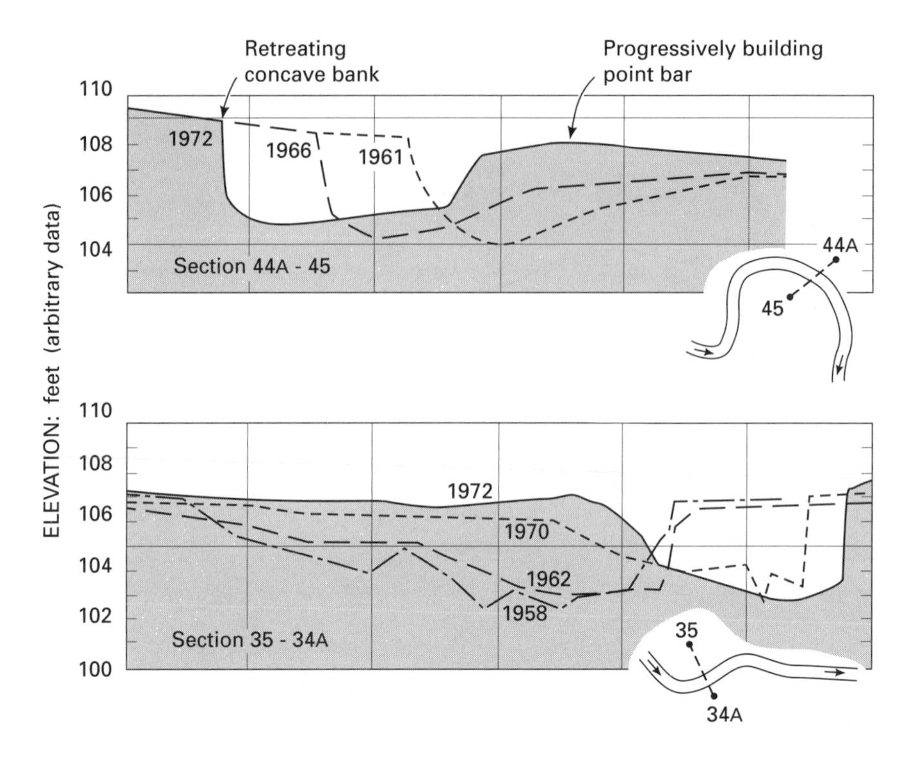


*Figure 1.4* Data obtained from successive surveys of Watts Branch near Rockville, Maryland, show lateral migration of a river channel by the building of a point bar into the stream and concurrent erosion of the opposite bank. Continuation of such point-bar building results in the development of a floodplain. The diagrams at the lower right indicate the positions of the cross sections relative to the channel bends. (From Leopold 1973.)

1.4). In some instances, deposition by overbank flow adds to the floodplain level. During a climatic regime when active aggradation is occurring, overland deposition can be a primary process of floodplain construction.

Only a few definitions are really necessary to an understanding of morphologic processes in rivers. This is one: A floodplain is a level area near a river channel, constructed by the river in the present climate and overflowed during moderate flow events.

Note the phrase "in the present climate," because a floodplain can be abandoned and at least partly destroyed when climate becomes drier. An abandoned floodplain is called a terrace.

## Channels and Climate

The consistency with which rivers of all sizes maintain the morphology typical of that climate is an indication that their channel is sensitive to the particular combination of discharge and load contributed from upstream. The river constructs and maintains its channel. The channel at any place is of such a size that the most sediment will be carried over a long period of time during those short periods when the flow is near bankfull.

The river channel responds quickly and sensitively to any change. Indeed, my own observations of channels in western states showed that streams in the semiarid areas changed from a state of erosion and instability during the first quarter of the twentieth century to a state of healing by vegetation in midcentury. Raw and unvegetated channels became stable and were gradually recolonized by vegetation beginning in about 1950. I found this to be the case in many western states. The change in upstream channels was apparently a response to the climatic shift at that time in the United States, England, and elsewhere. The British climatologist Hubert Lamb documented an increasing frequency and severity of storms after 1950 in the British Isles and northwest Europe. In the western United States the change brought on a slight cooling and a decrease in the number and frequency of intense short-lived rainfalls. Variability from place to place and from one season to another also increased.

This natural trend seems to have changed in the 1980s and 1990s, possibly because of the anthropogenic introduction of various gases into the atmosphere, and the poorly understood changes of ocean temperature in the equatorial region. The present outlook is for increasing air temperatures worldwide. The climatic changes of the past suggest that if the trend toward a warmer and more arid climate actually continues in the coming decades, the erosion of alluvial valleys seen in the thirteenth century, and again in the nineteenth, will be repeated in many of the semiarid areas of the planet where the rainfall is primarily of the thunderstorm type.

To understand how climate affects river channels, it is essential to perceive the primary difference between humid and arid locations in the semiarid parts of the globe. Figure 1.5 shows that the difference in the character of rainfall between a location with 14 inches annual rainfall (Santa Fe, New Mexico) and one with 8 inches (Las Cruces) is the number of small rains each year. The frequency of heavy rains is identical. Small rains foster vegetation and do not cause great discharges in channels.

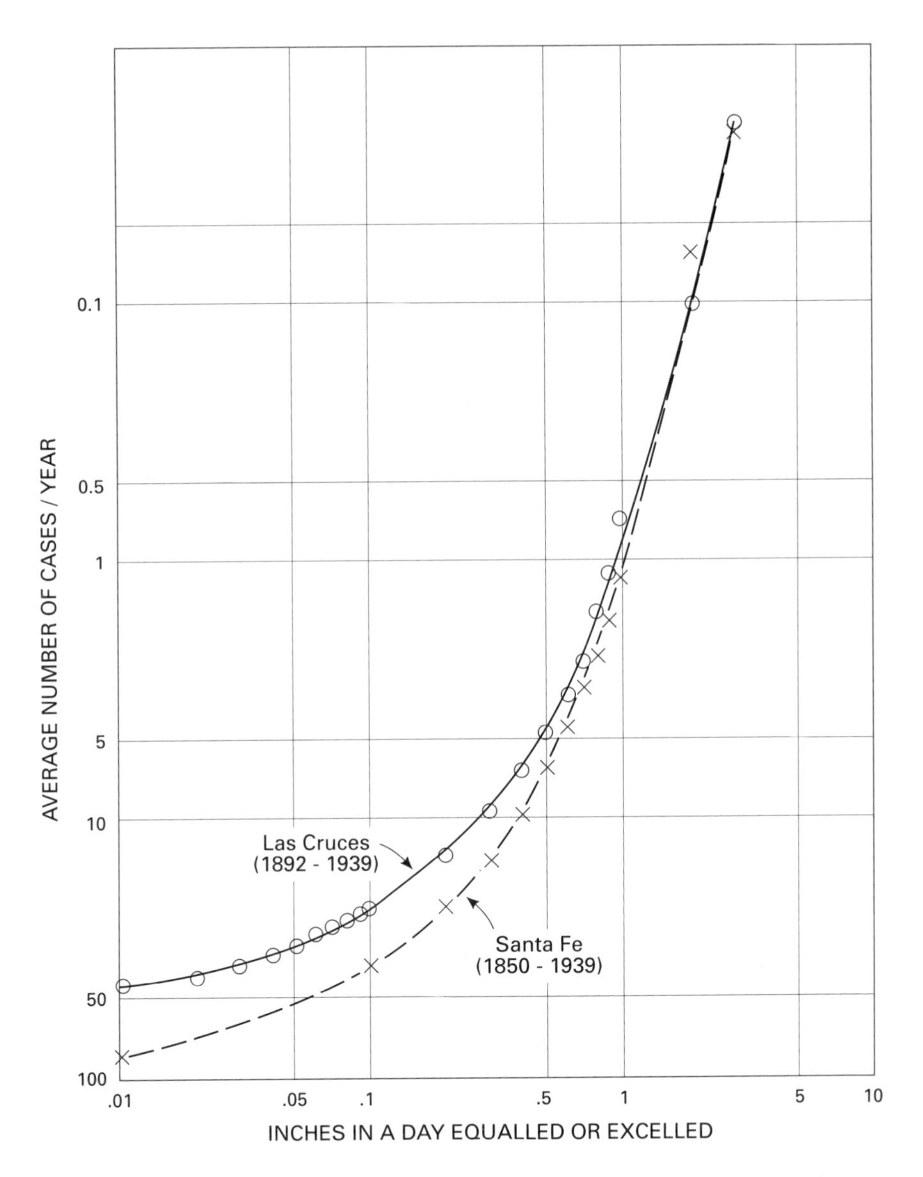


*Figure 1.5* Cumulative frequency curve of daily rains, Las Cruces and Santa Fe, New Mexico. (From Leopold 1951.)

Several important relations are apparent from this graph. Las Cruces gets only 60 percent of the mean annual rainfall of Santa Fe, yet the frequency with which the two stations receive large rains (more than an inch a day) is about the same. The difference in total annual rainfall is determined by the larger number of small rains in Santa Fe.

In a semiarid region the mean annual rainfall of a given station is higher than that of another station because of the larger number of rains, more especially of small rains. A corollary to this fact is that at a given station a large percentage of total annual fall is contributed by the small rains.

The period of valley erosion in the western states, 1880–1920, was not characterized by a change in annual rainfall but by a change in rainfall intensity. The period of gully erosion saw many heavy storms and few light rainfalls. It was a period that can be described as more arid than others.

These types of changes have occurred several times in the Holocene period, the 10,000 years since the retreat of Wisconsin ice. Gradual warming occurred for the first 4,000 or 5,000 years of the Holocene, culminating in the Altithermal period of temperatures higher than today. This period, in Europe called the Climatic Optimum, was characterized by warm temperatures and boreal vegetation. In the U.S. Southwest it was the end of a long warming period, and in geologic terms it was a time of valley aggradation or deposition. At the end of this depositional period, a calcareous soil developed and is now seen as paleosol, marked by deposition of calcium carbonate in the B horizon.

The geologic evidence leads to the generalization that valley alluviation or deposition occurs during periods of relative humidity, except perhaps in regions of very high precipitation. Erosion and valley evacuation or degradation take place in periods of climatic aridity, owing to the prevalence of sporadic heavy rains and the infrequency of small, light rainstorms. These changes profoundly influence river channels. During periods of aggradation, widespread deposition increases the elevation of the valley floor, resulting in a floodplain built at a relatively high elevation.

When such a period is followed by relative aridity, channels cut down; a previously constructed floodplain is not only abandoned but dissected, leaving only fragments standing above the valley floor. Terraces, the remnants of previous floodplains, are mute evidence of changes in previous conditions, either in climate or in tectonic activity.

Terraces stand above floodplains in many areas of the world and can seen nearly anywhere in the United States. I have studied such terraces

*Figure 1.6A* Three terraces in the alluvial valley of Salt Wells Creek, south of Rock Springs, Wyoming. Toward the back is the scarp of the high terrace, about 30 feet above the present creek; in the mid-foreground is the sage-covered middle terrace, 15 feet above the creek; at the front is the low terrace vegetated with tumbleweed, about 5 feet above the creek.

*Figure 1.6B* A terrace composed of materials of two different ages. The flat top, where growing plants can be seen, was once a floodplain. Later the stream lowered, cutting a wide channel. Subsequent climatic changes caused this channel to fill with the dark red silt seen in the center of the photograph. A more recent downcutting cut even deeper, exposing older material below the red silt. Rio Puerco near Gallup, New Mexico.

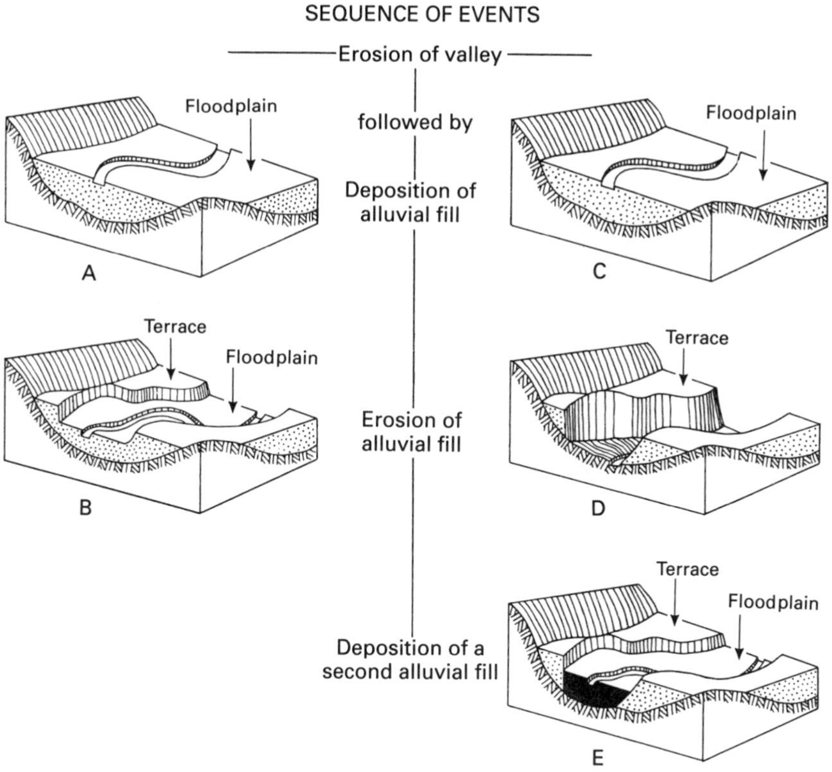

*Figure 1.7* The stages in development of a terrace. Two sequences of events leading to the same surface geometry are shown in diagrams A-B and C-D-E.

in New Mexico, California, Colorado, Wyoming, Maryland and elsewhere. Particularly on the channels of small rivers and creeks, three levels of terraces are readily apparent. Figure 1.6 shows typical examples.

The sequence of depositional and degradational events is depicted in Figure 1.7. The difference between a cut terrace and a fill terrace is that the former results from interrupted downcutting of a floodplain with no intermediate period of aggradation. That sequence is shown in diagrams A-B of the figure. If a period of aggradation follows the downcutting, that surface, when abandoned, is called a fill terrace. The filling process is shown in diagrams C-D-E of Figure 1.7. Such sequences of cutting and filling can lead to a variety of valley cross sections, as illustrated in Figure 1.8, which shows sequences that lead to no terrace, one terrace, or two

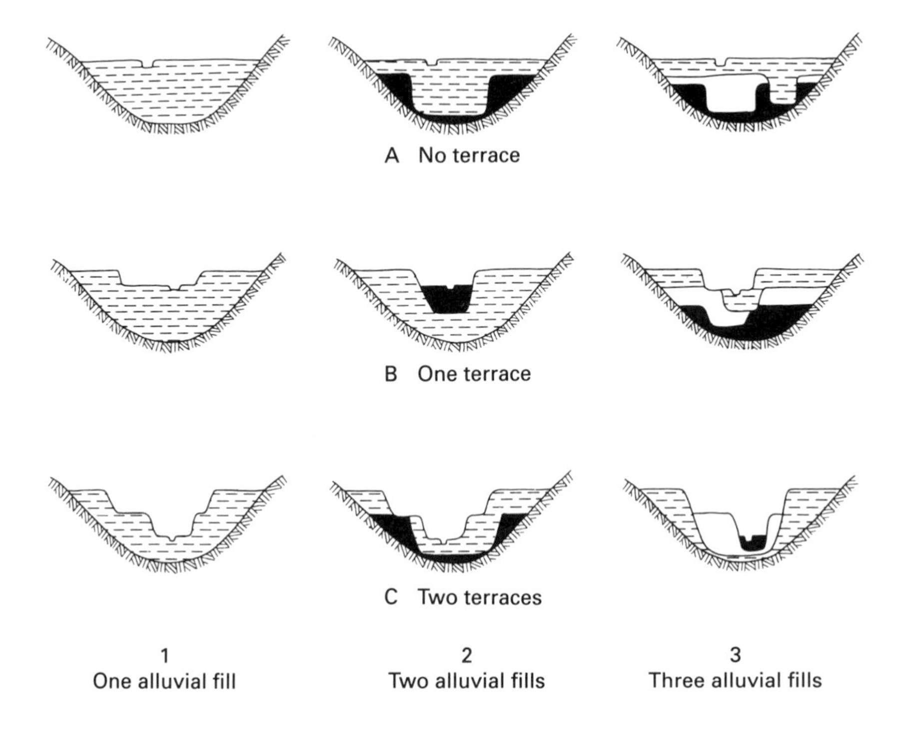


*Figure 1.8* Valley cross sections showing some of the possible stratigraphic relations in valley alluvium.

terraces. Actual examples from various parts of the western United States are shown in Figure 1.9.

The geologic evidence of terrace levels in many western valleys validates the theory that a period of valley aggradation occurred from the end of the Ice Age up to the Altithermal period, 4,000 to 6,000 years ago. During the arid conditions of the Altithermal, erosion carried away much of the accumulated valley fill, leaving a terrace standing 20 to 30 feet above the present channel in many valleys. There followed another period of aggradation that ended in 200 years of drought, approximately A.D. 1200–1400. In this dry period, again the erosion carried away previous deposits of valley alluvium. Humid and cool conditions dominated the continent in the well-known cold period called the Little Ice Age, ending in about 1860. A turn toward aridity caused widespread erosion of western valleys between 1880 and 1920. These alternate periods of erosion and deposition left indelible indicators of past climates, and

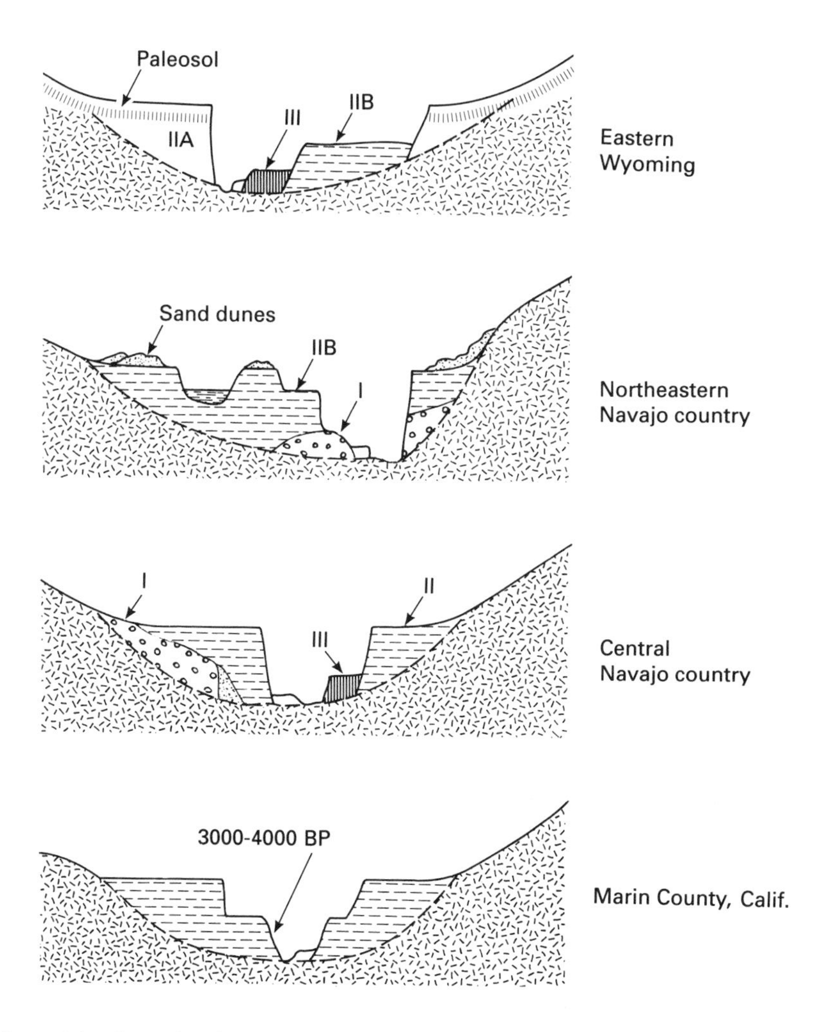


*Figure 1.9* Generalized cross sections of alluvial valleys in four areas of the western United States. In each, previous epicycles of erosion have cut down to bedrock, as has the modern gully. The stratigraphic units are briefly described in Table 1.2.

*Table 1.2* Alluvial chronology of western valleys, United States

| Period and name (after J. T. Hack) | Character and date | Inferred climate |
|---|---|---|
| Deposition I, Jeddito | Included extinct animals; Pleistocene | More arid at end of period |
| Erosion | Sand dunes locally; paleosol of more arid type | More arid |
| Deposition II, A and B, Tsegi | Often subdivided into parts; paleosols at end of first phase; second phase contained artifacts dated as late as A.D. 1200 | More humid |
| Erosion | 1200–1400 | Warmer; more arid |
| Deposition III, Naha | Ended with nineteenth-century erosion | More humid; colder near end |
| Modern gully erosion beginning about A.D. 1880 | Generally intensified by overgrazing | Summer rainfalls exceptionally intense |
| Initial aggradation or cessation of gully extension | Began 1940–1960 | Trend toward cooler; more precipitation in most but not all regions |

knowledge of the sequence provides some indication of what may be expected as climate changes in the next centuries.

The chronology of erosion and deposition illustrated in Figure 1.9 is shown in simplified form in Table 1.2.

The four generalized valley cross sections of Figure 1.9 typify wide areas in the states included in the figure—Wyoming, Arizona, and central California. In each area the erosion preceding Deposition I cut down to bedrock. The sediments of Deposition I contain a well-developed caliche horizon and in some places contain extinct fauna such as camel and extinct bison, indicating a late Pleistocene age (that is, prior to 10,000 B.P.). Most of that alluvial fill was eroded before the subsequent Deposition II. Some of the oldest portions of Deposition II are marked at the top by a calcium-carbonate-rich paleosol attributed to the more arid climate of the Altithermal period, about 5000 B.P. The younger alluvium of Deposition

II contains Paleoindian artifacts and pottery dated in the interval A.D. 950–1300.

All the sections have certain characteristics in common: the wide valley floor, or most extensive level, is underlain by Deposition II material. This is true in the California section (as well as in the others), where W. W. Haible, whose observations furnished the section shown, obtained material near the base of the major alluvial fill dated at 3000–4000 B.P.

In all the examples the modern gully, as well as most previous periods of downcutting, eroded down to bedrock at least in some places. The deep and steep-walled, or boxlike, character typified previous gullies as well as the modern gully. The high terrace, constituting in many places the valley flat, stands 15 to 30 feet above the present or recent channel bed; and a middle terrace, where one exists, stands 6 to 10 feet above the streambed.

Modern gullies began downcutting in the period 1880–1900 as a result of climatic factors, especially an increase in intensity of summer storms, and exceptionally heavy grazing by stock.

An example of Deposition II overlying bedrock is seen in Figure 1.10. An example of valley trenching in the 1880–1900 period is shown in Figure 1.11, on the Rio Puerco del Oeste at Manuelito, New Mexico.

The generalized sequence shown in Table 1.2 and illustrated in Figure 1.9 is common, but not universal. Interest in the chronology and its dating has spurred research that shows examples of other sequences, but substantiates the effect of changes in rainfall intensity. With R. C. Balling, Stephen G. Wells showed that in a given drainage basin the sequence of filling and erosion may differ among small headwater tributaries and trunk channels downstream. The erosion process is not necessarily contemporaneous in all parts of a drainage basin.

The terrace sequence in any valley is important because a stream impinging on and eroding a terrace deposit produces a large addition to the sediment load for each lateral unit of erosion. Terrace remnants are often lateral constraints on stream movement and thus control the width of valley floor that can be utilized in flood periods for amelioration of flood peaks by channel storage.

## Some Practical Insights Drawn from Alluvial History

The history of river cut and fill revealed in the stratigraphic relations in valley alluvium leads to some valuable insights into channel maintenance. The alluvial history shows that deposition leading to valley ag-

*Figure 1.10* William W. Emmett at a base of the 8-foot terrace in Arroyo de los Frijoles near Santa Fe, New Mexico. Note the bedrock near his feet, overlain by gravel, then 6 feet of silty alluvium. Deposition of the alluvium was in progress in 2200 B.P. in Deposition II and probably began at the end of the Altithermal, about 5000 B.P.

*Figure 1.11* The great trench of the Rio Puerco del Oeste at Manuelito, New Mexico, typical of arroyo cutting in southwestern valleys during the last decade of the nineteenth century.

gradation or alluviation is a slow process, whereas erosion is a rapid process. The valley fill accumulating in the early half of the Holocene took 4,000 to 5,000 years to fill western valleys with 30 to 100 feet of alluvium, a process of aggradation that ended about 5,000 years ago. In contrast, the deep gullies cut by erosion in pre-Columbian time took less than 200 years to evacuate a large part of the early Holocene fill. In the period 1880–1920, overgrazing and climatic change repeated the events of A.D. 1200–1400 in a period of less than 50 years.

In an effort to combat the erosion at the turn of the century in western states, government agencies including the Forest Service, the Soil Conservation Service, and experiment stations built thousands of small check dams in gullies. The results have been not only useless, but in some cases conducive to more erosion. Check dams can be useful only if a gully is deepening. Then they may provide a local base level to prevent further deepening. Check dams cannot store sediment because the volume to be stored is so small. The gradient of deposition behind a check dam is about half the gradient of the original valley, so the wedge of deposition extends upstream only a short distance. Check dams often fail by cutting around the dam; this lateral cutting enhances the erosion process and widens the gully in the vicinity of the failed dam.

As shown in Figure 1.10, many gullies immediately cut down to or near bedrock, so progressive deepening is not possible. Gullies cut by ancestral streams in the period A.D. 1200–1400 cut down to about the same depth as modern gullies.

Bank stabilization by vegetation is the best treatment of gullies in valley alluvium. It can often be facilitated by sloping the gully wall so that it no longer stands vertically. When such treatment is used, livestock should be kept out to protect the new vegetation.

It might be hoped that aggradation would fill the gullies cut during the 40 years between 1880 and 1920. Such filling has occurred in the geologic past, but the process is slow and takes many hundreds of years, and occurs only if climatic conditions are appropriate. At this point in time, the proper climate cannot be either forecast or influenced.

## Classes of Channels

Channels differ in shape depending not only on size of river but also on climatic-geologic setting. As indicated in Figure 1.1, the width increases downstream faster than the depth. Large rivers are very wide and may even resemble a lake. The Mississippi at high flow is 60 to 80 feet deep, but a mile or more wide.

If we compare rivers of the same size, those in a coastal plain setting such as in Alabama or Georgia are relatively deep, and they are muddy from the suspended sediment. In contrast, rivers in semiarid regions are relatively wide and shallow, examples being the upper Rio Grande in New Mexico and the Platte in Nebraska. Such rivers tend to be wide because the bed and banks are sandy and thus erodible, having only small amounts of fine-grained sediment load. These differences are also reflected in the slope or gradient, the size of material on the bed, the sinuosity or extent of meandering, and the nature of the bank material. Because a river channel can be characterized by a particular combination of these shapes and pattern parameters, a channel classification system is possible.

David L. Rosgen has proposed, tested, and explained a river classification system that is currently the most widely accepted manner of describing a channel. The classification is based on parameters of form and pattern but has the advantage of implying channel behavior. It also can indicate how restoration might be approached if a reach of river becomes aberrant or different from its normal condition. The Rosgen system describes an individual reach—that is, a short length of channel—a few hundred feet or a quarter of a mile. The system does not describe a whole drainage system. Under natural conditions a given river may vary in character and thus in class, even through short distances downstream, as a result of passage from one lithologic type to another, tributary entrance, or change in landscape character.

A river type according to Rosgen is defined by a particular combination of the following parameters: channel slope (gradient), bed material, ratio of width to depth, amount or degree of meandering as defined by the value of sinuosity, and degree of confinement or constraint to lateral movement. The classification system has seven types, A to G. In the simplified version considered here, each type has six subclasses that describe the size or coarseness of the bed material. Subclass 1 is bedrock, 2 is boulder, 3 is cobble, 4 is gravel, and so on. The total number of combinations is 42, but by far the largest number of channels fall in types A to D and subtypes 2 to 6, for a total of 20 most common field conditions. Figure 1.12 is an abbreviated explanation of the classification system and does not purport to include all possible types.

The preceding brief description of the Rosgen system cannot do it justice. It includes less common types omitted here. This summary merely indicates the variety of channel types that exist in nature and directs the reader to the Rosgen publication for a complete discussion of the implications.

## Riffles and Bars

There are characteristics of river channels that are so general that they must be recognized in any discussion of morphology. A straight or nonmeandering channel characteristically has an undulating bed and alternates along its length between deeps and shallows, spaced more or less regularly at a repeating distance of 5 to 7 widths. The same can be said of meandering channels, but this seems more to be expected because the pool or deep is associated with the bend, where there is an obvious tendency to erode the concave bank. The similarity in spacing of the riffles in both straight and meandering channels suggests that the mechanism which creates the tendency for meandering is present even in the straight channel.

The alternating pool and riffle arrangement is present in virtually all perennial channels in which the bed material is larger than coarse sand,

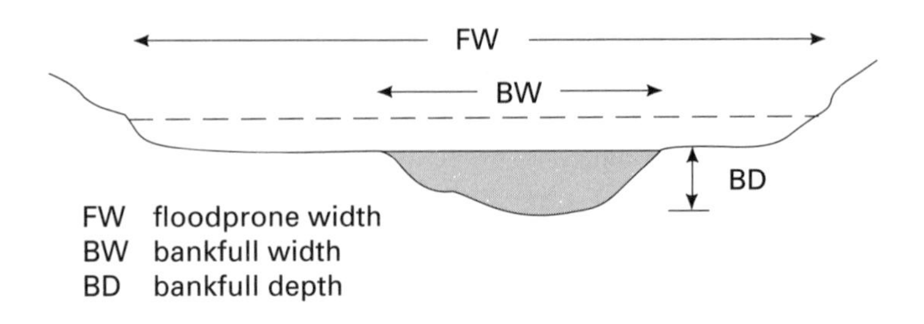


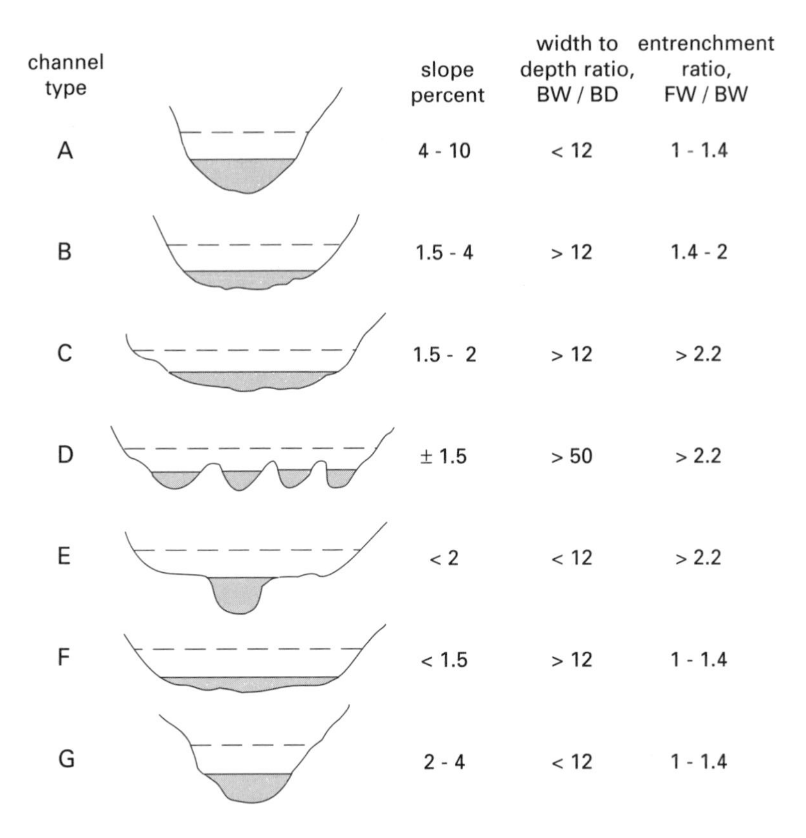


Subtype number based on channel bed material

| 1 | 2 | 3 | 4 | 5 | 6 |
|---|---|---|---|---|---|
| bedrock | boulder | cobble | gravel | sand | clay |

*Figure 1.12* The Rosgen system of channel classification.

but it appears to be most characteristic of gravel-bed streams—whether the gravel is the size of a pea or of a human hand. There appears to be a latent tendency for the development of pools and riffles even in boulder-bed channels.

Another longitudinal morphology exists in very steep channels in mountainous areas, particularly where the bed material consists of boulders and large rocks. In step-pool morphology the steps are often nearly vertical, the pools short and either deep or shallow. The spacing of the steps is much shorter than that of the pool-riffle channel, often 2 to 4 widths, but far less regular than the riffle spacing. The step-pool is also caused by hydraulic factors—that is, it is a natural phenomenon resulting from the transport and deposition of rocks by flood discharges. The causative mechanisms are not well known, especially the reason for the observed spacing.

A diagrammatic sketch of the plan and profile of the pool-riffle sequence and its relation to alternate bars is shown in Figure 4.14. The flow over a bar at high stage involves water actually being forced upward to rise over an obstruction, because the riffle is a mound on the streambed. At low flow some water sinks into the obstructing bar and flows through the gravel to emerge from the bed on the downstream end of the riffle. This sinking of water into the gravel is part of the reason that trout and salmon thrash their redd, or nest, out of the upstream part of a riffle; the downward flow into the gravel keeps the fish eggs from washing downstream.

Measurement of lengths of individual pools and riffles is not only a matter of judgment but is subject to considerable variation along a particular reach. A riffle bar in Seneca Creek at Dawsonville, Maryland, drainage area 100 square miles, is shown in Figure 1.13. A prominent riffle may be a bar adjoining one bank and sloping off to a deep hole at the opposite bank; it may be a central bar not flanked by deep holes; or it may be a low, off-center mound. In Seneca Creek the average length of one repeating distance is 324 feet, which is 5.1 times the mean channel width. The average length of pool in Seneca Creek is 1.6 times the length of riffle. The comparable figure for Watts Branch, a smaller stream a few miles to the south, is 1.1.

At low flow the water surface over a pool and riffle sequence tends to consist of alternating flat reaches of low gradient and steeper reaches often involving white water. This appearance of smooth water over the pool and riffles over the bar—terms well known to trout fishermen—led me to use these terms in describing the feature.

As the water rises during flood, the difference in appearance of the

*Figure 1.13* A riffle bar on Seneca Creek near Dawsonville, Maryland, looking downstream at low flow. The head of the next pool can be seen in the background.

water over pool and riffle tends to disappear. At sufficiently high flow, about bankfull, the longitudinal profile of the water surface tends to become less stepped. Still, some difference in slope over pool and riffle remains. The riffle is then said to be "drowned out," a process that apparently occurs at smaller discharge in a meandering river than in an otherwise comparable straight reach of channel. The significance of the stepped or nonlinear profile at bankfull stage will be discussed when we compare it to the profile of a meandering reach.

In a study of the effect of diversion and realignment of certain gravel streams in Scotland on their ability to maintain trout, Tom Stuart noted that newly constructed streambeds dredged by a dragline were of uniform depth and without pools and riffles. With the aim of producing the usual pool and riffle sequence, in one river he directed the operator of the dragline to leave piles of gravel on the streambed at intervals appropriate to riffles—that is, 5 to 7 widths apart. After a few flood seasons these piles had smoothed out and in all respects appeared natural for a pool and riffle sequence. Moreover, the riffles so formed have been stable over a number of years.

Pole Creek, a mountain stream that has incised itself into a moraine of

Wisconsin age near Pinedale, Wyoming, has a coarse gravel bed derived from the moraine. Through this reach the stream averages 80 feet wide and 3 to 4 feet deep at bankfull stage. It exhibits alternating deeps and shallows, which in form are typical pools and riffles, but their spacing is variable and not clearly related to any function of width.

In many of the pools in Pole Creek boulders were conspicuously absent and the bed material was fine enough to be counted by measuring individual pebbles. To obtain a quantitative measurement of the concentration of boulders in the rapids or riffles, I counted the number of boulders equal to or greater than 3 feet in diameter in sample reaches of pool and riffle. The average number per 100 square feet of stream was zero in the pools sampled. In the nearby riffles the averages were 0.18, 0.27, 0.38, and 0.65.

Median grain diameter in pools varied between 0.04 and 0.4 foot, and local channel gradient from 0.002 to 0.013. The rapids, by contrast, were composed principally of boulders, which were measured individually in place on a sampling grid. The comparative median diameter was 1 to 2 feet, and the average slope through the rapids was 0.02. This same sampling method was applied to boulders seen on the surface of the moraine into which the stream was incised. The average number of boulders per 100 square feet on the moraine was 0.24.

Although these measurements are crude, they tend to support the conclusion that in Pole Creek the pools have a relative dearth of large boulders compared with the source material, and that boulders have been concentrated in the riffles by stream action. Thus, boulders must have been swept out of incipient pools and collected in incipient riffles.

At Seneca Creek, Maryland, I painted the individual pieces of gravel (0.25 to 6 inches in diameter) lying at the surface of a gravel bar during low flow when the bar was exposed. During subsequent high flows all the painted particles progressively moved, but the bar itself was the same height and topography as before. Some of the painted pebbles were found on the next riffle downstream. In these studies the movement of gravel of medium size on the riffle requires a discharge that fills the channel about three-quarters full (depth equal to 0.75 bankfull depth), which has a recurrence interval of about one year.

In gravel-bed channels during periods of observation extending up to 7 years, we found no indication that the bars comprising riffles move downstream with time. Movement of gravel bars or riffles appears to be relatively slow.

One of the requirements for the existence of pools and riffles in non-meandering streams is apparently some degree of heterogeneity of bed-material size. Channels that carry uniform sand or uniform silt have little tendency to form pools and riffles.

Alternate deeps and shallows also occur in rivers that are incised into deep canyons. The Colorado River in the Grand Canyon is one example. This river, as well as many others flowing between rock-walled cliffs, are noted for their rapids where there is a local steep gradient of the water surface. But rapids belie the existence of numerous deeps not apparent on the water surface. Figure 1.14 gives a profile of two reaches of the Colorado River upstream of the Grand Canyon.

In the Grand Canyon the steep rapids sections are spaced an average of 1.6 miles apart. Their locations are dictated by tributary entrances and faults. But local deeps are much more numerous, and this periodicity is probably due to the same factors that cause riffles and pools in more common river types. In the reaches illustrated in Figure 1.14, the average spacing of deeps in the 7-mile reach 187–194 miles, is 2,500 feet or 11.2 channel widths. In the reach of miles 199–205, the spacing is 2,100 feet or 9.6 channel widths. Although these spacings are larger than the expected 5 to 7 channel widths in successive pools of many streams, the occurrence of these alterations does not depend on the rapids and is dictated by quite different factors. It is my conjecture that the deeps and shallows of the Grand Canyon result from the same basic causes as pools and riffles in nonincised rivers, but in rock-walled canyons the channels are influenced by additional factors.

## The Coarse Surface Layer

The extensive study of rivers worldwide in the last several decades has shown that most gravel-bed channels have larger cobbles or pebbles at the surface of the bed than in the layer immediately below the surface. This layer of coarse surface material has been called armoring or pavement. For example, Paul Komar reports that Oak Creek, a mountain river in Oregon, has a surface layer of 5 centimeters median size (50 percent of the material equal or finer), with the subsurface or next lower level, the subpavement, at 1.8 centimeters.

It is generally assumed that armoring results from the winnowing away of fine particles to leave a lag deposit of the coarse fraction at the surface. This is probably the process that causes surface pavement immediately below a dam where a channel is exposed to the discharge of clear water, the sediment having been trapped above the dam. Where pavement occurs in channels carrying the natural sediment load, other factors

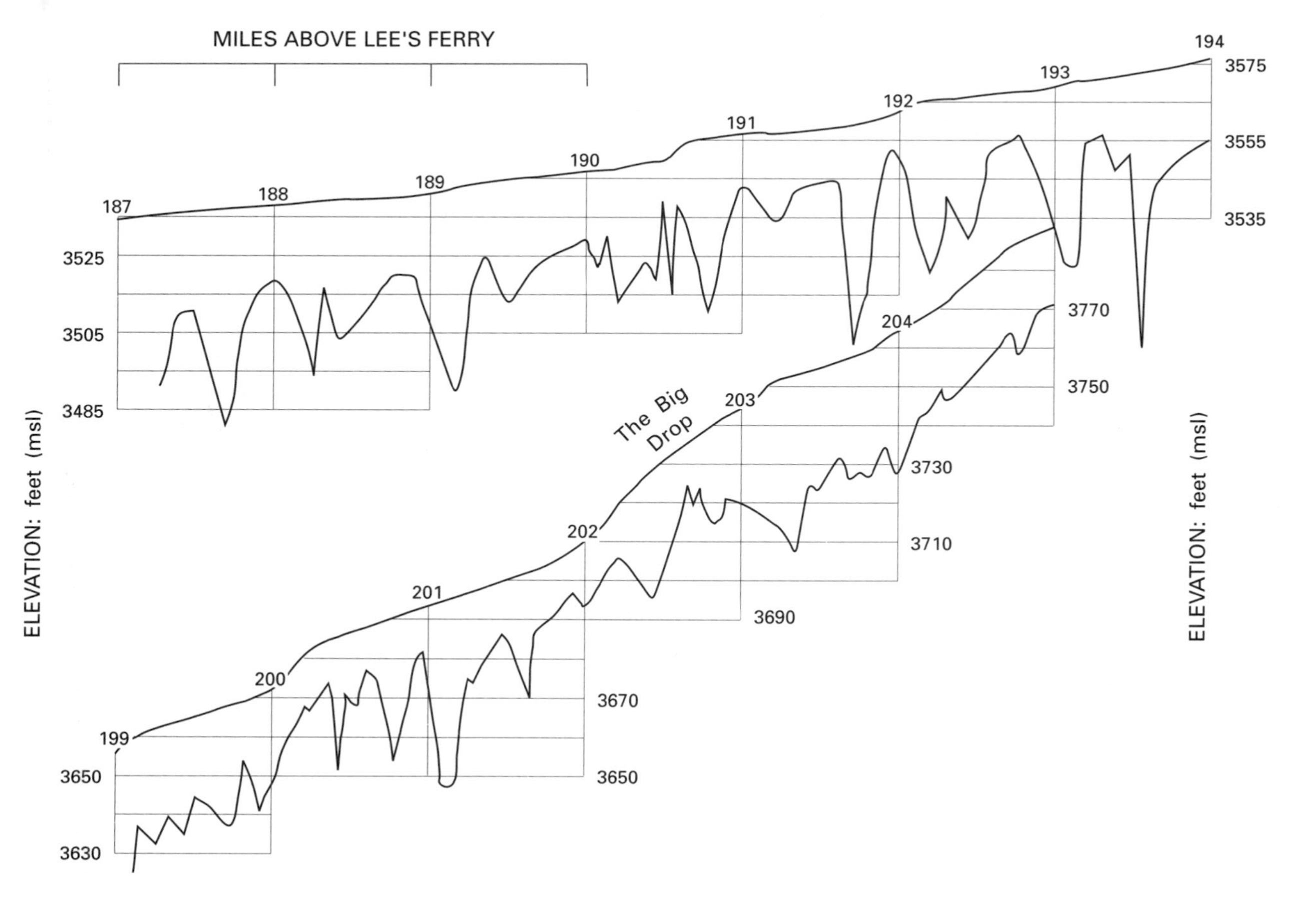


*Figure 1.14* Profile of the water surface and bed of the Colorado River in Cataract Canyon, Utah. The depths were measured by the author during the U.S. Geological Survey expedition of June 1967 at a discharge of 38,900 cfs.

must be considered. At least in some relatively uniform environments, as in a laboratory flume, an increasing discharge flowing over a bed of heterogeneous size does not put in motion the smallest particles, then the somewhat larger particles, and finally the largest. Rather, when motion begins, nearly all sizes move at the same time.

Natural channels are characterized by their nonuniformity in topography and distribution of sediment sizes. T. E. Lisle and M. A. Madaj emphasized local variability in both the direction of sediment transport and the magnitude of the flow-induced stress. They found that degree of armoring is different in aggrading and degrading reaches. Armored locations appeared to be depleted of fine material rather than being enriched by coarse particles.

Various aspects of armoring are under study by investigators and remain a subject needing clarification. Another process, the action of dispersive stress, can lead to concentration of larger particles at the surface of a streambed. An example of the action of dispersive stress appears in sandy ephemeral channels viewed in the usual dry state. Pools and riffles are generally absent, though careful observation or detailed mapping discloses an analogous feature—thin surface accumulations of coarse material in the form of gravel bars. The distribution of these bars is strikingly reminiscent of the occurrence of pools and riffles in gravelly perennial streams, for they tend to be spaced at 5 to 7 widths along the channel length and remain there with only minor change from year to year. In these bars by far the majority of the cobbles are at or very near the surface; the sand below is quite free of rocks and cobbles. Such gravel bars, then, are mere surface features that we presume are caused by the same general process that accounts for riffles and bars in perennial rivers.

That large rocks accumulate at the surface of sandy ephemeral washes is particularly surprising in view of the fact that the channel bed scours at high flow and fills again to approximately the same level when the high flow ceases. You can observe the same phenomenon in the kitchen: place white flour in a cake pan and add some whole wheat flour. When you shake the two together, they will not mix. Rather, the shaking will separate the coarse whole wheat grain from the fine-grained white flour, with the larger grains accumulating at the surface.

In rotating-drum experiments conducted by various people to test the rate of abrasion of particles, the large particles were generally on top of the mixture. Ralph A. Bagnold explains this phenomenon as the effect of the intergranular dispersive stress. The knocking of particles against one another increases as the square of the particle diameter; hence differential stress on the larger particles may be enough to force them to the surface, where the dispersive stress is zero.

The same phenomenon occurs in dry granular material that flows under gravity. When a truck dumps dry gravel in a pile, the largest particles come to the surface, roll down the face of the conical pile, and tend to segregate themselves at the base.

The full significance of this phenomenon is not known. The concentration of the largest movable particles near the surface of the streambed seems to occur in a variety of channels in quasi-equilibrium. Whether or not the phenomenon contributes to the armoring observed in gravel rivers is unknown.

CHAPTER TWO

# River Measurement

## Need for River Discharge Data

The United States is unique in that more of its citizens are supplied with potable water than those of any other country. There is still a rural population that obtains water from individual wells, but even small towns have a water supply system. About 80 percent of the water used in this country comes from surface sources, the rest from groundwater. Nearly all public supplies are treated, with the treatments ranging from simple chlorination to full-scale filtration and chemical treatment. Furthermore, all public surface-water sources have been developed subsequent to hydrologic analysis of available streamflow data, thanks to far-sighted engineers.

Analysis of streamflow data can give good estimates of the types of information needed for water development and management. These include the amount of water available, the frequency of its deficiencies and excesses, the volume of storage needed for specific conditions, the frequency and magnitude of floods, the chemical and biological quality of supplies, and the duration of various low-flow condition.

All of these purposes are served by a network of river gaging stations and the published data obtained by their operation. The importance of data publication cannot be overemphasized; fortunately, there is essentially no proprietary withholding of river flow data. Credit for this laudable state of affairs is due the United States Geological Survey, one of the most important data collection agencies in the world. In the last part of the nineteenth century the engineers of this remarkable organization not only saw the need for a widespread uniform network of observation stations, they developed standard operating procedures, organized a training program for hydrographers, and instituted a publication program for producing and disseminating the data obtained.

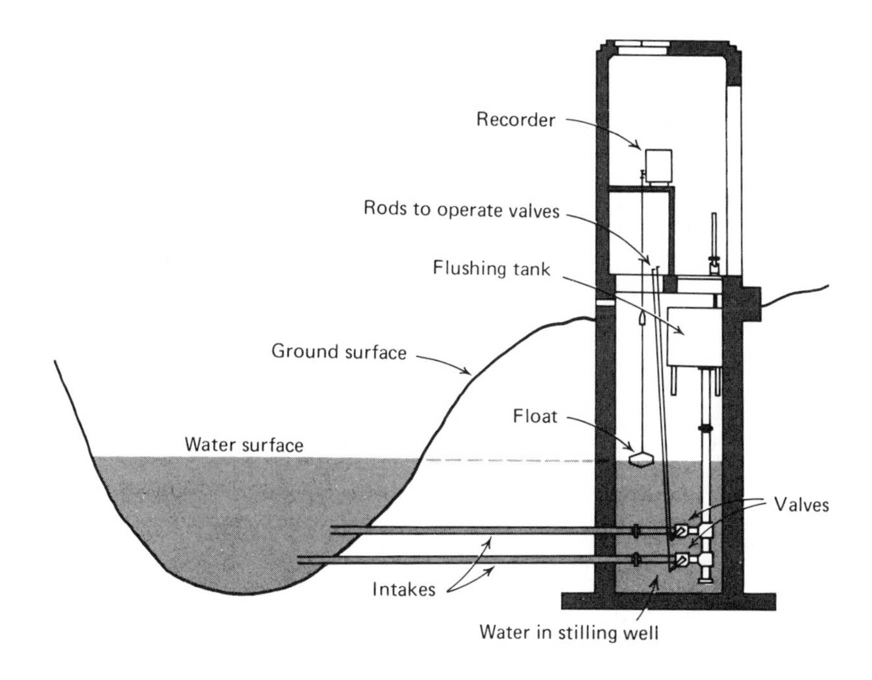


*Figure 2.1* Diagram of a gaging station, showing the relation of the water in the stilling well to the river.

## The Gaging Station

Driving along roads in city or country, beside a stream channel you will often see a vertical round corrugated tube with a conical roof. There are other types of gage structures as well—square brick ones, tall tubes, or square wood houses. The contents are virtually the same. The gaging station holds a device that records the changing height of water in the adjacent channel. The vertical housing is a water reservoir connected by a pipe to the stream, and the water level in the little house goes up and down in concert with the rise and fall of level in the channel. As shown in Figure 2.1, a float on the water surface within the house is connected to a device whereby the water level is recorded as a function of time. The recorder may be an ink trace on a paper chart moved by a mechanical clock. It may be a paper tape on which the water level information is punched, and it may also include radio or telegraph transmission to some distant office. There are other ingenious devices for countering the effect of ice in winter, for substituting water pressure for the usual float, and

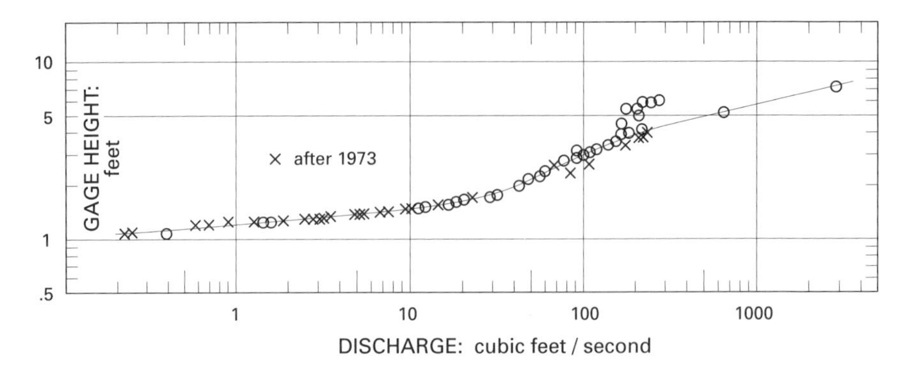


*Figure 2.2* The discharge rating curve at Watts Branch, Maryland. The observations that deviate from the main curve near a discharge of 250 cfs were influenced by accumulations of organic debris at the gaging station near bankfull stage.

for obtaining samples of the water for later analysis. The essential item present in all stations is the recorder of water surface levels.

Water surface level is not equivalent to discharge, however. The cross-sectional area of flowing water and its mean velocity must be known. These are obtained separately, near the recording gage, from a bridge, from a trolley suspended from a cable, or by wading at low flow. The cross section and velocity are obtained simultaneously by measurement of depth and velocity when a current meter is lowered into the water.

In the United States the standard current meter has cups like a wind anemometer, whereas in Europe a propeller-type rotor is employed. In both situations the velocity is measured at about 20 to 30 positions across the channel and the discharge is computed for each position.

Associated with every gaging station is a gage plate or vertical scale on which the water level is measured. The height of water on this gage plate is called stage or gage height. When the total discharge for the cross section is computed, its value is plotted against the gage height. Successive measurements of stage and discharge are plotted in what is called the discharge rating curve (Figure 2.2). The measurements at this gage are very consistent over a period of nearly 26 years, except for 7 points close to bankfull stage at discharges between 180 and 250 cfs. These aberrant values were caused by logs and brush clogging the concrete measurement section.

Each plotted point in Figure 2.2 is a measurement of velocity and gage height, and thus discharge. At most gaging stations a measurement is made about once a month—except during storm periods, when the busy hydrographer makes as many measurements as he can at many gaging stations. Over a period of time the rating curve is developed, with the hope that it will include a wide range of discharge conditions.

The mean velocity of rivers in flood varies from 4 to 10 feet per second. The mean velocity attained in large rivers tends to be slightly higher than that in small rivers. There are, of course, many local situations where, owing to constrictions or rapids, velocity attains greater values. The figures cited above include a large majority of river channels in reaches that have no unusual features. For rivers of moderate size (2 to 100 square miles of drainage area), the flow at bankfull stage will ordinarily have a mean velocity on the order of 4 feet per second. If one had to make a guess without any measurement data, that figure would be a usable approximation.

The U.S. Geological Survey has analyzed individual velocity measurements made by current meter at the point of maximum velocity in river cross sections. The data were from routine measurements at 48 gaging stations on 27 large rivers throughout the country. A frequency table of 2,950 maximum values was compiled. Analysis showed the mean to be 4.84 feet per second, the median 4.11, and the mode 2.76 feet per second. Data on the Mississippi River constituted 13 percent of the sample and had a median value of 8.0 feet per second.

Less than 1 percent of the total measurements exceeded 13 feet per second. The highest velocity known to have been recorded with a current meter by the U.S. Geological Survey was 22.4 feet per second in a rock-bound section of the Potomac River at Chain Bridge near Washington, D.C., on May 14, 1932. Velocities of 30 feet per second (20 miles per hour) have been reported but were not measured by current meter. No greater values are known.

## The Role of the Hydrographer

Each state now has a district hydrologist, whose office is usually located in the state capital. There may be several subdistricts that act as offices for portions of the state, because field personnel must travel to gaging locations within those subdistricts. Until the last few decades the hydrographer was a graduate engineer, but now individuals with technical training are employed. Hydrographers continue to bear a great responsibility, because the measurements they obtain are the basis for design and construction of projects costing millions of dollars. During flood periods these individuals are often at great risk under conditions of rain, snow, flood, or darkness. During storm periods the measurements made at high discharge are the unusual but highly important events contributing to the rating curve. In flood times where brush, trees, and even houses are floating down a high-velocity river, to be the sole occupant of a swaying open sled or cable car that hangs from a cable stretched across the channel is perilous. Yet these demanding conditions are common for all the hydrographic staff and are among the most important circumstances for data-collection purposes.

## Discharge Records

In quieter times these persons are rotated into central offices to check each and every detail of the field notes taken during measurement duty. They also contribute to the compilation of the recorded data for publication and wide distribution.

A summary of each field discharge measurement is compiled on a form crucial to the study of rivers. This summary form, the 9-207, includes for each river measurement the date, width, cross-sectional area, mean velocity, gage height, discharge, and name of hydrographer. These forms are not published but can be obtained for any gaging station by application to the Geological Survey district office in the state.

The public record of streamflow is published by the Geological Survey in annual volumes. These volumes can be found in the government documents section of all large libraries, under the title *Surface Water Records* for each state. They can also be retrieved from computer data bases. Each currently operating gaging station is represented in the annual volume, and the main tabulation consists of the value of mean discharge for each day of the year at that gage. Additional descriptive data include geographic location of the gage, the drainage area, a brief history of the gage, the periods of record taking in the past, notes on diversion, and reservoir storage upstream of the gage. Also specified are the momentary peak discharges experienced during the year, data that are valuable for studying flood frequency.

With the advent of computers, modern record collection and publication has been much improved. Analyses formerly made by intensive hand computation are now routinely made and in many states are published in compact form. A typical example is the book titled *Streamflow Characteristics for California,* compiled by the U.S. Geological Survey. Especially useful are published values for the duration curve of each station, which indicate the percentage of time given discharge values are equaled or exceeded at the station. For many states the annual volumes now include summary data on suspended sediment and chemical quality of water for stations where such measurements are made.

Other highly valuable analyses have been published for various river basins in the United States. The series of volumes entitled *Flood Frequency* appear in water supply papers 1671–1688 of the U.S. Geological Survey.

The first gaging station was established on the Rio Grande in 1895, so the longest record is just short of 100 years. The number of gaging stations has gradually increased; today there are 7,590 operating stations in the United States. An early gage is shown in Figure 2.3.

The length of time to maintain a gaging station has always been debatable, because the number of possible gaging sites is so large. It could be argued that no one place deserves operation forever while other locations go unmeasured. Network design for attainment of maximum usefulness within the constraints of budget and personnel has long been a preoccupation of the Geological Survey. No perfect solution is possible. When the record of a specific station is long enough so that it can be correlated with other stations, and thus an estimate of flow is available by statistical means, then that station may be discontinued. The matter is complicated and will not be treated in detail here. But it is important to realize that many river gaging stations are so affected by the works of man that they do not provide a representation of the hydrology of the river basin. The flow of most rivers in the United States is now outflow from a reservoir, and because water is held over from one season to another the discharge record is not a record of the natural flow.

The Geological Survey has in place a network of chemical quality sampling carried out at some gaging stations, and a developing program of assessing biologic materials at selected stations. Somewhat similar sampling is done by the Environmental Protection Agency, but those programs are outside the scope of this book.

Some physical measurements not made by these agencies are needed at gaging stations that would be useful for monitoring long-term changes as well as for describing the morphology of the river channel. Extensions of present descriptions of the gaging site, they need be made only once, or perhaps once every 20 years. They include a permanently monumented cross section of the channel, a description of bed material made by the pebble counting method, a survey of the longitudinal profile of bed and water surface in the reach of channel, and the determination of bankfull stage by a standard uniform method. We call such a procedure a channel geometry survey (Figure 2.4).

In addition to the discharge and suspended sediment network, a small

*Figure 2.3* A patriarch of the gaging network: Rio Grande at Wason, below Creede, Colorado. The station was established in 1907 as a staff gage. The house seen here was built in 1910. A recorder was installed, and a cable was built at the gage house. When the station was discontinued in 1954, Lisle Alsdough of Del Norte moved the cable to another location.

net of bedload measurement sites is needed to include representative river channels in various climatic and geologic regions. Until recent decades such a network was impossible because no sampling equipment was available. But we now have a tested, easily available sampler called the Helley-Smith, which is simple to operate and practical for field use.

*Figure 2.4* M. Gordon Wolman surveying on a typical point bar. Seneca Creek, Maryland.

## The Bench-Mark Stations

In mid-century the Water Resources Division of the U.S. Geological Survey established a small number of gaging stations carefully located in basins protected from human activities. These sites included national and state parks, monuments, and other small areas without roads, clear-cutting of forests, or mining. The purpose of these stations is to provide a minimum number of stream locations that reflect the natural hydrologic reality, and to record the effects of any climatic changes that occur. It is hoped that these few stations will be operated without interruption for a very long time.

Fortunately, successive administrators of the Geological Survey have realized the importance of records uninfluenced by man-made alterations of the drainage basin. The number of such bench-mark stations remains small, only 57, but has not decreased even in times of budget stringency.

## Backyard Hydrology

You can learn a great deal about hydrology by engaging in the interesting and rewarding activity of direct observation. Hydrology is a discipline that requires hands-on collection of data. Observation, the very essence of the hydrologic business, can be as simple as placing a thermometer on the back porch and reading it every morning. Of higher value and sophistication is buying a plastic rain gage for a few dollars and putting it on the front lawn, to be inspected each time it rains. A rain gage can also be made from a coffee can. In this case, buy a baby bottle from the dime store; most such bottles have gradations in cubic centimeters. You can pour the rainwater from the can into the bottle, record the volume in cubic centimeters, then convert the volume into inches of rain.

A still more sophisticated procedure is to place a staff gage in the nearest creek, gully, or channel and read the gage at certain times. If you are fortunate enough to live near a creek or have your office near one, it might be possible to place the gage where it can be seen from your window and read through binoculars. You can make a staff gage out of any old piece of wood, lettering or burning into it a scale of gage height in feet or centimeters.

Once you are reading a staff gage, you are into the crux of hydrologic science, because it is then necessary to construct a rating curve and measure stream velocity. Obviously this requires that a cross section be surveyed, a bench mark established, and velocity measurements made at various stages of flow. The easiest way to measure velocity is by floats, and the best float is an orange peel. It has just the right specific gravity to float nicely at the surface, it is brightly colored and thus easily seen, and it is readily available.

Choose a fairly straight reach of channel. Measure a distance along the bank of at least one width—but two or three widths would be better. Mark the beginning and end with stakes or colored markers. Pace along the bank or measure with a tape the distance between the markers. A stride (right footprint to right footprint) is for most people close to 5 feet.

You will use at least five floats, so have a few extra. Throw one upstream of the first marker and start counting seconds as it passes the marker. Measure time in seconds with a stopwatch, with the second hand of your wristwatch, or by counting out loud—one, and two, and three . . .

Note the time at which the float passes the lower marker and write that time in the field book. Throw your floats at various distances from the bank, so that you will measure velocity in different places in the channel width. Average the times in seconds. Divide the distance by the time in seconds to get a mean surface velocity in feet per second. Multiply this by 0.8 to get the mean velocity in the cross section.

This velocity must be multiplied by a cross-sectional area to get the discharge. To get the cross-sectional area, take a rule or meterstick or rod marked in appropriate units, preferably tenths of feet, and at uniform distances measure the depth of the water as you wade across. Take no less than 10 measurements; 15 would be better. The depth should be in even centimeters or to 0.05 foot. That is, readings might be 0.05, 0.55, 0.65, 0.80, 1.15, and so forth in feet; in centimeters they might be 6, 10, 14, and the like.

Measure the water surface width with care, using a tape or pacing carefully. Average the depths, multiply by the width to get the area in square feet. Then multiply by the mean velocity to get the discharge in cfs. Record all the figures as you compute them, writing them in your notebook not on scraps of paper.

Record the discharge with the staff gage reading taken at the same time. The gage height plotted as ordinate against the discharge plotted as abscissa on double-log (log-log) paper is the beginning of the discharge rating curve. For double log paper the line through the plotted points usually has a slope of about 0.34.

An excellent way to begin this kind of field observation is to make some preparations in advance of rainstorms. Install a staff gage, survey the cross section by wading, then wait for a storm. When rain appears imminent, put the plastic rain gage out on the lawn in an open space. From under an umbrella, watch the water at the staff gage. Read the rain gage every 10 minutes. When the water begins to rise, read the staff gage every 2 minutes. Keep careful note of the clock time of every reading. If you are lucky enough to get a burst of rain and a resulting rise and fall of the river, a great deal can be learned from even a single storm. Your observed hydrograph can be plotted and analyzed as discussed in the next chapter.

(I observe the staff gage in my yard by looking at it through a telescope. During a storm I step outside under my umbrella every 10 minutes to read my plastic rain gage.)

Individual observation sharpens awareness of the basic processes in nature that provide us with the necessities of life. Our personal measurements can add important details to the knowledge of our resource base.

CHAPTER THREE

# Down the Channel System

## The Hydrograph

Rivers drain water from the continents to the oceans and are the principal routes of transport for the products of weathering. Gravity provides the force by which both excess water and movable debris are brought from higher to lower elevations.

In accomplishing this transfer, the water that flows off the land toward the oceans forms and maintains a highly organized system of physical and hydraulic features. So complex are the interrelations that to focus on any single portion tends to make one lose sight of other equally important features. In the natural environment, the interrelationships make it difficult to visualize all of the system simultaneously. Yet it is precisely these interrelationships that constitute the most distinctive and pervasive characteristic of natural systems including rivers.

One reason why the interrelationships are difficult to visualize in rivers is that large variations occur through short periods of time. A storm that lasts a few hours or a few days produces runoff that appears in the river channel and subsequently drains away. The water level rises as the storm flow arrives at a given point and falls as it passes downstream. This rise and fall may be graphed as a function of time. Each location in a river system displays characteristic features in such a graph.

Discharge is defined as volume per unit of time. It is usually expressed as cubic meters per second (cms) or cubic feet per second (cfs). A plot of stream discharge as a function of time is, by definition, a hydrograph. An important aspect of hydrology involves the analysis of hydrographs to derive quantitative characteristics of the basin and its channels. A given basin will characteristically produce nearly the same hydrograph from different storms of equal magnitude and distribution. Hydrograph analysis, the use of the unit hydrograph, and analysis of rainfall-runoff relations are detailed in standard texts on hydrology. Of concern in this volume are the principles important to the fluvial geomorphologist, the student of rivers.

The morphology of channels involves more than rainfall-runoff relations. But an understanding of those relations must begin with the hydrology of a basin and the runoff in its channels. An example of such interrelations is presented to illustrate several important characteristics of a runoff hydrograph. In Figure 3.1 is a plot of rainfall and the resulting hydrograph measured at the U.S. Department of Agriculture Walnut Gulch Experiment Station, Tombstone, Arizona, on August 18, 1961. The basin is called W-3, its drainage area is 8.99 square kilometers (3.47 square miles), and its mean gradient from headwater to gaging station is 0.019. The vegetational cover is grass-woodland.

A graph of the rate of rainfall as a function of time is called the hyetograph. Note that the scale for plotting rainfall intensity is different than that used for runoff rate because the runoff rate is always smaller than the rate of the rainfall causing the runoff. It is smaller because of the infiltration that takes up some of the precipitation and because of the storage during the process of runoff generation.

In this example the maximum rainfall rate was 3.3 inches per hour (84 mm/hr), whereas the maximum runoff rate was 0.31 inch per hour (7.9 mm/hr).

The runoff began some time after the beginning of rainfall, 18 minutes in this case. Local depressions in the ground must be filled before overland flow can begin (depression storage). There is also a delay known as lag between the center of mass of rainfall and the center of mass of runoff. The former occurred at about 9:45 and the latter at 10:28, a lag of 43 minutes. The lag time is a hydrologic quantity of great significance.

Early in the history of hydrograph analysis it was recognized that the time distribution of accumulated percentage of runoff volume, called the distribution graph, was a useful way to compare hydrographs. W. B. Langbein found that if the time scale of the distribution graph was expressed in terms of lag rather than hours, the resulting dimensionless graph fitted most hydrographs for large as well as for small basins. His graph is shown in Figure 3.2.

The graph is presented in two forms. The S-shaped line is the usual distribution graph showing the percentage of runoff accumulated with passage of time. Time in this graph is expressed not in hours but in lags—that is, the time period between the center of mass of the rainfall

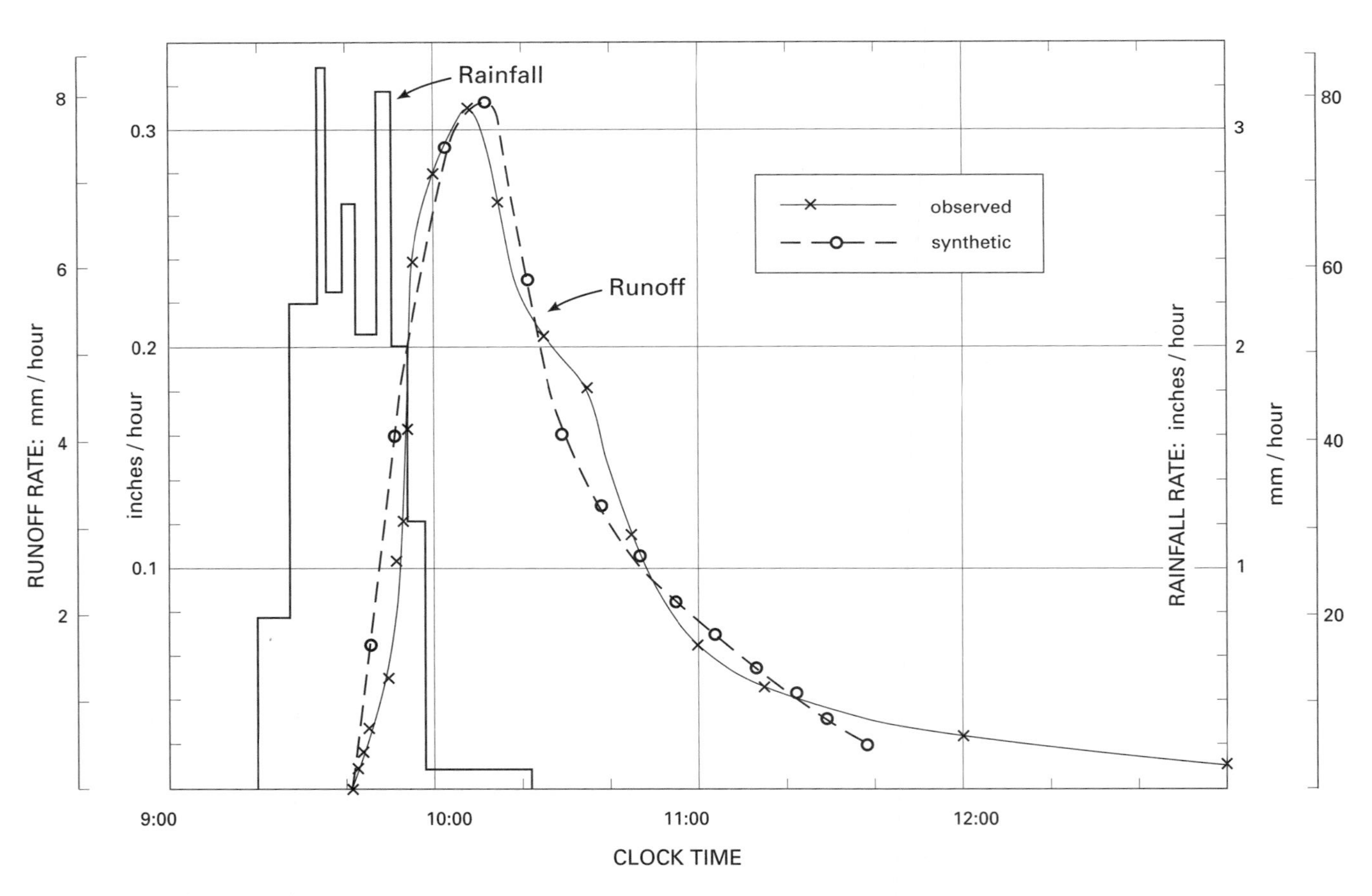


*Figure 3.1* A hyetograph of rainfall (rain intensity as a function of time) and the resulting runoff hydrograph for a storm of August 18, 1961, at the W-3 basin of Walnut Gulch Experiment Station, Tombstone, Arizona. (Data from U.S. Department of Agriculture.)

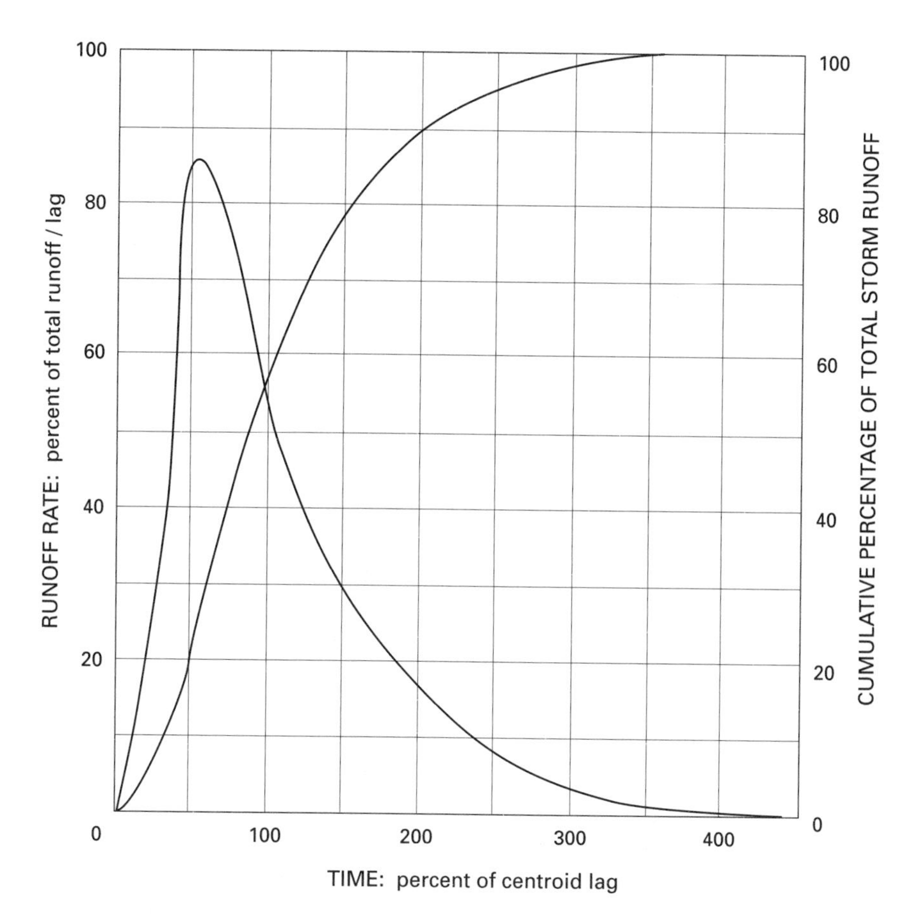


*Figure 3.2* Summation graph and derived unit hydrograph. (From Langbein 1940.)

and the center of mass of the runoff. The peaked curve is the slope of the distribution graph and is in the form of a hydrograph. On that graph several important traits of hydrography are shown. Practically all of the water has run off after 3.5 times the lag; that is, nearly all runoff ends after a time equal to 3.5 lags. In the Tombstone example runoff began at 9:42 A.M. and after 3.5 lags or 150 minutes (at 12:12 P.M.) the runoff rate was small, although it did not cease until about 4.8 lags. Figure 3.2 shows that the peak rate of runoff occurs about 0.55 lag after the beginning of runoff. In the Tombstone example the peak occurred at 10:08, or about 26 minutes, 0.60 lag, after runoff began.

Note the way the Langbein hydrograph of Figure 3.2 is labeled. To construct a graph of the slope of the distribution curve, the slope must be expressed as percentage of runoff for some time period. A logical choice is one lag. Where the distribution curve is steepest, thus where the hydrograph peaks, the slope of the tangent to the distribution graph extended to a period of time equal to one lag gives a runoff rate of 87 percent of total runoff per lag.

The distribution curve can be used to compute a synthetic hydrograph that will agree closely with the observed hydrograph. The computation is described in Leopold 1991. The computed synthetic hydrograph using a lag of 43 minutes (0.72 hour) is shown as the dashed line in Figure 3.1.

## Average Discharge

The average discharge is defined as that flow rate which, if continued every day of a year, would yield the observed annual volume of water. The average discharge usually fills a channel to about one-third the channel depth, and this flow rate is equaled or exceeded about 25 percent of the days in a year. That is, the river flows at a discharge less than average about 75 percent of the time. This figure varies among rivers between 60 and 75 percent.

Eight river channels in the upper Green River, Wyoming, were analyzed to determine the percentage of depth of channel filled when the discharge is equal to its mean annual value. The average was 43 percent. A study of 21 streams in the west central part of California showed that the mean annual discharge filled the channel to 0.28 of its bankfull depth on the average.

The discharges that are less than the average value contribute about 25 percent of the total yearly volume of runoff; those discharges less than half the average value contribute about 15 percent of the total volume. In most rivers these low values derive from the emergence of groundwater. Average discharge is, of course, important in analysis of water supply, but the average depends on the particular period of record over which the average was compiled. Serious errors in water management have been made by too heavy reliance on the average runoff value. A much more useful analysis is a frequency study of the whole array of available data so that the probability of various levels of excess or deficiency may be considered.

Because average values of discharge depend on the drainage area, rivers of different sizes are not comparable using average values. Comparison is facilitated by expressing the average as discharge per square mile of area or in inches of runoff. One inch of runoff from one square mile is 2.32 million cubic feet or 53 acre-feet. This would be produced by 0.074 cfs flowing continuously for one year.

A map of the United States showing average annual runoff may be found in the *National Atlas of the United States,* published by the U.S. Geological Survey in 1970.

## Channel Storage

When you go outside to water the garden and turn on the faucet, water does not immediately come out at the far end of the hose. The hose must be filled with water before the far end begins to flow. The volume of water needed to fill the hose is called channel storage. By the same token, when you are finished and turn off the faucet, water flows out of the hose at a rapidly decreasing rate. The water in channel storage is draining out.

The river acts in a similar manner. Storm water inflow at some upstream point must at least partly fill the volume of the channel in order for outflow to feel the storm inflow. The time distribution of the discharge as measured at the downstream location changes.

Published works in engineering hydrology usually concentrate on dams, reservoirs, and other works on rivers large enough to be represented adequately in the river measurement network. These are basins having drainage areas of 25 square miles or larger. Environmentalists and even planners are often involved in the hydrology of creeks and small streams. They may be faced with assessing the impact of development on a basin of a square mile or less. For such small basins, river measurement stations are few; seldom are there two gaging stations in tandem along the length of such a small stream. Therefore hydrograph changes along the channel cannot be ascertained by analysis of changes between successive measuring points. Procedures applicable to such analyses are available in principle but not in specific example. In Chapter 2 a field procedure was outlined that would result in a hydrograph of discharge for a single storm event. Such a hydrograph developed by the most simple means—a stick for a staff gage, a measuring tape, orange peel for floats, and graph paper—can be just as satisfactory for many purposes as the W-3 basin of the USDA Experiment Station. Most interested observers who are not professional hydrologists are more likely to be dealing with hydrographs developed by simple means, not by instrumented basins such as W-3.

Once field observations are in hand, procedures explaining how the observed hydrograph of a small basin affects downstream locations are not available in the literature. The hydrograph will change shape and peak discharge as channel storage affects it in its downstream travel. A method of computing the effect follows.

Imagine water flowing from a faucet into a tub that has a drainpipe in the bottom. No appreciable outflow occurs down the drain until some water has accumulated in the tub to build up a depth over the drain orifice. Thus, in the early stages of putting water into the tub, the rate of inflow is greater than the rate of outflow. If you then turn off the faucet and the inflow ceases, the outflow down the drain continues until the water temporarily stored in the tub has drained out. During this time the outflow exceeds the inflow. Total outflow must equal total inflow. The relationships in any increment of time are expressed by the storage equation that states, "Outflow equals inflow plus or minus the rate of change of storage."

Think of a reach of channel as if it were a reservoir, or tub. The flow into the upper end of the reach is the inflow, the outflow is the water passing the downstream cross section, and the volume of the channel in between is the reservoir. The channel reach acts as if it were indeed a reservoir having the same relation of outflow rate to water level or storage volume, as if the long reach of channel were a lake just upstream of the outflow point. The storage volume in the channel thus acts like the bathtub storage. Channel storage in a river alters the outflow hydrograph, as would a reservoir.

The same kind of storage or reservoir action translated the hyetograph (precipitation versus time) into a hydrograph (stream flow versus time) that has a more rounded shape and smaller peak. Similar changes occur as the inflow hydrograph is compared with the outflow graph at the upper and lower ends of a reach of river.

In the left-hand portion of Figure 3.3, a storm hyetograph of precipitation as a function of time is represented by the tall rectangle. The precipitation started abruptly, lasted a short period of time at a constant intensity (inches per hour) and ended abruptly. The sketch of the discharge in an upstream channel reach as a function of time is the hydrograph (labeled "Discharge at 2"). It is drawn so that the area enclosed under the curve is smaller than the area under the precipitation rectangle. These areas represent volumes of water (flow rate multiplied by time equals volume). The difference in volume of precipitation and the consequent volume of runoff is what has been infiltrated into the soil during the time surface runoff collected and has become channel flow. Note also that the runoff hydrograph began slightly later than the beginning of the precipitation.

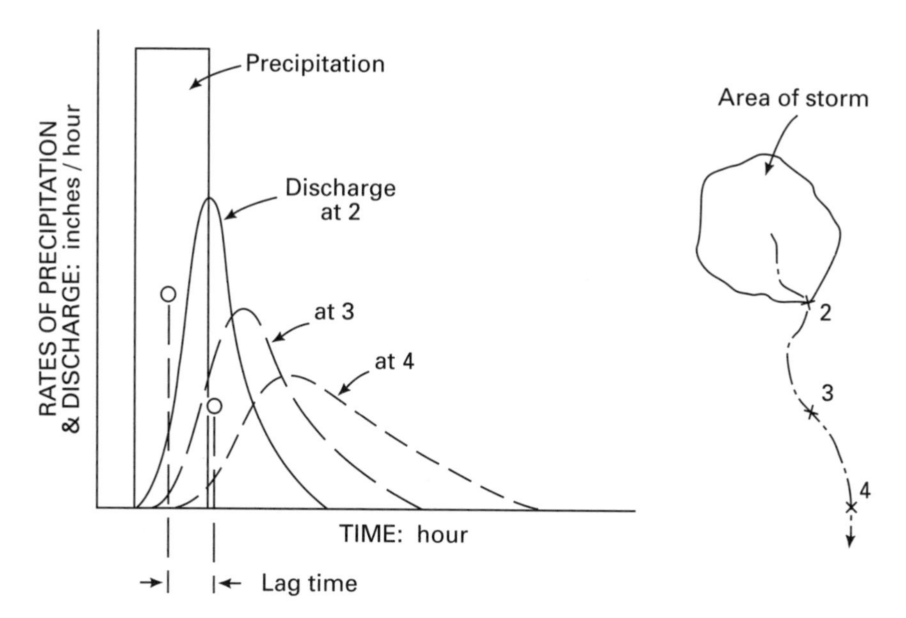


*Figure 3.3* A diagram of the rainfall hyetograph (rectangular area) and the resulting hydrograph at location 2. The small open circles have been placed at the center of the mass of rainfall and runoff; the time difference between these circles is the centroid lag. The storm rain fell on the circular basin above location 2, and the flood wave moved downchannel to 3 and 4. Hydrographs at locations 3 and 4 are shown.

Let the right-hand diagram represent a plan or map of the area under consideration. The balloon-shaped area received the precipitation. Location 2 is the point on the channel where hydrograph 2 was observed.

The figure shows that no rain fell in the zone between points 2 and 3, so all the discharge experienced at 3 flowed earlier past point 2. Because it takes time to flow that distance, the beginning of flow at 3 is later in time than at 2, and at point 4 it is even later.

The first part of the hydrograph (discharge is rising) is called the rising limb, and that part after the peak is the recession limb. The point on the hydrograph where the curvature of the recession limb changes—that is, the point of inflection—indicates the time when inflow ceases. All of the recession limb of the hydrograph later than the point of inflection is water draining out from channel storage. Therefore, the volume of water in the channel for various values of outflow rate can be compiled by measuring the area under the recession limb at different times. For each time chosen, the outflow rate is shown by the ordinate value of the hydrograph.

The point of inflection of the recession limb marks the transition from that part of the hydrograph maintained by inflow into channels and that portion representing drainage from channel storage. This is true for small basins, but may not be true for large ones. The reason is that large basins are likely to have appreciable inflow from groundwater, which alters the base flow. In addition, numerous tributaries may enter a long reach, complicating the form of the hydrograph. Simple hydrographs such as the one in Figure 3.3 are caused by discrete storms having a sharply defined beginning and end. Storms over large areas usually have complicated patterns of time and area, resulting in complex hydrographs.

## Flood Routing

The process of determining the timing and shape of a flood wave at downstream points is called flood routing.

In natural channel networks, a storm produces a runoff hydrograph in each small basin. At the junction of tributaries these respective hydrographs augment one another, and thus the flow below the junction is the sum of the rates of flow of the tributaries at the junction point. One such tributary may be on the rising stage while the joining one may be at peak or on the recession limb. The magnitude of a flood downstream depends strongly on the coincidence or lack of coincidence of the contributing basins.

Along the length of a drainage system, channel size and thus channel storage increase downstream. The flood peak from a given storm decreases downstream through the action of channel storage. The strongest tendency for overflow or for flood conditions is immediately downstream of that part of the basin experiencing the greatest runoff. Therefore, flooding of out-of-channel banks usually occurs just downstream from the storm location. The size, distribution, duration, and placement of a storm within a basin materially affect the location and distance along the major valley where overflow and the resulting flood damage will occur.

In a small headwater tributary a severe flood may peak within a few tens of minutes of the time of heaviest rain, and the peak may last only a few minutes before receding. In great rivers such as the Mississippi, the flow takes weeks to build up to the peak and high water may last a month or more.

For a simple problem in which only one hydrograph is available rather than two hydrographs from separate gaging stations, it can be assumed with little probability of error that channel storage acts as a reservoir. Therefore we can use a reservoir storage analysis.

Channel routing involves successive solutions to the storage equation:

outflow = inflow ± rate of change of storage

The computation illustrated follows the procedure outlined in Dunne and Leopold (1978, p. 353). The equation to be solved at each chosen time period is

$$\frac{S_2}{\Delta t} + \frac{O_2}{2} = \frac{S_1}{\Delta t} - \frac{O_1}{2} + \frac{I_1 + I_2}{2}$$

where $S$ is storage volume, $I$ is rate of inflow, $O$ is rate of outflow, subscripts are 1 at beginning of period and 2 at end of period, and $\Delta t$ is the duration of each step in the computation.

The storage characteristics will be estimated from the recession limb of an observed hydrograph. The problem posed is to compute the hydrograph at points downstream from the measuring station on basin W-3, introduced earlier.

Begin by reading the simultaneous values of discharge and time for the whole hydrograph plotted in Figure 3.1. In Table 3.1 the time and observed discharge are shown in columns 1 and 2, read at intervals of 5 minutes. The 5-minute period was chosen to give a reasonable number of points to represent the full hydrograph.

Column 2 is observed discharge in inches per hour, but because it is a reading for 5 minutes the value is also a volume of water equal to inches per hour for 5 minutes. In inches it is equal to the value in column 2 divided by 12, for there are 12 units of 5 minutes in an hour. We wish to accumulate the volume of runoff in the recession limb.

In the period later than noon, the tail of the hydrograph constituted 0.015 inch. So by accumulating the volumes in column 2 in inches and adding 0.015 inch, column 3 is the accumulated volume in inches for 5-minute intervals backward from noon. For example, at 12:00 the runoff rate was 0.022 inch per hour which in 5 minutes is a volume of 0.022 divided by 12, or 0.002 plus the amount 0.015 that was in storage after 12:00, giving 0.017 inch. At 11:45 the rate was 0.025 inch per hour or 0.002 inch, which added to the previous accumulation of 0.021 inch gives 0.023 inch.

Now compute the values of the outflow relations. Choose a time (distance) for the routing that is the length of channel down to the outflow

*Table 3.1* Discharge and volume of observed runoff at Tombstone W-3 on August 17, 1968

| 1<br>Time<br>(A.M.) | 2<br>Observed<br>discharge<br>(in/hr) | 3<br>Accumulated<br>storage (in) |
|---|---|---|
| 9:40 | 0 | |
| 9:45 | 0.025 | |
| 9:50 | .050 | |
| 9:55 | .190 | |
| 10:00 | .280 | |
| 10:05 | .300 | |
| 10:10 | .310 | |
| 10:15 | .275 | |
| 10:20 | .245 | 0.158 |
| 10:25 | .205 | .137 |
| 10:30 | .195 | .120 |
| 10:35 | .185 | .104 |
| 10:40 | .150 | .089 |
| 10:45 | .117 | .076 |
| 10:50 | .098 | .066 |
| 10:55 | .082 | .058 |
| 11:00 | .065 | .051 |
| 11:05 | .049 | .046 |
| 11:10 | .041 | .042 |
| 11:15 | .038 | .038 |
| 11:20 | .034 | .035 |
| 11:25 | .032 | .033 |
| 11:30 | .030 | .030 |
| 11:35 | .028 | .027 |
| 11:40 | .026 | .025 |
| 11:45 | .025 | .023 |
| 11:50 | .024 | .021 |
| 11:55 | .023 | .019 |
| 12:00 | .022 | .017 |

*Table 3.2* Storage outflow relations at Tombstone W-3 for time period 20 minutes (= 0.33 hr) on August 17, 1968

| 1<br>Outflow<br>$O$<br>(in/hr) | 2<br><br>$S$<br>(in) | 3<br>$\frac{S}{\Delta t}$<br>(in/hr) | 4<br>$\frac{S}{\Delta t} - \frac{O}{2}$<br>(in/hr) | 5<br>$\frac{S}{\Delta t} + \frac{O}{2}$<br>(in/hr) |
|---|---|---|---|---|
| 0.245 | 0.158 | 0.527 | 0.404 | 0.772 |
| .205 | .137 | .457 | .355 | .560 |
| .195 | .120 | .400 | .303 | .498 |
| .185 | .104 | .347 | .255 | .440 |
| .150 | .089 | .297 | .222 | .372 |
| .117 | .076 | .253 | .195 | .312 |
| .098 | .066 | .220 | .171 | .269 |
| .082 | .058 | | | |
| .065 | .051 | .170 | .138 | .203 |
| .049 | .046 | | | |
| .041 | .042 | .140 | .120 | .161 |
| .038 | .038 | | | |
| .034 | .035 | .117 | .102 | .151 |
| .032 | .033 | | | |
| .030 | .030 | .100 | .085 | .115 |
| .028 | .027 | | | |
| .026 | .023 | .077 | .064 | .090 |
| .025 | .023 | | | |
| .024 | .021 | .070 | .058 | .082 |
| .023 | .019 | | | |
| .022 | .017 | .057 | .046 | .068 |

station for which the computation is being made. The routing envisions the hydrograph downstream by 20 minutes from the assumed inflow point. To estimate the distance along the channel, we can use the approximation that the flood wave proceeds downstream at about half the speed of the water particles.

In this example 20 minutes was chosen, or 0.33 hour. Table 3.2 repeats in columns 1 and 2 the flow rate and associated storage from columns 2 and 3 of Table 3.1. Then the third column is the storage from column 2 divided by the routing time, 0.33 hour. The units are inches per hour. Column 4 shows the result of subtracting half the outflow rate in column 1 from the storage rate in column 3, as indicated in the storage equation. Column 5 is similar, except that half the flow rate is added to the storage rate.

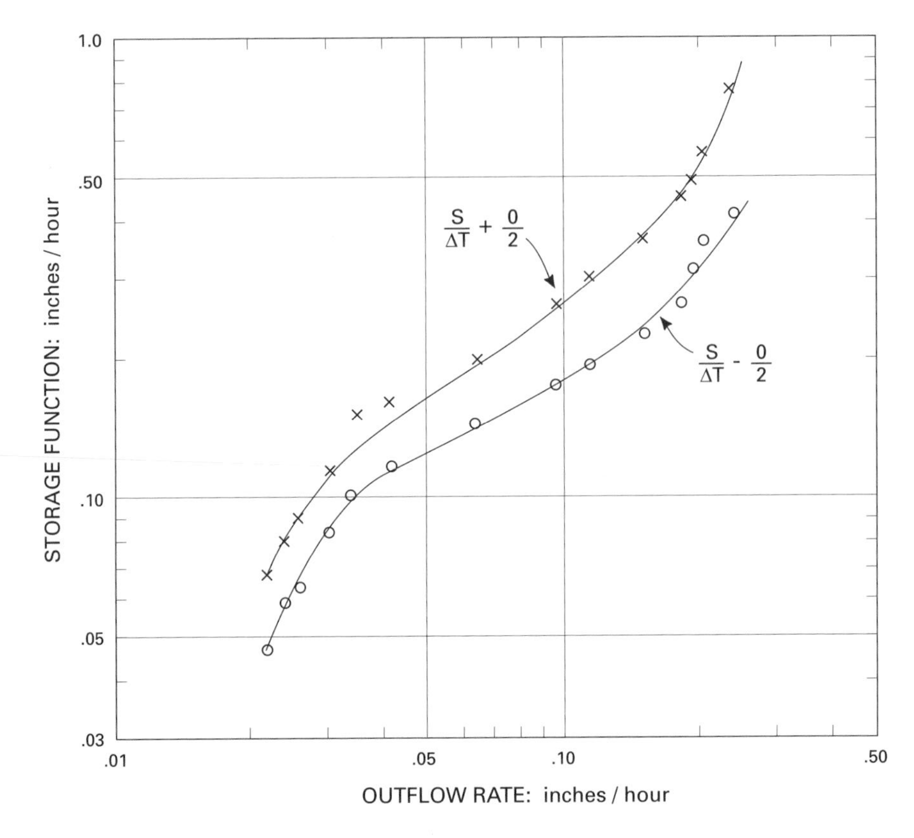


*Figure 3.4* Storage function plotted against outflow rate, computed from the recession graph of the storm of August 18, 1961, on basin W-3.

The outflow rate of column 1 is now plotted against the two storage factors of columns 4 and 5, and the curves are shown in Figure 3.4.

The routing is carried out in Table 3.3. The hydrograph data observed are in columns 1 and 2. Column 3 shows the inflow occurring between the beginning of the time period and the end, a period of 20 minutes in this example. For convenience the computation is made at 10-minute intervals, a matter of choice. The first value of the end of the period is chosen as 9:50, but 20 minutes earlier the inflow was zero. So the value in column 3 at the time 9:50 is half of 0.050. This small inflow goes to storage, as will be seen. At 10:00 the average inflow is the average between 0.280 and zero, or 0.140, which entered in the abscissa of Figure 3.4 gives a value on the ordinate of $(S/\Delta t - O/2)$, or 0.220. Add this number to the inflow of 0.025 and get 0.360 which, entered into the ordinate of the other curve of Figure 3.4, on the abscissa gives a value of

*Table 3.3* Routing increments of 20 minutes (0.33 hr) at Tombstone W-3 on August 17, 1968

| 1 | 2 | 3 | 4 | 5 | 6 | 7 | 8 | 9 | 10 |
|---|---|---|---|---|---|---|---|---|---|
| | | Routed 20 minutes | | | | Routed 40 minutes | | | |
| Time (A.M.) | Inflow in/hr | Average inflow $\frac{I_1+I_2}{2}$ | At begin $\frac{S}{\Delta t}-\frac{O}{2}$ | At end $\frac{S}{\Delta t}+\frac{O}{2}$ | O (in/hr) | Average inflow $\frac{I_1+I_2}{2}$ | At begin $\frac{S}{\Delta t}-\frac{O}{2}$ | At end $\frac{S}{\Delta t}+\frac{O}{2}$ | O (in/hr) |
| 9:40 | 0 | | | | | | | | |
| 9:45 | .025 | | | | | | | | |
| 9:50 | .050 | 0.025 | 0.060 | 0.060 | 0.085 | 0.012 | 0.015 | 0.027 | 0.025 |
| 9:55 | .190 | | | | | | | | |
| 10:00 | .280 | .140 | .220 | .360 | .150 | .075 | .145 | .220 | .075 |
| 10:05 | .300 | | | | | | | | |
| 10:10 | .310 | .180 | .275 | .455 | .180 | .115 | .190 | .305 | .118 |
| 10:15 | .275 | | | | | | | | |
| 10:20 | .245 | .263 | .420 | .683 | .240 | .195 | .300 | .495 | .195 |
| 10:25 | .205 | | | | | | | | |
| 10:30 | .195 | .253 | .400 | .653 | .230 | .205 | .315 | .520 | .200 |
| 10:35 | .185 | | | | | | | | |
| 10:40 | .150 | .198 | .300 | .498 | .195 | .218 | .330 | .548 | .205 |
| 10:45 | .117 | | | | | | | | |
| 10:50 | .098 | .147 | .225 | .372 | .150 | .190 | .290 | .480 | .190 |
| 10:55 | .082 | | | | | | | | |
| 11:00 | .065 | .108 | .185 | .293 | .110 | .153 | .235 | .388 | .160 |
| 11:05 | .049 | | | | | | | | |
| 11:10 | .041 | .070 | .140 | .210 | .072 | .111 | .180 | .281 | .105 |
| 11:15 | .038 | | | | | | | | |
| 11:20 | .034 | .050 | .123 | .173 | .054 | .082 | .155 | .237 | .086 |
| 11:25 | .032 | | | | | | | | |
| 11:30 | .030 | .036 | .103 | .139 | .037 | .055 | .128 | .183 | .058 |
| 11:35 | .028 | | | | | | | | |
| 11:40 | .026 | .030 | .083 | .113 | .030 | .042 | .115 | .157 | .045 |
| 11:45 | .025 | | | | | | | | |
| 11:50 | .024 | .027 | .060 | .087 | .026 | .032 | .088 | .120 | .032 |
| 11:55 | .023 | | | | | | | | |
| 12:00 | .022 | .024 | .056 | .080 | .024 | .027 | .068 | .095 | .027 |

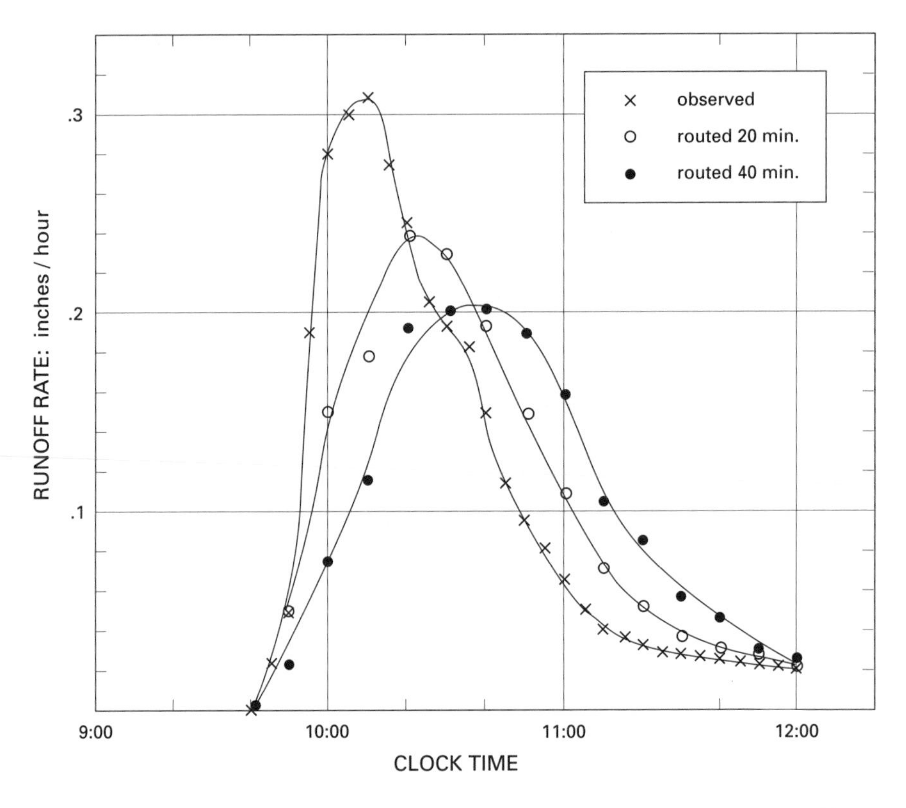


*Figure 3.5* Routing the observed storm on the W-3 basin, shown in Figure 3.1. The routing shows two hydrographs 20 minutes apart, which represent the relation of discharge to time 0.4 and 0.8 mile downstream from the original point of measurement.

outflow of 0.150, to be plotted as the outflow of the routed hydrograph at 10:00 on Figure 3.5.

At the next time 10 minutes later, at 10:10, the average inflow between 9:50 and 10:10 is (0.050 + 0.310) ÷ 2 = 0.180. This value entered on the abscissa as flow rate in Figure 3.4 gives a storage function of 0.275, which added to 0.180 gives 0.455. At the ordinate of 0.455 of storage function read an outflow rate of 0.180, which is plotted in Figure 3.5 at time 10:10.

At the next time 10 minutes later, 10:20, the average inflow is (0.245 + 0.280) ÷ 2, or 0.263. Enter this value on the abscissa of Figure 3.4 and read the value of $(S/\Delta t - O/2)$ as 0.420, which added to 0.263 gives a storage function of 0.683. Enter that value in the ordinate of Figure 3.4 and read on $(S/\Delta t + O/2)$ an abscissa value of 0.240, which is plotted on Figure 3.5 at time 10:20. It can be seen that the values of time in column 1 are plotted against the outflow rates in column 6 to produce the outflow hydrograph of Figure 3.5.

To route for the next downstream point, later by 20 minutes, values of outflow of column 6 are now inflow for the new hydrograph. At the end of the computation, time in column 1 is plotted against outflow in column 10.

The routing envisions the hydrograph that would be observed at a location downstream at a distance of 20 minutes from the assumed inflow point. Using the approximation that the flood wave proceeds downstream at about half the speed of the water particles, and assuming that the water velocity in a storm flow is 3 to 4 feet per second, the flood wave may proceed at about 1.75 feet per second, and in 20 minutes has moved $1.75 \times 60 \times 20 = 2{,}100$ feet or 0.40 mile in the 20 minutes.

This procedure allows a quantitative estimate of how a flood wave moves downstream, being continually ameliorated by the reservoir action of channel storage. Visualization of this process of hydrograph alteration as it passes down the channel is essential to an understanding of channel dynamics. A river channel maintains its form even though it is constantly subjected to variation in discharge as weather goes from dry to wet and back to dry. The form that it takes and the processes of adjustment within the channel are the focus of the science of fluvial geomorphology.

## CHAPTER FOUR

# Meanders and Bars

### Meandering, the Predominant Pattern

Channel pattern is the term to describe how a river looks from above, as seen for example from an airplane. Many geomorphologists, myself included, have written about channel patterns as if there were three types: meandering, straight, and braided. Meandering channels can be highly convoluted or merely sinuous but maintain a single thread in curves having definite geometric shape. Straight channels are sinuous but apparently random in the occurrence of bends. Braided channels are those with multiple streams separated by bars or islands.

My concept of pattern was changed, however, by the results of a survey I conducted to locate geographically the occurrence of various channel patterns. In a light airplane Herbert Skibitzke and I flew nearly 2,000 air miles over all the river valleys in western Wyoming, northern Utah, and southeastern Idaho, mapping channel pattern.

We were surprised to see that about 90 percent of all valley length has meandering stream channels. Braided channels were rare and confined to a few localities. The only valleys with braided channels occurred near the Lost River area of southeastern Idaho and in west-flowing tributaries of the Snake River north of Jackson, Wyoming. There are a few channels with well-vegetated islands that appear stable; some reaches of the Snake River are of this sort. These reaches are similar to the meandering Mississippi River on the eastern border of the state of Iowa. But most river channels are meandering or sinuous in all climates and on all continents.

Not only is meandering the predominant pattern, it exhibits in its details the role that energy utilization and dissipation play in governing channel morphology. These aspects include:

1. The relation of meander wave length to channel width and its curvature

2. The relation of radius of curvature to energy loss by friction
3. The relation of the shape of the meander curve to the distribution of energy utilization
4. The relation of energy utilization to distance along the meandering channel.

Each of these will be illustrated in the pages that follow.

## The Open System in Steady State

A reach of river is a transporting machine. Potential energy at the upstream end is progressively changed to the kinetic form along the channel and the kinetic energy is transformed to heat, doing some work during the transition. The kinetic energy is utilized to carry some sediment and accomplish some erosion as it is changed to heat through turbulence and friction at the boundaries. This transformation of energy from potential form through kinetic form and finally into work and heat characterizes all transporting machines—steam engines, automobiles, airplanes, and others. The potential energy in man-made machines is usually derived from coal or petroleum, both forms of energy storage. In rivers the potential energy is in the form of topographic elevation above the ultimate base level of the ocean.

Most of these examples are in a class called open systems which have available a continual source of potential energy. Thus, an ongoing process of transformation and utilization is possible as long as the store of potential energy is present. The river system is a classic case, because precipitation at high elevation provides a continuing supply of the potential energy of elevation. The utilization of the energy as it is transformed to heat conforms to the laws of physics, including conservation of energy, conservation of mass, and Boyle's law.

One of the characteristics of such an open system is its tendency toward two conditions: that of minimum work and that of uniform distribution of work or energy utilization. These two conditions cannot be simultaneously satisfied, so the result is a compromise.

One of the characteristics of compromise is minimization of variance, a condition statistically known as the most probable state. The compromise toward the most probable state is exhibited in channel meanders. Meandering is thus an especially appropriate morphologic form to demonstrate the physical results of the tendencies characteristic of an open system approximating a steady state.

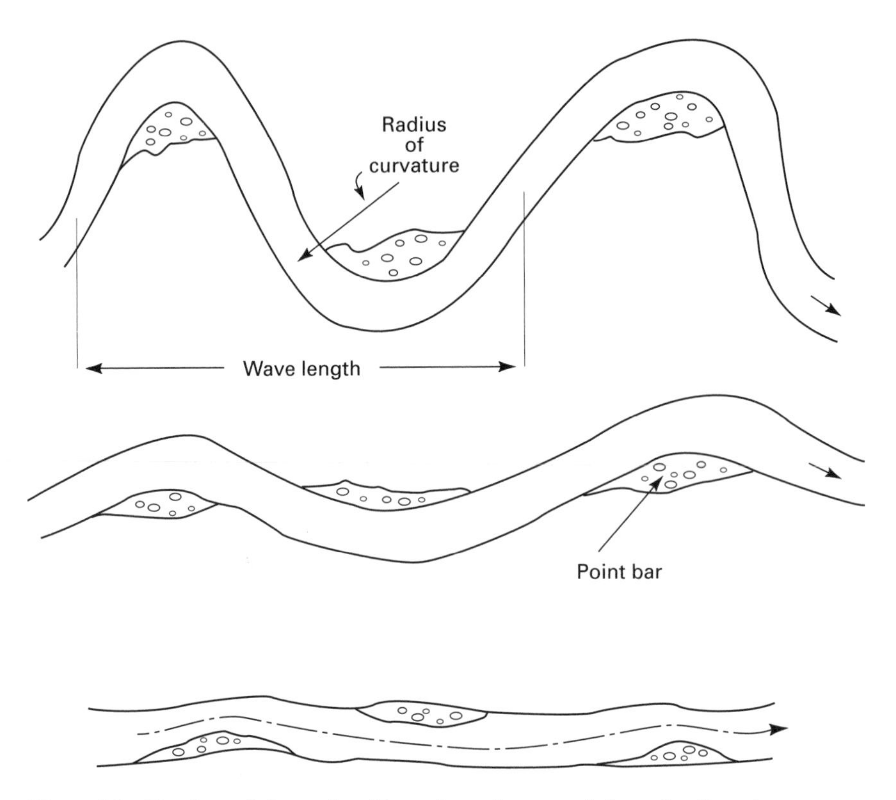


*Figure 4.1* Sketches of channels with various degrees of sinuosity. Even in a channel that is nearly straight, alternate bars lead the thalweg in a sinuous path.

## Width and Wavelength

River channels are seldom straight except for short distances. Even in a straight reach the thread of deepest water, called the thalweg, tends to wander in a sinuous course having a wavelength that relates to channel width as in fully developed meander curves (Figure 4.1).

There is in channels of all sizes a remarkable relationship among the wavelength, channel width, and radius of curvature. The wavelength averages about 11 times the channel width and nearly always is between 10 and 14 channel widths, as seen in Figure 4.2. The radius of curvature of the central portion of a channel bend averages about one-fifth of the wave length. These two relations mean that the radius of a curve is about 2.3 times the channel width.

So closely do channel curves tend to adhere to these dimensions that the forms of meanders of large rivers look very much like the forms of small streams. So close is the resemblance that unless a scale is provided,

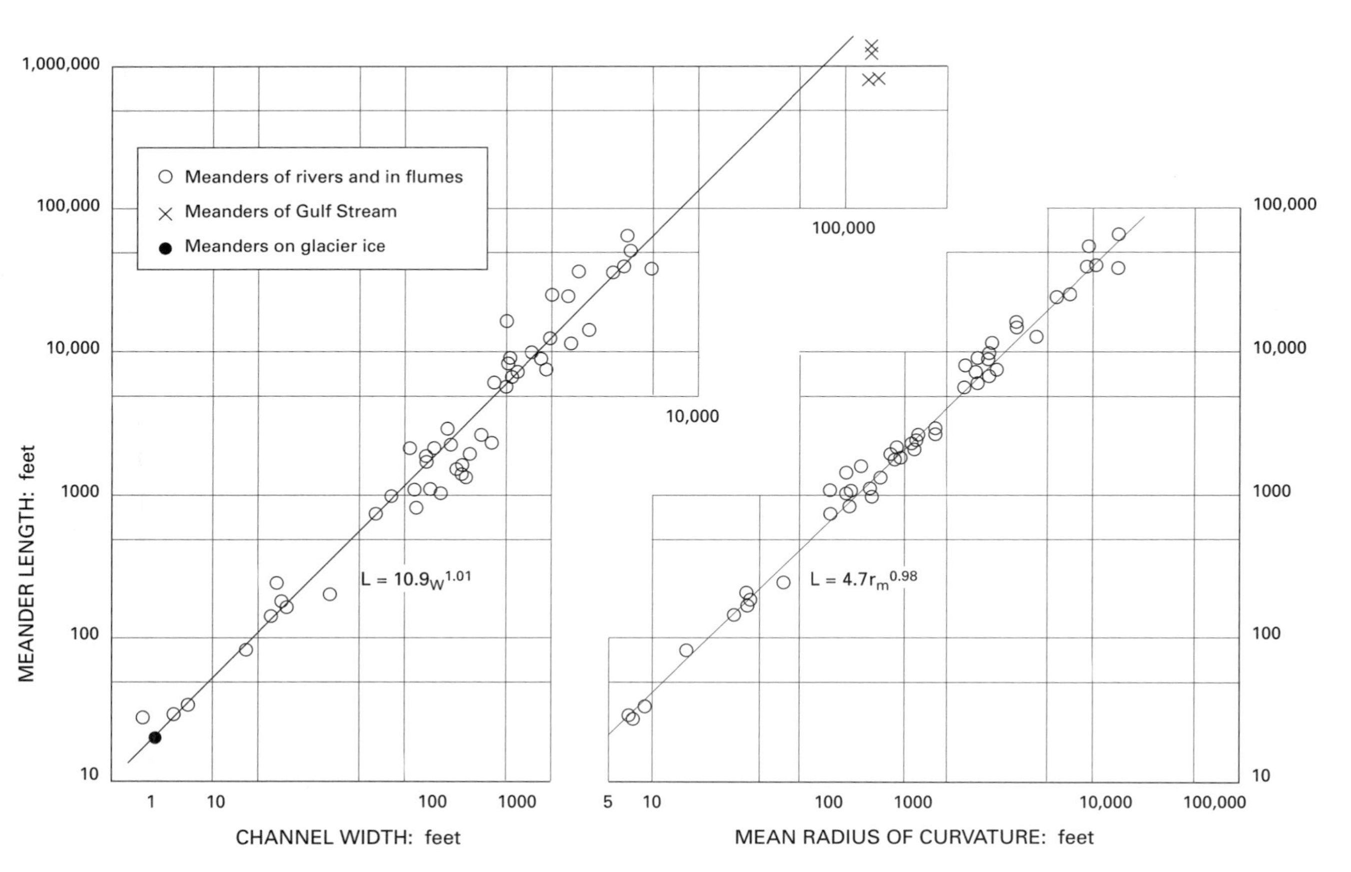


*Figure 4.2* Relations between meander length and channel width, and between meander length and mean radius of curvature.

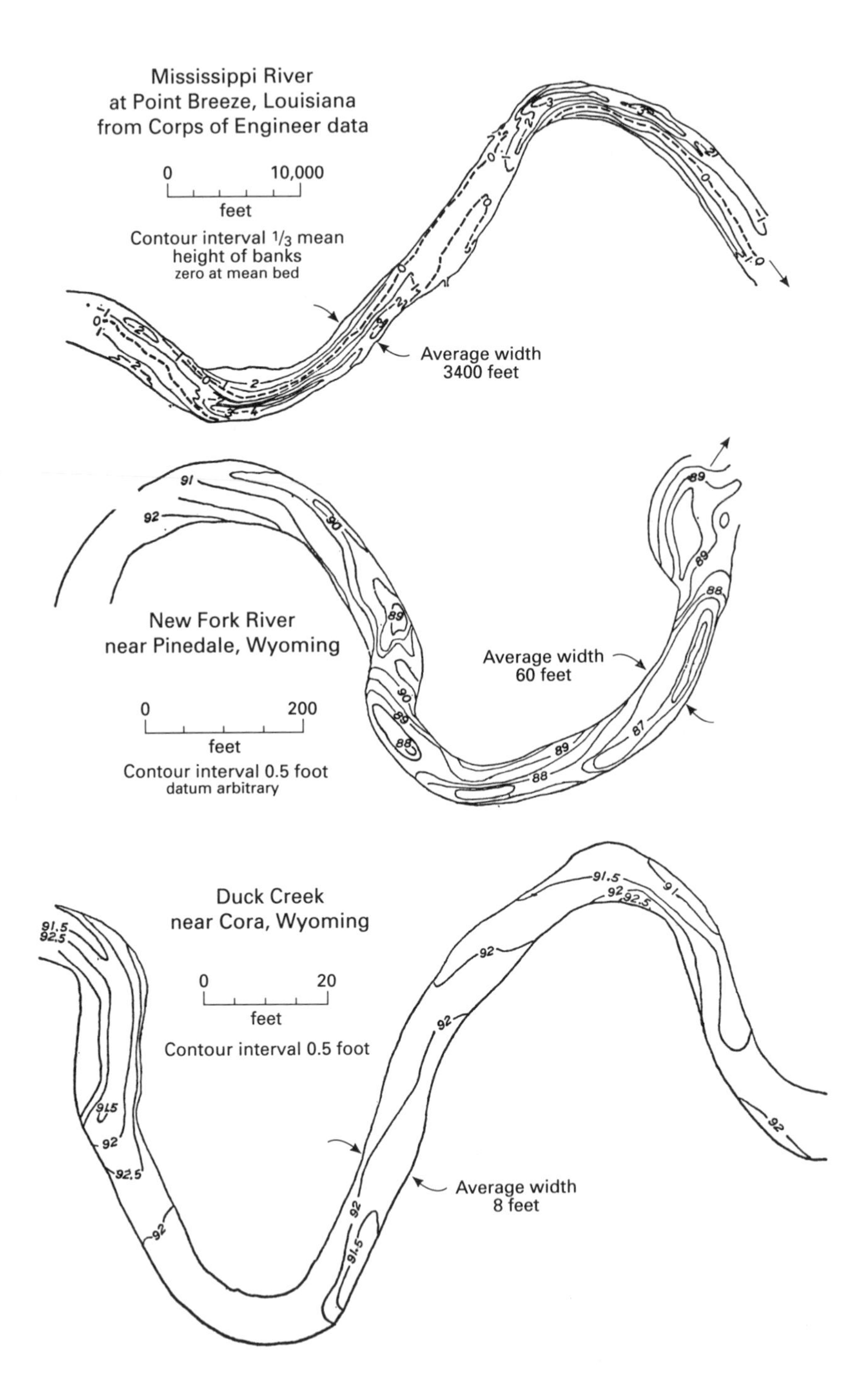


*Figure 4.3* Planimetric map of a meander bend on each of three rivers that vary greatly in size; they are scaled so that the meander lengths are equal on the printed page. (After Leopold 1962.)

*Figure 4.4* A meandering channel melted into the ice of the Aletsch Glacier, Switzerland.

the size of a river seen in an aerial photo cannot be judged. This is demonstrated in Figure 4.3, where three rivers of different size are drawn so that the wavelength is the same on the page.

In Figure 4.2, the width as a function of wavelength includes data from very small to very large rivers. The smallest shown is a river I measured on a supraglacial channel of Dinwoody Glacier, Wind River Mountains, Wyoming. The beautifully developed meanders of the channels on glacier ice, shown in Figure 4.4, I photographed on the Aletsch Glacier, Switzerland. The measurements of the Gulf Stream meanders were made from maps by F. C. Fuglister and by H. Stommel. Modern instrumentation has permitted even more detailed patterns including the cut-off rings, an example of which is shown in Figure 4.5, adapted from maps of Allan R. Robinson and colleagues.

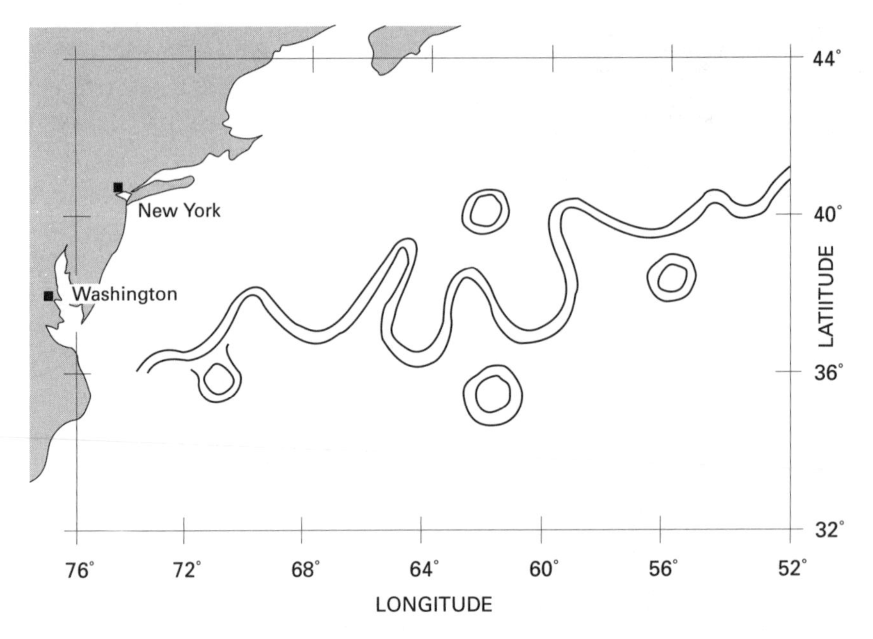


*Figure 4.5* Meanders of the Gulf Stream current off the east coast of the United States. Some bends have pulled away and formed circular rings. (After Robinson et al. 1989.)

The patterns visible on glacial ice and in the Gulf Stream show that the existence of meanders does not depend on sediment load or even on fixed boundaries. The phenomenon is purely hydrodynamic. The meander channels in the Gulf Stream flow are expressions of variations in temperature that reach to depths of at least 1,000 meters (3,000 feet), accompanied by variations in flow velocity. Velocities of flow vary from zero to 0.60 meters per second (2 ft/sec), as shown by Robinson and associates (1989).

I observed what appears to be a comparable phenomenon where the flow of the Gunnison River enters the still water of the Morrow Point Reservoir in the Black Canyon of Gunnison National Monument, Colorado. The entering water assumes a meandering path clearly seen from the cliffs above the reservoir (Figure 4.6). It is probable that the releases from the Blue Mesa Reservoir upstream came from the cool depths of the lake and entered the warm surface water of the lake downstream. This is another example of a meandering streamline unconfined by any fixed boundary.

The interrelation of width, radius, and wavelength is only the first of several shape characteristics that show a tendency toward energy conser-

*Figure 4.6* The meandering stream line of cool river water entering the warm water at the upstream end of the narrow Point Reservoir, Black Canyon, Gunnison, Colorado. The meandering path is unconfined by any channel banks.

vation. The relation of channel width to radius of curvature has its counterpart in the hydraulics of pipes. The head or pressure difference between two locations along a pipe system depends on the friction or resistance offered by that length of pipe. The resistance increases considerably if the pipe is not straight; if a bend of 90 degrees is introduced, the resistance is increased still more. It has been demonstrated that for a bend in a pipe the resistance or head loss depends on the ratio of bend radius to pipe diameter. The resistance is least if this ratio is 2.3. Clearly, the mean ratio of radius to width in a meander is the configuration that provides the least energy loss from friction. Recalling that energy is expended in bed friction and bank erosion, and that erosion is closely dependent on radius of bend, the erosion process reaches an equilibrium with channel shape when erosion is minimum and thus energy loss also is minimum.

## Effect on Flow Resistance

The increased resistance to flow caused by elbows in a pipe system has been well studied. An energy gradient is required to force water to flow around a bend, and the amount of energy expended is measured by the required loss of head. For the same reason, river curves cause energy loss that is reflected in the water surface slope. Another type of resistance to flow is the rugosity (roughness) of the channel bed, often approximated by the size of the relatively coarse bed material, $D_{84}$. (This term means that 84 percent of the rocks on the bed are smaller.) Other forms are the roughness of the channel banks and the roughness caused by bars, riffles, and other undulations of the riverbed. Hydraulicians have been unsuccessful in differentiating the effects of these various sources of flow resistance. At least river curves are sufficiently uniform and repetitive that experiments are possible, keeping other forms of resistance constant but varying the curvature of the channel. We carried out such experiments in a laboratory flume. The bed material, bank character, and cross-sectional shape and size were kept constant, but curves of different amplitude were introduced. Figure 4.7 shows one curved channel used. Our experiments proved that the frictional loss due to channel curvature is much larger than previously supposed. For purposes of the present discussion it can be said that curves in a channel introduce at least as large a resistance to flow as all other forms of roughness. This statement is in keeping with the fact that the average relation of radius of curvature to width is the one that minimizes the resistance.

## Shape of the Meander Curve

As the consistency of the ratio of the radius of curvature to width became evident from many examples, we sought a theoretical concept that would illuminate the meander geometry. Knowing that the sharpness of the

*Figure 4.7* One of the sinuous channels constructed in the laboratory to study the effect of channel curves on flow resistance.

curve is related to the flow resistance, we studied the angles of deviation from the downstream direction. The deviation angle measured at different distances along the meander channel varied as a sine function of the distance. We called it a sine-generated curve and found that it has the equation:

$$\theta = \omega \sin \frac{S}{M} 2\pi$$

where $\theta$ is the angle between the direction measured at a given point along the curve and the mean downstream direction, $\omega$ is the maximum

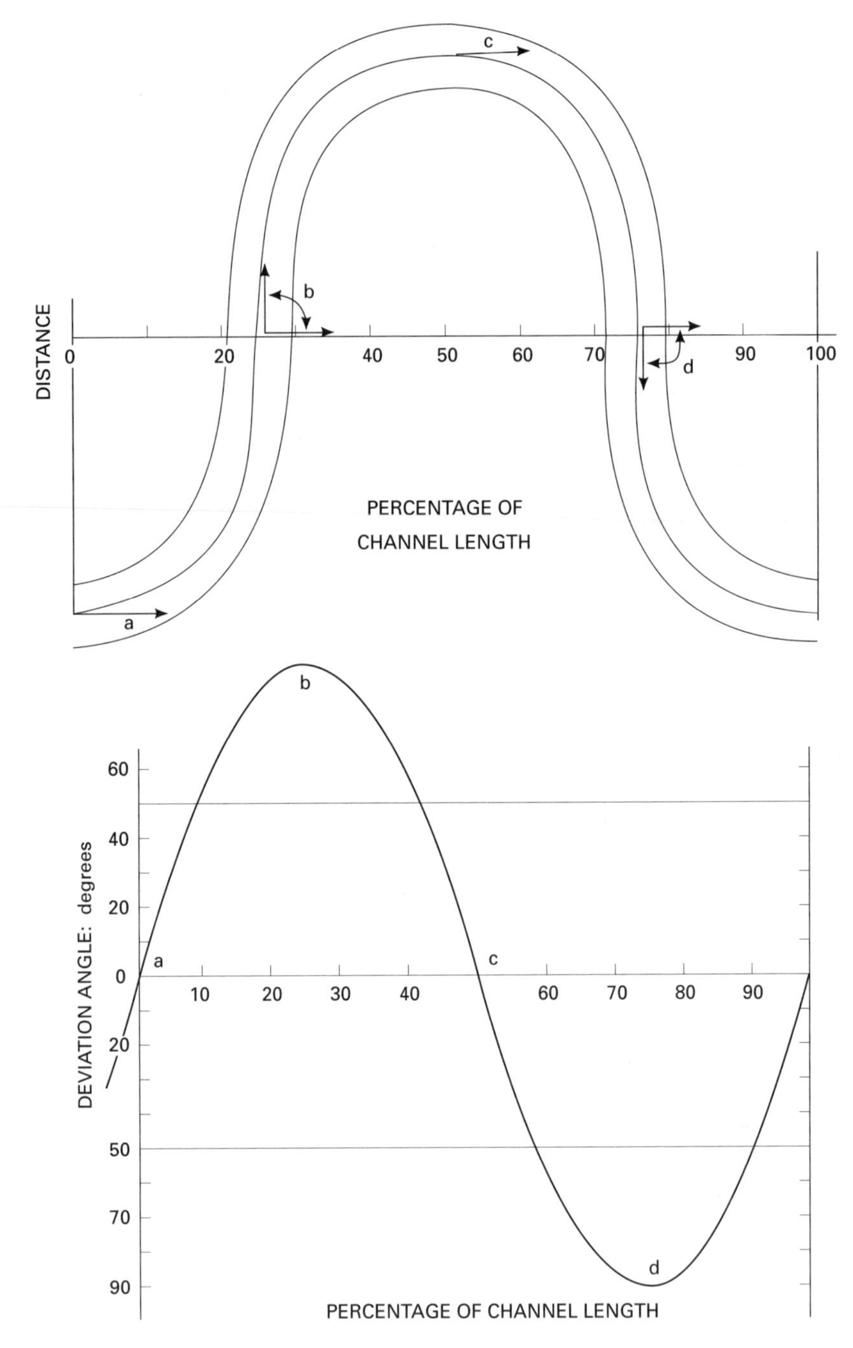


*Figure 4.8* A curve typical of a meandering channel, drawn so that the maximum deviation angle from the downvalley direction is 90 degrees *(upper diagram)*. Below is the plot of the angle of deviation as a function of distance along the curved path. At position *a* on this sine curve the deviation angle is zero, for the flow direction is downvalley. At position *b*, the direction is perpendicular to the downvalley direction and the deviation angle is 90 degrees.

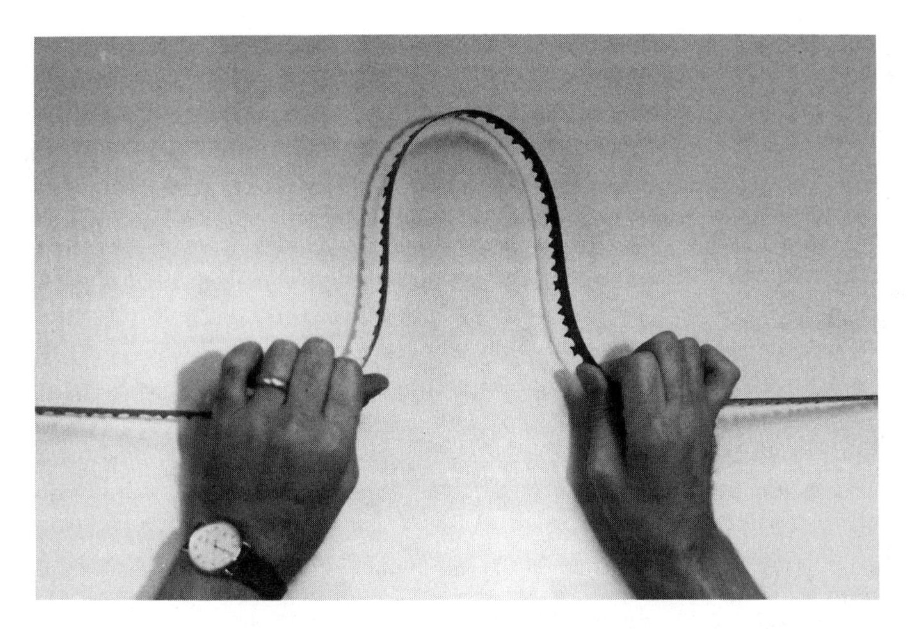

*Figure 4.9* A steel spring forced into a bend takes the form of a sine-generated curve that distributes stress uniformly along the curve from zero stress in the unbent section to maximum stress in the apex of the bend.

angle of deviation or the maximum value of $\theta$, $S$ is the distance along the path, and $M$ is the total path distance in a unit wavelength (repetition distance).

A typical meander curve is shown in the upper diagram of Figure 4.8. The lower graph is a plot of the deviation angles measured on the meander curve. Note that the beginning of the meander curve is chosen as the apex of a bend, designated *a*, where the local channel direction is downvalley and its position on the lower curve has a deviation angle of zero.

At a distance along the channel equal to one-quarter wavelength, at *b*, the deviation angle is to the left and has a maximum value. At the top or apex of the meander bend, at *c*, one-half wavelength, the deviation angle is again zero. Downstream from the halfway point, at *d*, the deviation is to the right with a maximum value at the two-thirds point.

The lower graph of Figure 4.8, derived from the upper graph, is a sine curve. Thus the deviation angle is a sine function of the distance along the meander.

The sine-generated curve has an interesting property. The shape represents the most uniform distribution of change along the curve. This feature is most easily demonstrated by the bending of a blade of spring steel, as shown in Figure 4.9. Where the hands hold the blade, there is

*Figure 4.10* A train carrying steel railroad tracks 700 feet long crashed near Greenville, South Carolina. The wreck contorted the rails into a series of sine-generated curves. (Photo by United Press International.)

no bending so there is no stress of tension or compression. The stress increases to a maximum at the apex of the curve, where the outside of the blade is in maximum tension and the inside in maximum compression. Stress increases uniformly from zero to a maximum. The curve assumed by the blade is a sine-generated curve.

A spectacular example of sine-generated curves is provided by the configuration of steel railroad tracks that were contorted in a railroad accident near Greenville, South Carolina (Figure 4.10). The rails were continuous lengths of steel measuring 700 feet, forced by the collision into a foreshortened form. The response was a sine-generated shape that distributed the bending stress in a uniformly progressive manner.

The sine-generated curve is different from a half-circle, or a parabola, or a sine curve. In Figure 4.11 three curves are compared, drawn so that each has the same total length along the curve and the same wavelength. They are a sine curve, segments of a circle, and a sine-generated curve. The shape of the sine-generated curve is drawn so that the maximum

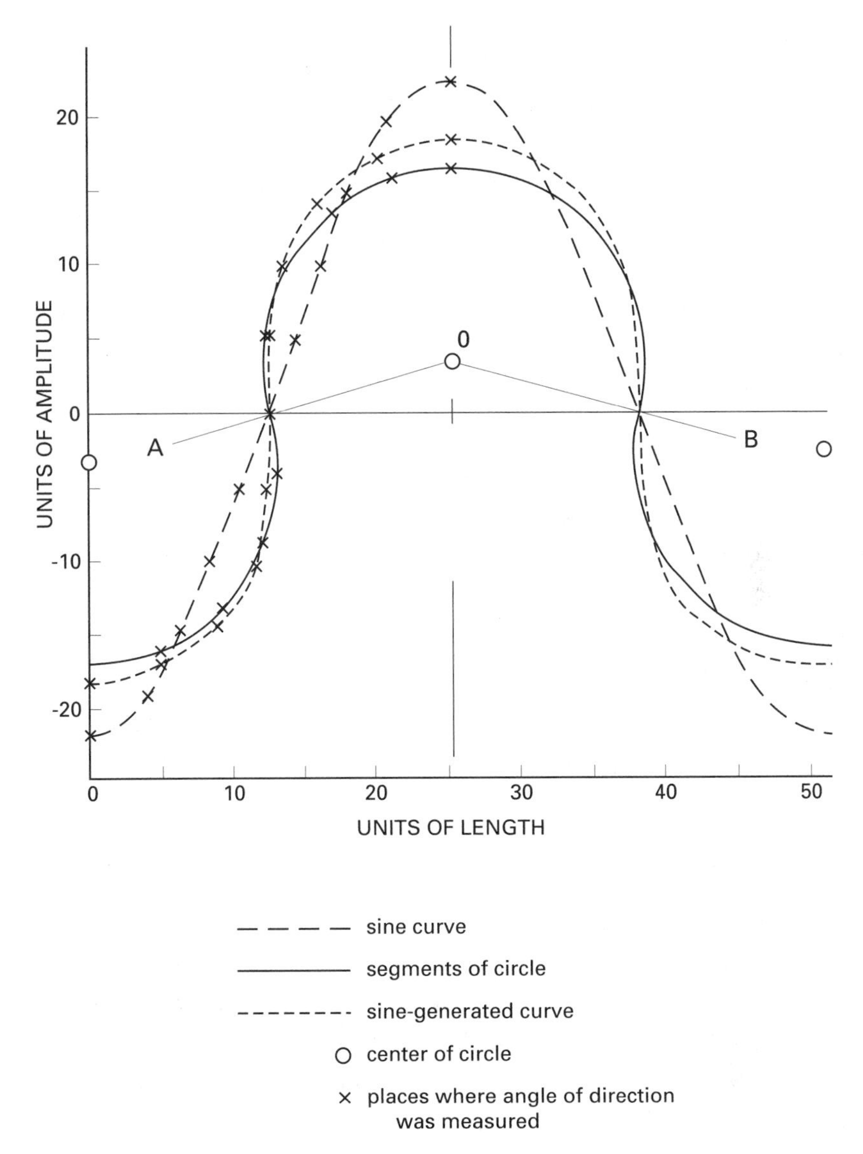


*Figure 4.11* Three curves of same wavelength (51 units) and path length (100 units): a sine curve, a circle, and sine-generated curve.

deflection angle is 90 degrees. All the curves have a curve length of 100 units and a wavelength of 51 units.

At equal distances along the path of each curve, the angles of deviation from the downvalley direction were measured by protractor; they are tabulated in Table 4.1. These deviations are then squared and the sums of squares are computed. The sum of the squares of deviations is called the variance. In this case the deviations are the angles from the downvalley direction at points along the curve.

The sum of the squares, or the variance, is slightly smaller for the sine-generated curve than for the sine curve or circular arc, as can be seen in Table 4.1. The relationship with the smallest squared variance is the most probable, and in this case also has a physical meaning. The work done by flow in erosion of the banks is dependent on the force required to turn the current in a new direction as the water flows around a bend. Therefore, minimization of the sum of the turning angles will tend to minimize the total work of erosion on the bank. Still, as in other aspects of the river form, this minimization is in opposition to the uniformity represented by equal angles of deflection, provided the curve were in the form of a circle. It seems that minimized variance (least work) is approached at the cost of less uniformity.

The difference in the sums of the squared deviations of the three types of curves is not large, but the trials that have been computed show a consistently smaller value for the sine-generated curve. Despite its size, this difference must be effective, for meander curves in nature are definitely not circular and closely fit the sine-generated form. As in all river data, there is considerable heterogeneity or scatter, and the fit with the sine-generated form holds only for uniform river curves. Chance differences in bank erodibility and bed form cause random differences in the sums. But a considerable scatter is both expected and observed, as in other applications of the concept of the most probable.

The sine-generated curve has two properties:

1. Compared with other curves of comparable path length and wavelength, variance from the mean downstream direction is minimum.
2. Its shape represents the average path of random walks of given length between two fixed points.

These properties imply that the most probable path is the average path of a random walk within the constraint of a given length of path between two points. The characteristics are quite in keeping with the concept that the most probable state is also the state of minimum variance. Thus, by

*Table 4.1* The deviation of several curves from the downvalley direction, and the squares of these deviations. Curves are of equal length and equal wavelength.

| Units of distance along curve | Deviation of path from downvalley direction (degrees) | | | Square of deviation | | |
|---|---|---|---|---|---|---|
| | Sine curve | Circle | Sine-generated curve | Sine curve | Circle | Sine-generated curve |
| 0 | 0 | 0 | 0 | 0 | 0 | 0 |
| 5 | 53 | 23 | 24 | 2,809 | 529 | 576 |
| 10 | 62 | 46 | 47 | 3,844 | 2,116 | 2,209 |
| 15 | 68 | 69 | 67 | 4,624 | 4,761 | 4,489 |
| 20 | 69 | 92 | 90 | 4,761 | 8,464 | 7,056 |
| 25 | 69 | 98 | 90 | 4,761 | 9,604 | 8,100 |
| 30 | 69 | 92 | 84 | 4,761 | 8,464 | 7,056 |
| 35 | 68 | 69 | 67 | 4,624 | 4,761 | 4,489 |
| 40 | 62 | 46 | 47 | 3,844 | 2,116 | 2,209 |
| 45 | 53 | 23 | 24 | 2,809 | 529 | 576 |
| 50 | 0 | 0 | 0 | 0 | 0 | 0 |
| | | | Sum: | 36,837 | 41,344 | 36,760 |

deposition and erosion, the river assumes a pattern that is both the most probable and the one having the smallest sum of the squares of deviations.

## The Pool-Riffle Sequence

In nearly any discussion of channel shape and pattern, reference to the alternating deeps and shallows is inescapable. In sand-bedded channels this alternation of deeps and shallows is absent, but resistance is offered by dunes and ripples. Even in a straight reach, bars tend to form on alternate sides of the channel and the thalweg (the deepest part of a channel) takes a sinuous course skirting these bars. Figure 1.3 showed point bars occurring on alternate sides of a channel. Often the deepest location in a channel is off the end of a point bar. If in time the channel becomes more sinuous, the bank opposite the point bar becomes the apex of a bend and the points of inflection between successive apexes become the shallow zones or riffles.

The successive shallows and deeps that make up the pool-riffle sequence provide a necessary part of the total hydraulic resistance to flow. The three-dimensional form of channel curves and the concurrent undulation of the channel bed occur in many kinds of channels. Channel bends are mirrored in valleys, even when the valleys cut through mountain ranges. The shape characteristics of meanders are seen in ephemeral as well as perennial channels. For all these reasons, a plausible theory to explain the meander phenomena is of major importance to river science.

Consideration of the motion of particles in a mixture leads to the hypothesis that an alternation of groups and deficits does not result from curvature in the paths of movement. In other words, the alternation of deeps and shallows in rivers is independent of channel bends or meanders. Indeed, the sequence of deeps and shallows seems to dictate certain aspects of meanders, causing river curves to accommodate to the pool-riffle sequence.

The first salient fact is that the riffle bar is an accumulation of objects that may range in size from fine gravel to large cobbles. Pools and riffles do not occur in channels whose bed consists of sand.

The rocks or particles making up a riffle bar move at times during high flow, but the riffle bar itself moves little if at all. Particles that are carried away from the bar surface are generally replaced by others arriving from upstream. That units such as rocks move into and out of a concentration of units is the concept of a kinematic wave. In everyday experience, a concentration of cars stopped at a red light remains a group of cars, even though the individuals constituting it continually change as some depart and others arrive. Equally commonplace is the concentration of cars on a highway. The groups of automobiles are called platoons, and these are separated along the highway by distances essentially free of cars. Our driving experience confirms the general truth.

The kinematic wave in the highway is such a close analogue to the movement of particles on the gravel streambed that its operation sheds light on the river case. Units such as cars flowing along a path are not randomly distributed, but tend to bunch into groups because of interaction among the units. Even when the introduction of cars into the stream from many lateral entrances is random, any initial random distribution quickly assumes the character of platoons separated by gaps with few or no cars. The interaction among cars that leads to this grouping derives from our driving habits. One important lesson, taught in every driving school, is that the higher the speed, the longer should be the distance separating each car from the car ahead. We drivers instinctively adhere to this rule in the way we apply brake and accelerator.

The result of this behavior is that the forward speed of moving cars decreases as the distance between cars is smaller. When cars are literally bumper to bumper, the forward motion is zero. All cars are stopped. The interdependence of the spacing of cars and their speed is expressed as a mathematical relation widely used by highway planners and road designers.

We have simulated the grouping phenomenon by simple games in which the movement of chips on a game board is governed by a random process such as the throw of dice or the turn of a card. If the game rules specifying the ability to move forward depend on spacing or presence of chips ahead in the line, the game quickly becomes a series of groups or platoons separated by distances without any chips.

The platoons and the spaces between them resemble the pool and riffle sequence in that a riffle is a concentration of more rocks per unit of length than is a pool. It was our hypothesis that the pool-riffle sequence is a kinematic wave train resulting from the interaction of rocks moving in the downchannel flow. Whereas platoon generation on the highway resulted from driver reaction to concentration, rock particle grouping on bars was the result of interference that caused slower net speed when concentration of rock particles increased. Obstruction due to hiding of one rock behind another, or protection of one rock from the shearing action of the ambient water by close contact with another rock, suggests a logical mechanism causing interaction. But advancement of the theory did not appear persuasive without field testing and data collection.

## Interaction of Rocks

Rocks in a gravel bar do not know any rules of driving behavior, but they are sensitive to the tendency of small rocks to hide behind larger ones or nestle in pockets between them. The relation of larger to smaller rocks has been treated in detail in the literature on sediment movement. In some transport equations there is a specific graph that describes a "hiding factor."

It is well known that in a heterogenous gravel, one containing a variety of particle sizes, transport does not occur in the manner experienced in unigranular material on the bed of a flume. In flumes it was shown by the experiments of Shields that initial motion of a particle of given size required a particular shear stress or force of the moving water; the larger the particle, the larger the force required.

Were this the process on a gravel streambed, one would expect the smallest particles to be carried away at lower flows, then successively larger particles moved with increasing discharge, until the winnowing left only the largest grains to be moved by the highest flows. This sequence is not observed. Because of the tendency for hiding and interaction, winnowing occurs primarily where there is little or no incoming sediment, as in the case of clear water released from a dam.

When motion of particles on a gravel bar begins, it appears that individual particles are dislodged and moved—and they may be of different sizes. When one or more particles of a given size move, many others of the same size do not move. At sufficiently high discharge or stress, all particles at the surface, regardless of size, seem to be mobile.

In view of the effect of interaction in other kinematic examples, John P. Miller and I devised an experiment aimed at obtaining quantitative information on the effects of interaction of rocks. We reasoned that the closer rocks are to one another, the larger would be the discharge required to move them. Rocks closely spaced would tend to require more discharge and therefore would move less often. As a result, their downstream mean speed would be less than that of rocks spaced far from one another. This would be analogous to the close spacing of cars resulting in their low speed down the road.

The experiment was carried out in an ephemeral channel near Santa Fe, New Mexico. Though mostly sand, the channel had some cobbles from 20 millimeters to more than 100 millimeters in size. We collected many of these rounded rocks, classified them by size, and painted them a bright color. Each rock was also measured and weighed. The weight in grams was painted on each rock for identification. The rocks were arranged in groups, each including 24 rocks of 6 sizes, and placed in a 3-mile stretch of the dry streambed. In each group the rocks were spaced at a chosen distance, from 1 diameter apart to more than 8 diameters apart. Because of the identification number, we knew where each rock had been placed. About 700 rocks were situated on the channel bed and its tributaries at any one time.

This channel had flowing water only during summer thunderstorms, which occurred about three times a summer. After each storm we or our associates walked the 3-mile length of the channel, looking for rocks that had moved, recording distance moved, and noting other characteristics of the event. The upper half of Figure 4.12 shows a typical rock group on the streambed.

Our project extended over 7 years and there were about 20 storms sufficient to cause flow. Of course, in any storm many rocks did not move, either because the flow missed the part of the channel where they were placed or because the flow depth was insufficient to cause movement. All rocks recovered after movement were returned to their original position. About 80 percent of all rocks moved were found in the channel reach searched. The lower half of Figure 4.12 shows a rock group after storm flow. Including all rocks inspected after storms, moved or unmoved, about 14,000 rocks were observed. Nonmovement was of course a significant piece of information.

The results are summarized in Figure 4.13. The graph is constructed to show what discharge would move all rocks of a given size, spaced a certain distance apart. If we read horizontally across the graph, for a given discharge value an increasing size of rock is moved as the spacing between rocks increases. If we read on any vertical line, for the same size rock to move requires increased discharge as the rocks are spaced closer together. The line marked 8 diameters spacing is the largest value needed. Rocks spaced farther apart than 8 diameters were uninfluenced by other rocks.

These findings confirm the hypothesis that closely spaced rocks are moved less frequently than rocks far apart, thus travel downstream at a relatively slower speed. We reason that because of the proven interaction,

*Figure 4.12* *Above,* 24 painted rocks of 6 sizes that were placed in a streambed near Santa Fe as part of an experiment to evaluate the effect of rock spacing on propensity to be moved by the flow. *Below,* one of the rock groups after having been subjected to storm flow.

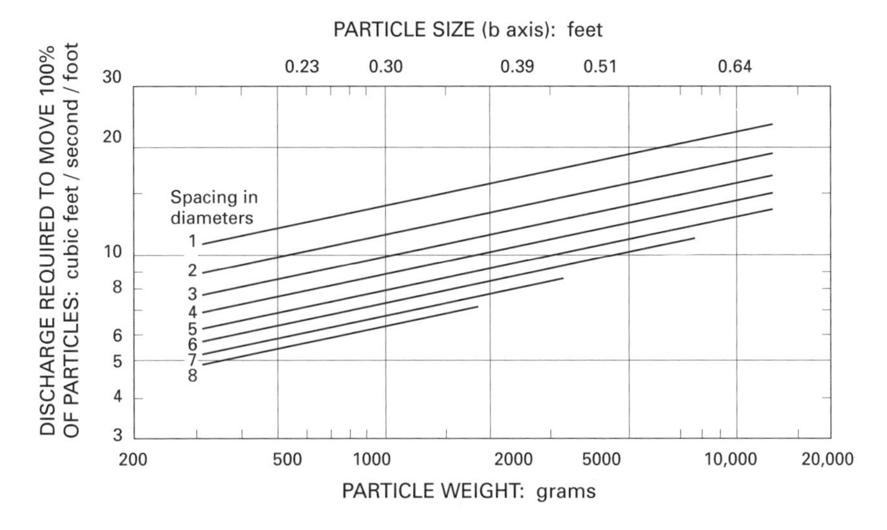


*Figure 4.13* The empirical relationship of rock size, rock spacing, and discharge required to move the rocks in the Santa Fe experiment.

rocks introduced at random into a channel cannot be transported in a random arrangement but must bunch up in platoons, which form the riffle bars. Unexplained is the fact that the spacing of bars is directly related to channel width (5 to 7 channel widths for a large number of rivers of all sizes).

## Channel Curvature and Water-Surface Profile

The interaction of grains, clasts, and rocks proceeds regardless of other channel factors, and to the results of this process the channel must adjust. One of its adjustments is in the rate of energy expenditure as water flows over pool and bar. The pool tends to be deeper than water over the riffle bar and offers less resistance to flow at low discharges. Therefore, at low flow the gradient of the water surface is small over the pool, whereas over the riffle bar it cascades down at a steep slope. The gradient of the water surface is a direct measure of the rate of loss of potential energy due to the frictional resisting forces.

At higher discharges when the water is deeper, the slope over the riffle bar decreases or flattens to some extent, whereas the gradient over the

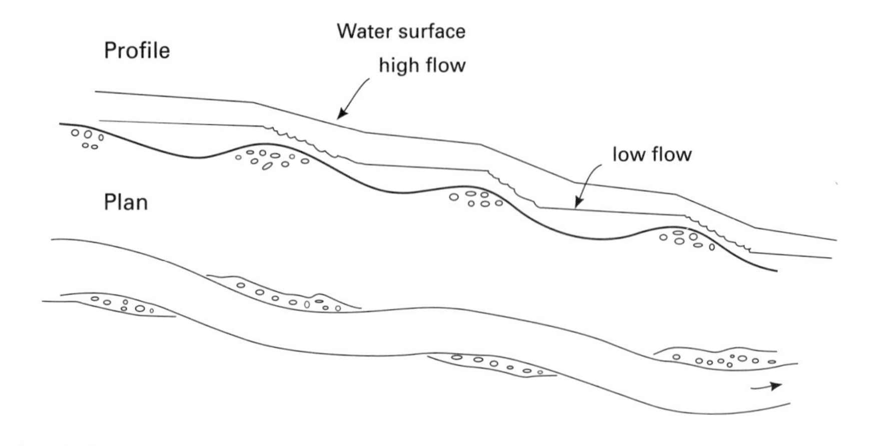


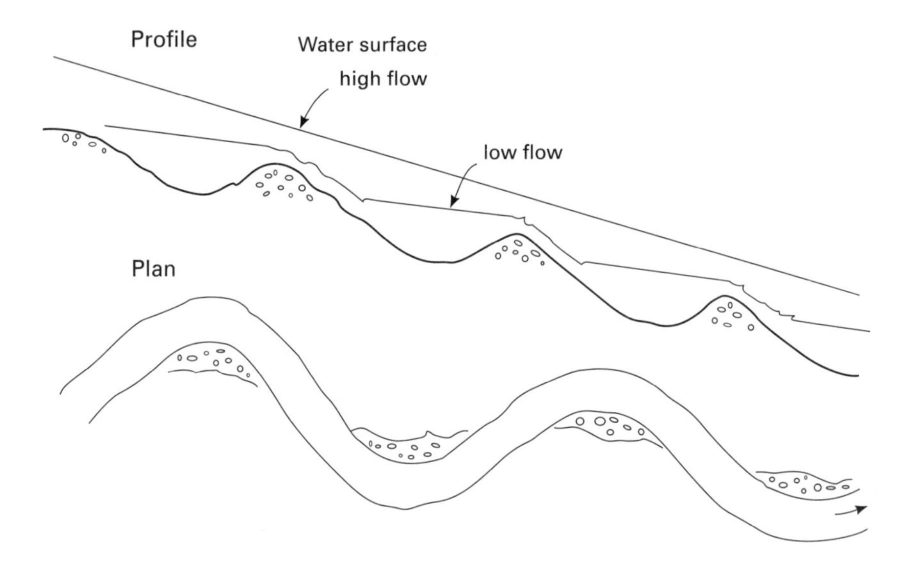


*Figure 4.14* Bed and water-surface profiles of a meandering and a straight reach of channel at low flow and at high flow.

pool increases. The result, as shown in Figure 4.14, is a somewhat subdued steepness over the riffle, which still retains some of its stepped character. At high flow over the meander, the water profile is straight without steps.

Physical systems in which energy is utilized tend toward minimum work, that is, minimum energy expenditure. This principle has already been noted in that the sine-generated shape minimizes the total variance of direction and thus is a shape of minimum erosion, or minimum work. The other example previously discussed is the nearly constant ratio of radius of curvature of a bend to channel width. This ratio provides minimum energy loss, as shown in pipes.

An opposing tendency is also present, the tendency for uniformity in the rate of energy expenditure as water flows along the channel. When uniformity of energy loss cannot be achieved, the tendency promotes the most uniform change of stress. We saw this characteristic in the shape of the bent steel spring (Figure 4.9) that took the form of a sine-generated curve. Stress changed uniformly in the steel from no stress in the straight or unbent portion to maximum bending stress at the apex of the curve.

The conditions of minimum work and uniform distribution of work are opposing and cannot simultaneously be fulfilled. These tendencies appear in many aspects of rivers and explain many features of river mechanics.

In the present context, the stepped profile of the pool and riffle sequence of a straight channel represents a variation in rate of energy expenditure along the channel length, from steep to flat to steep to flat. There is a tendency to smooth out these steps and straighten the energy grade line (the water-surface profile). One way this can be accomplished, as seen in Figure 4.14, is by steepening the gradient in the pool or deep section. Inserting curvature into the pool creates an additional form of resistance. We know that curvature in pipe or channel increases energy loss. The result of the curvature is to steepen the gradient over the deep pool and thus to attain a nearly straight profile at high flow. The straight water-surface profile at high flow is readily evident on the Popo Agie River, Wyoming (Figure 4.15).

As expected, to attain a more nearly straight profile by introducing curves or meanders, the whole profile steepens or more total energy is consumed. The river takes a form such that some departure from minimum work is the cost of a closer approximation to uniform work done along the channel. The comparison of the profiles of straight and curved reaches in Figure 4.14 shows the overall steeper profile of both bed and

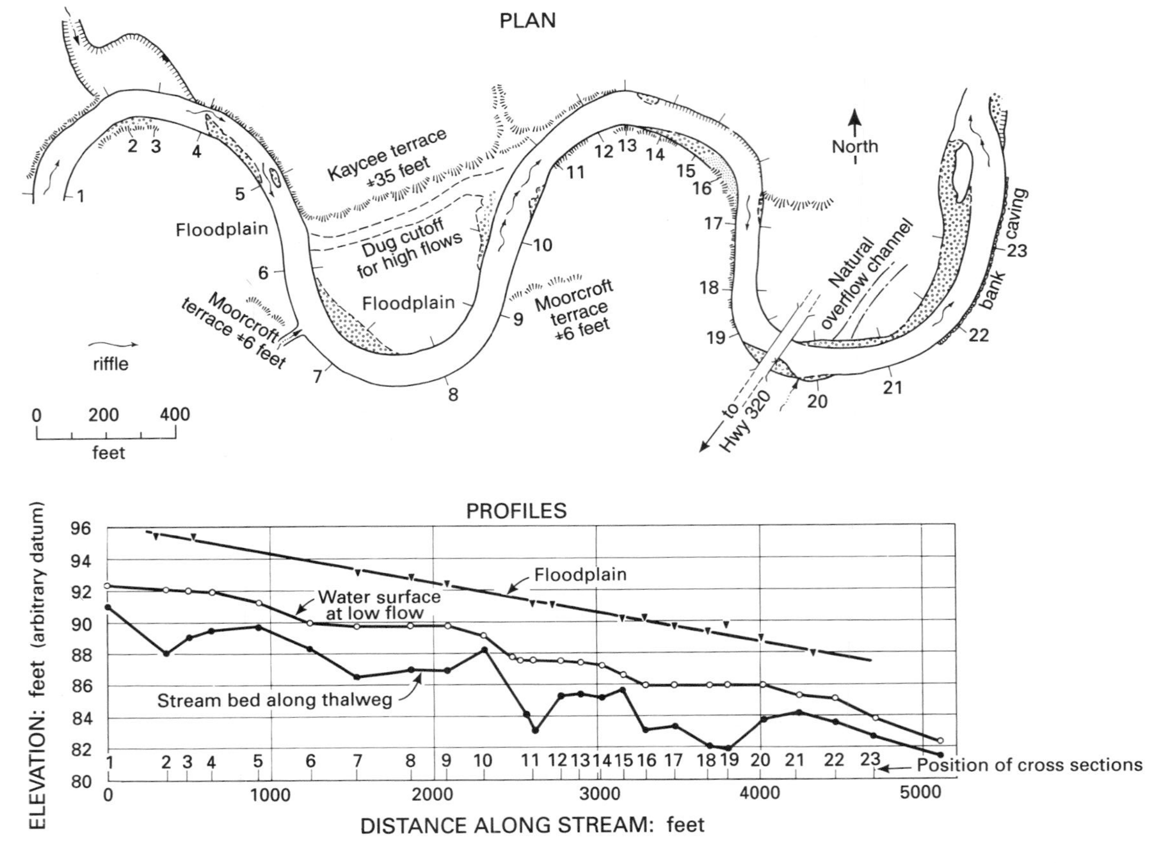


*Figure 4.15* Surveyed water-surface profiles of a meandering reach of the Popo Agie River near Hudson, Wyoming. At high flow the profile is straight, with no remaining evidence of shallows or deeps.

*Figure 4.16* Meanders of the East Fork, near Boulder, Wyoming. The bedload trap was on this river.

water surface in the meander reach than in the straight reach of the same river. This feature has been demonstrated in actual river measurements.

The ubiquity of the meander pattern is typified by Figure 4.16. The heterogeneous composition of bed particles leads to interaction among all sizes. As a result, the bed material does not lie uniformly distributed along a channel but bunches up in bars or riffles, which represent concentrations separated by pools or relative deficiencies in grains. These accumulations or channel bars tend to deflect the thread of flowing water toward a bank, with consequent erosion and initiation of a bend. Therefore, the shape, cross section, and flow within the channel take on the character of meanders. These characteristics are apparent in channels of all sizes (Figure 4.17).

The facts presented lead to a definite hypothesis of meander formation and development that may be called the *theory of minimum variance.* The processes of erosion and deposition are carried out by forces of shear closely associated with the distribution of velocity. If any local point is an area of unusually large shear, erosion occurs. A local lowering in shear is a place of reduced transport or deposition. The end result is an averaging of the anomalies and an approach to the most probable state, or the locus of minimum variance. When many adjustable factors are op-

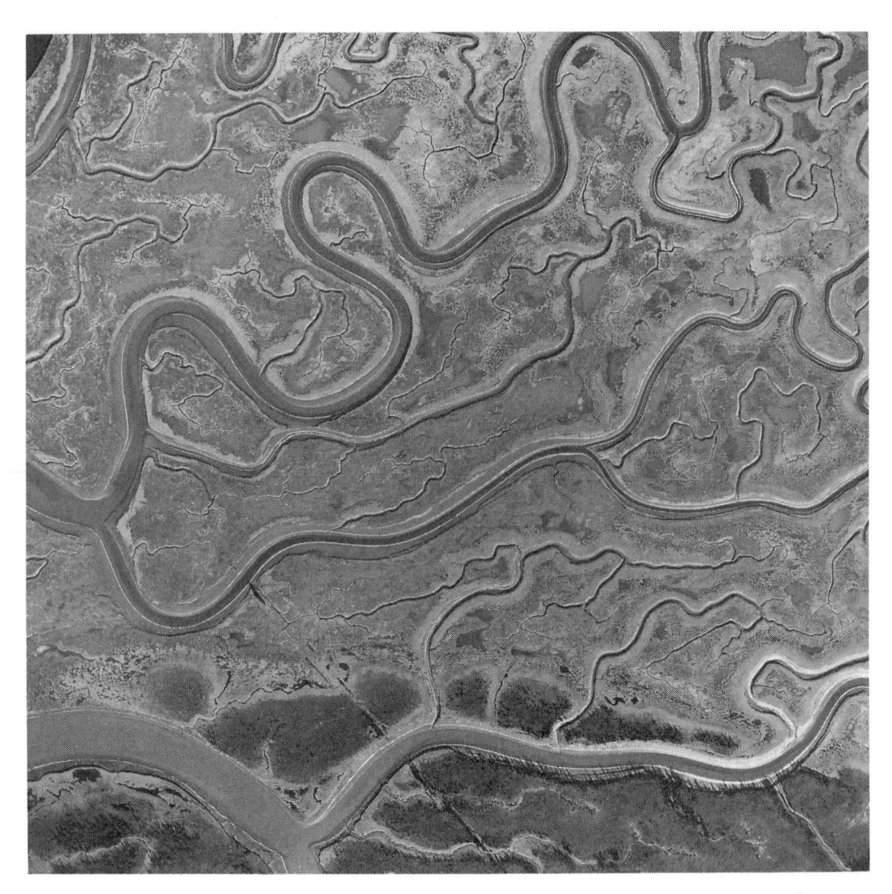

*Figure 4.17* The floodplain near the mouth of the Copper River as it enters the Gulf of Alaska in the vicinity of Cordova, Alaska.

erative, as in rivers, minimum variance is a term that includes the variance of several factors simultaneously. All factors participate, and no one factor takes a large fraction of the total adjustment.

Meanders represent a most probable configuration that is a compromise between minimum total work and uniform distribution of power expenditure.

CHAPTER FIVE

# Distribution of Discharge in Space and Time

## Relation of Average Flow to Drainage Area

The amount of water discharged over time past a point on a river depends on the contributing drainage area. The relation is not linear because precipitation during a given storm covers only a limited area. For any storm the average intensity decreases with increased area and so does the average storm rainfall. Even in great storms that cover vast areas, the inverse relation between intensity and area holds. This effect, especially marked in thunderstorm precipitation, is illustrated in Figure 5.1 for several storms I studied in New Mexico and Arizona. It is interesting that the slope of these area-depth curves for desert thunderstorms agrees with the curves for much more powerful storms in Texas that stemmed from different meteorological conditions. It might be expected, then, that the floods produced in rivers would follow a similar pattern; that is, that the discharge per square mile would decrease as the size of the drainage basin increases. Still, the addition of tributaries as the drainage area gets larger would make the flood discharges of large basins much greater than those of small basins. The general relation of discharge to drainage areas of different sizes warrants discussion.

A list of gaging stations representing the humid eastern United States was chosen in the drainage basin of the Susquehanna River in Pennsylvania, where the annual precipitation is about 40 inches per year. The mean annual river discharge and the maximum flow of record for the selected stations are plotted against drainage area in Figure 5.2.

The salient feature of the graph is that the annual flow is about 1 cfs

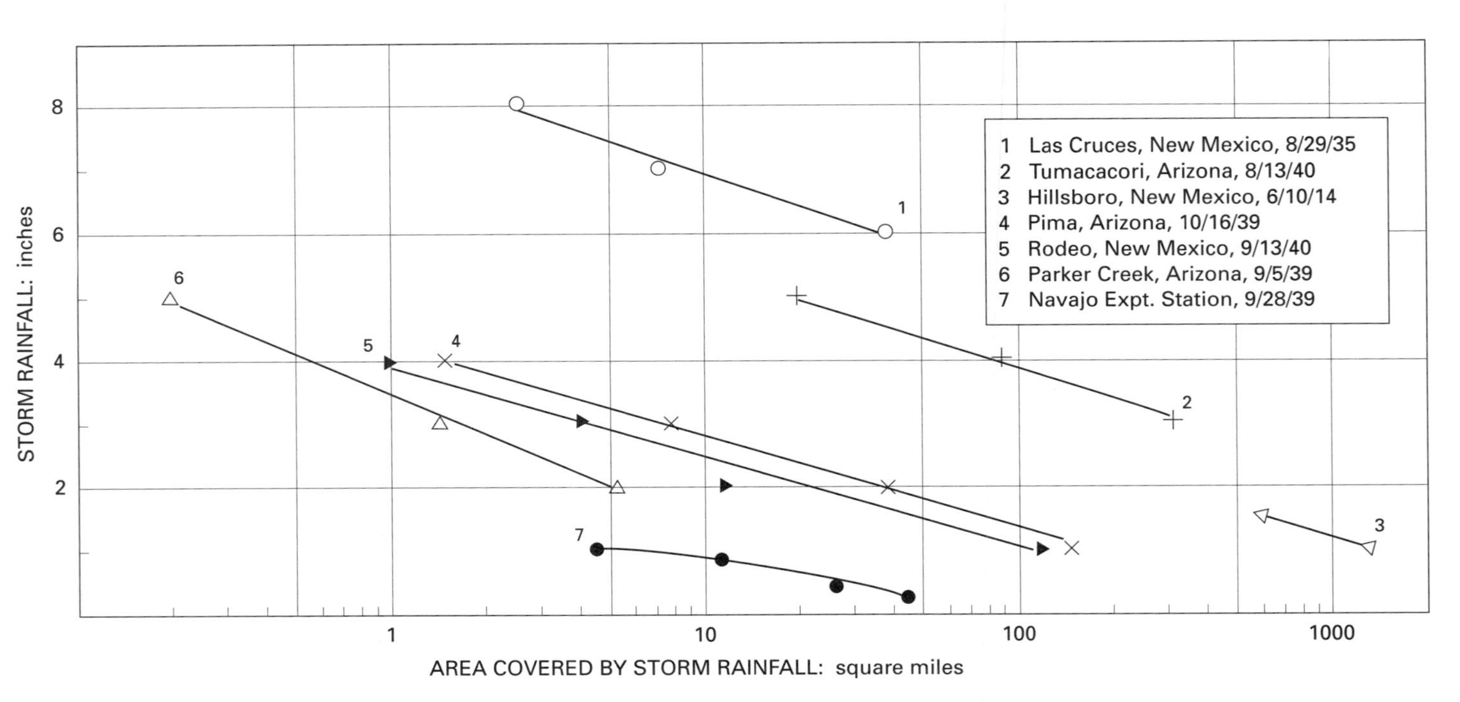


*Figure 5.1* Depth-area curves for local thunderstorms in New Mexico and Arizona. (After Leopold 1942.)

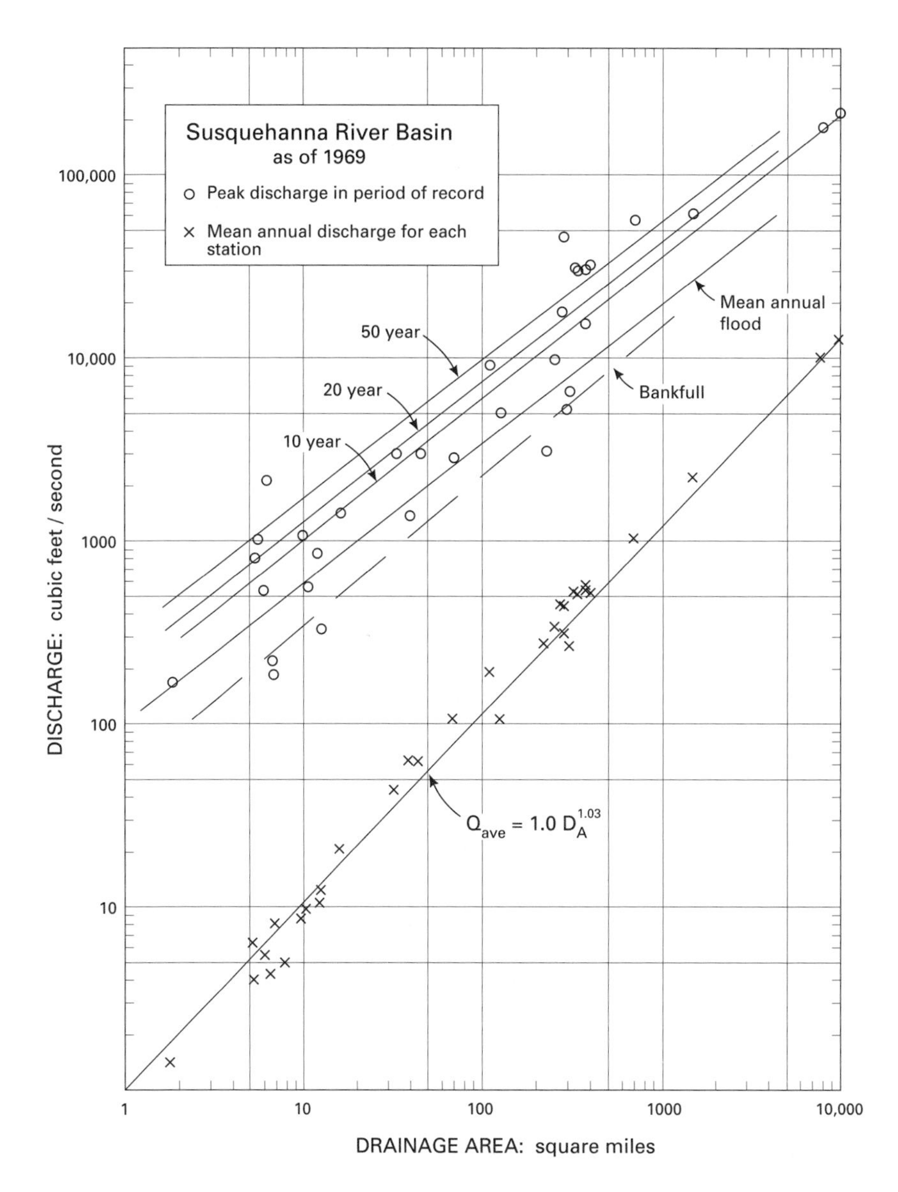


*Figure 5.2* Mean annual discharge and peak discharge of record, as functions of drainage area at 34 gaging stations, Susquehanna River Basin, Pennsylvania. Regional flood-frequency curves are also shown.

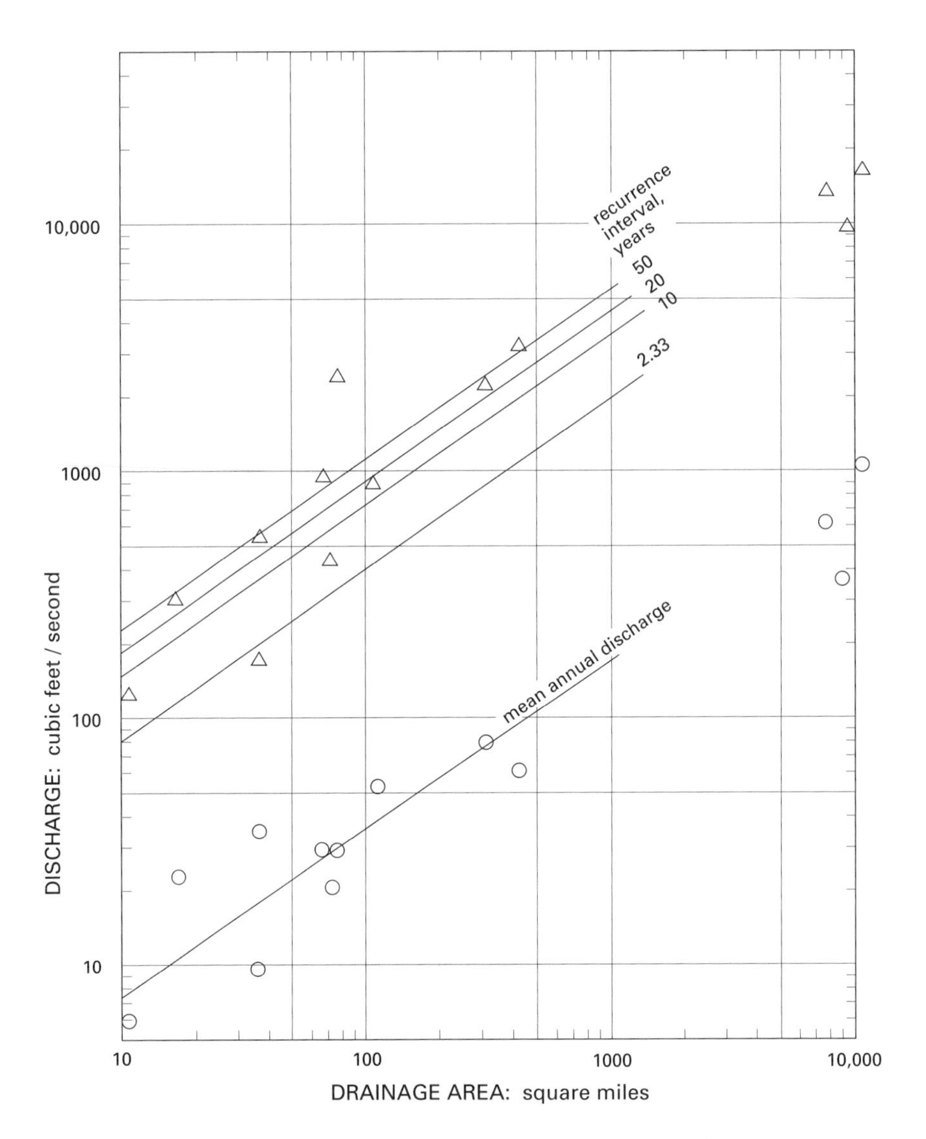


*Figure 5.3* Mean annual discharge as a function of drainage area for some semiarid locations, Rio Grande Basin, Colorado and New Mexico. For each station the highest discharge of record is also shown.

per square mile for drainages of all sizes. The line fitted by eye through the points has the equation

$$Q_{\text{average}} = 1.0\, D_A^{1.03}$$

where $Q_{\text{average}}$ is the mean annual discharge in cfs and $A$ is the drainage area in square miles. The exponent close to unity implies that mean discharge increases in direct proportion to drainage area. In contrast, flood or peak flows increase with drainage area only to the power 0.7 or 0.8. Peak flow increases less rapidly with area than does mean discharge.

The peak flow of record at different stations depends, of course, on the particular storm experience during the years of record. Therefore, the recurrence intervals of the peak flow data plotted in Figure 5.2 vary because stations with long records are more likely to have experienced great floods than stations with more abbreviated records. To express flood flow of different recurrence intervals, the regional flood-frequency curves for the Susquehanna Basin area are plotted against drainage area in Figure 5.2 for a few selected recurrence intervals. These are typical relationships for humid climates in that the mean annual flow is related to drainage area in nearly linear fashion. The same is true of some mountainous areas in semiarid regions as well, but the discharge per square mile is different from that in higher rainfall zones.

Figure 5.3 is a similar diagram representing gaging stations on and at the foot of the Sangre de Cristo Mountains of New Mexico and Colorado. The area is generally semiarid, but mountain masses occur, the crests of which may experience 25 to 30 inches of rainfall annually. The majority of the basins receive 14 to 20 inches annually. In the humid regions, heavy storms cover thousands of square miles, whereas in the semiarid areas widespread storms are of low intensity and generate much less total rainfall.

In the humid East the average flow is about 1 cfs per square mile, or 10 cfs for 10 square miles. In the semiarid areas exemplified by Figure 5.3, the average flow is about 8 cfs for 10 square miles, but variation among stations is large. The principal difference is in the slope of the graph relating mean annual flow to drainage area. For small basins in the mountain country of the semiarid zone, the discharge is slightly less than 1 cfs per square mile, but much less for large basins. At 1,000 square miles the annual flow is 170 cfs, in contrast to 1,000 cfs for that basin size in the East.

The difference is easily explained. The mountainous areas, though

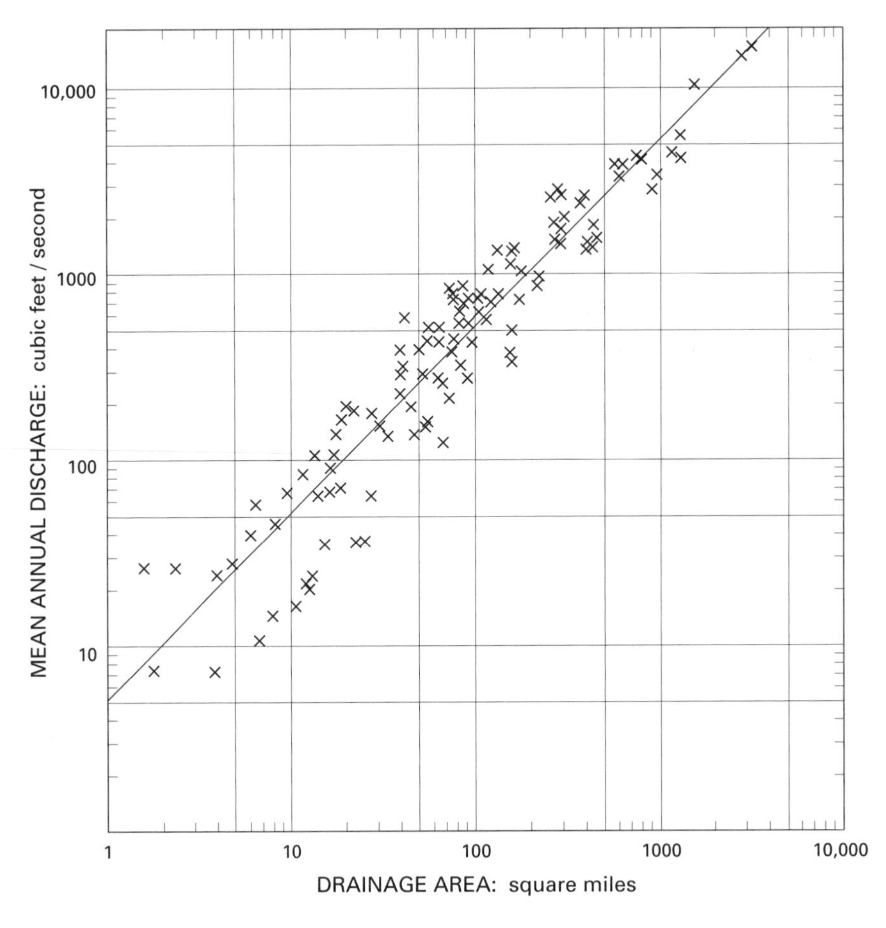


*Figure 5.4* Mean annual discharge as a function of drainage area for the Western Cascade Basins. The average is 5 cfs per square mile.

massive, are merely prominences or islands in a much larger area with low rainfall. Therefore, any basin of large size incorporates a large proportion of area with low water yield. In the high rainfall region of the northwestern United States exemplified by the Western Cascade Basins, the mean annual discharge is directly related to drainage area as in the eastern states, but at a value much higher (Figure 5.4). The annual discharge averages 5 cfs per square mile, rather than the 1 cfs per square mile in the East.

The topography and climate of northern and central California vary strongly even over short distances. The Coast Range concentrates precipi-

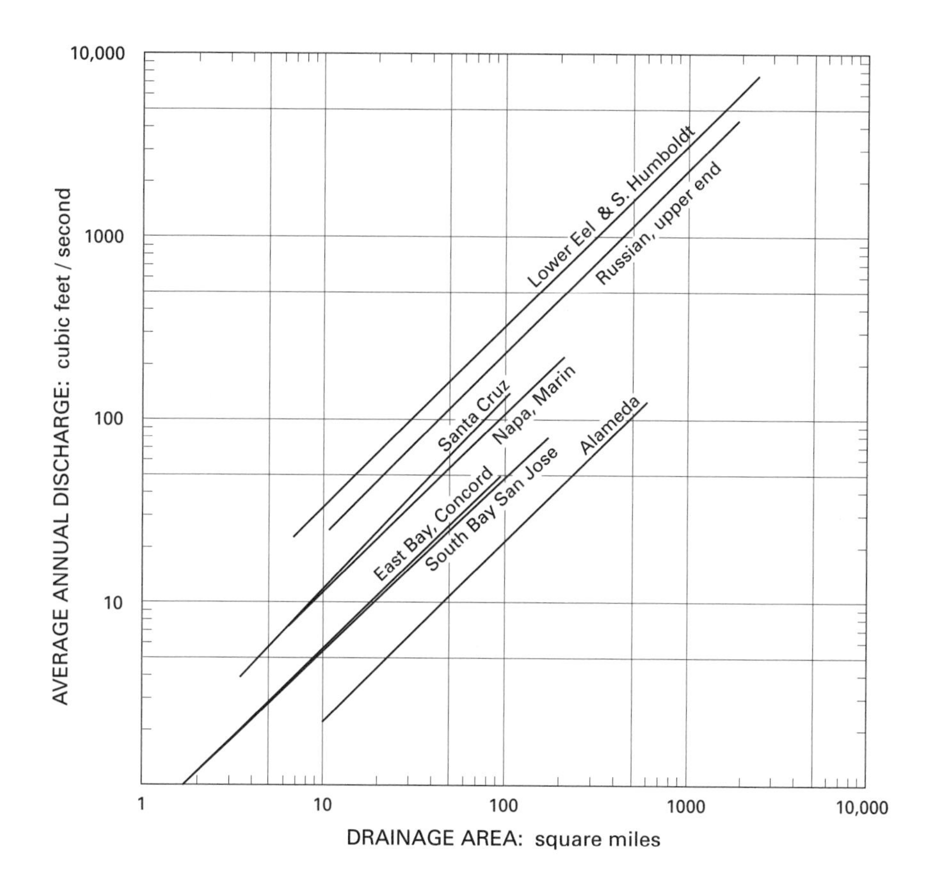


*Figure 5.5* Average annual discharge as a function of drainage area for various basins in central and northern California, showing the effect of differences in topography and latitude.

tation on the western border of the state and tends to shield inland valleys. As a result, the relation of drainage area to average discharge varies from the Eel-Humboldt region to the Alameda-San Jose area, which is both inland and south. This variation can be seen in the curves of Figure 5.5. For a 10-square-mile basin the average discharge varies from 30 cfs in the north to 2 cfs in the south. Thus, in the estimation of average discharge from basin areas, strict attention must be given to regional differences due to topography and orographic shielding.

In summary, mean annual discharge is nearly proportional to drainage area, but flood discharge per square mile decreases with an increase in basin area. The reason is that average storm precipitation is smaller over large areas than over small areas.

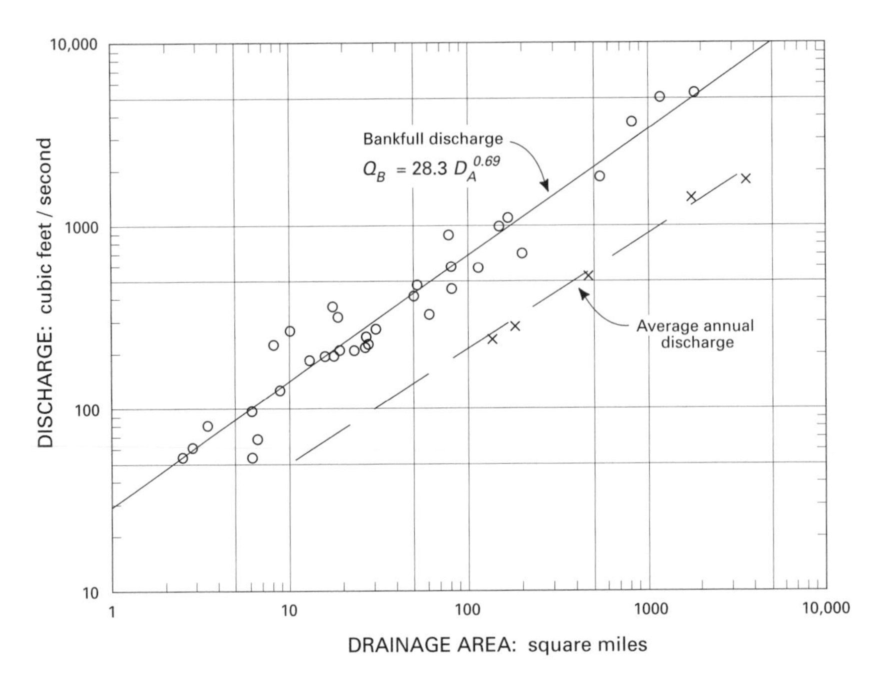


*Figure 5.6* Bankfull discharge as a function of drainage area for the upper Salmon River area, Idaho. (After Emmett 1975.) Also shown is the relation of mean annual discharge to drainage area.

## Bankfull Discharge

The level of the floodplain is the elevation of the top of channel banks. When the channel is flowing full, the water surface is at floodplain level and the flow rate is the bankfull discharge. This has morphologic as well as hydrologic significance, for the bankfull discharge is considered to be the channel-forming or effective discharge. It is an empirical fact that, for most streams, the bankfull discharge is the flow that has a recurrence interval of approximately 1.5 years in the annual flood series (defined in Chapter 7). For example, the floodplain elevation for Seneca Creek at Dawsonville, Maryland, has a discharge value of 2,000 cfs corresponding to the recurrence interval of 1.5 years. The gage height of bankfull condition is 6.6 feet.

At this location, where the drainage area is 101 square miles, the ratio of bankfull to average discharge is 20:1, for the average discharge is 1 cfs per square mile. This value is typical of the ratios for basins of moderate size. The ratio increases as the basin size decreases. For example, Watts Branch near Rockville in the same county, drainage area 3.7 square miles, has a bankfull discharge of 220 cfs and an average discharge of 1 cfs per square mile, so the ratio of bankfull to average discharge is about 60:1.

For a series of previously ungaged river locations in the Salmon River Basin in Idaho, William W. Emmett determined bankfull stage from field surveys. His relation of bankfull discharge to drainage area is presented in Figure 5.6. From gaging station data in the same area a line has been drawn to show the relation of average discharge to drainage area.

In hydrology, few relations among variables are perfectly correlated. Despite scatter in graphical plotting, it is both common and useful to draw average or smooth curves through scattered data and utilize them to compare regions or localities. Figure 5.7 shows an average relation developed from data that have scatter. But in this case, as in many others, the spread of data through four orders of magnitude gives credence to the average relation.

The relation of bankfull discharge to drainage area for various regions is shown in Figure 5.7. Bankfull discharge for a given drainage area is remarkably similar among basins in the Salmon River of Idaho, the upper Green River of Wyoming, and the Yukon River area of Alaska. There is a similarity in this relation of the San Francisco Bay region, with 30 inches of annual precipitation, and wetter areas such as the West Cascades and Puget Lowlands in Washington, and the Great Smoky Mountain area of North Carolina and Tennessee.

## Flow Duration

It is useful for many purposes to know the percentage of days in a year when given flows occur. In more general terms, the percentage of time during which specified discharges are equaled or exceeded may be compiled in the form of a flow duration curve. This is a cumulative frequency curve of flow quantities without regard to chronology of occurrence. Such a curve may express daily, weekly, or monthly average flows. Its most common form is the percentage of days in a year the mean daily flow is equaled or exceeded. The mean daily flow on a given day is the mean of the varying flow rates occurring during that day. This is the quantity published for each day at a gaging station in the volumes titled *Surface Water Supply of the United States*, compiled by the U.S. Geological Survey and now on CD-ROM. From these data the number of days having averaged-flow values in different categories of size are tallied.

The shape of the flow duration curve reflects the ability of the basin to store water temporarily in the ground and to release it later as a contri-

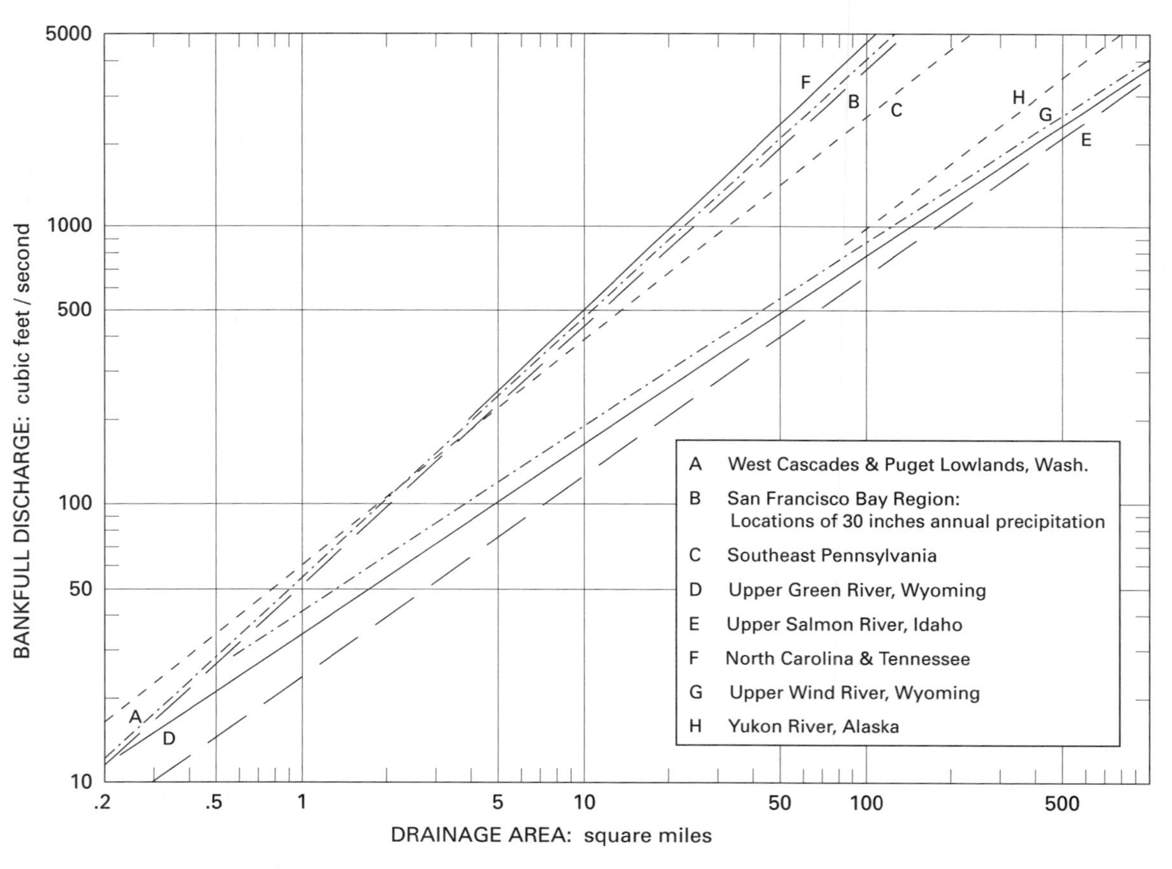


*Figure 5.7* Bankfull discharge as a function of drainage area for several regions in the United States.

bution from groundwater. In nonstorm periods the low flow may represent perched or near-surface storage and thus may not be an indication of the full amount of ground storage. Stream flow during nonstorm periods is stored groundwater that drains out into the channel.

If the amount of ground storage above channel bed level is small, as in impervious or poorly jointed rock, the amount of water stored is limited and the low-flow end of the duration curve will be steep. If, by contrast, large amounts of water are in storage, the low flow will be sustained over longer periods of time and the duration curve will be flat in its lower end. A steeply sloping duration curve is characteristic of a highly variable stream, the flow of which is primarily from direct storm runoff. The flat or low-sloped duration curve is typical of a stream draining a basin with high ground-storage capacity that sustains or equalizes the flow.

Examples of flow duration curves for various stations in Idaho are shown in Figure 5.8. All the curves except that for the Salmon River near Obsidian are generally parallel despite large differences in actual discharge values for any given percentage of time. The shapes of these curves show that at most stations the discharge does not differ much 60 percent of the time. For example, the discharge equal to or exceeded 40 percent of the days in a year is only moderately larger than that equal to or exceeded 99 percent of the time.

This similarity in shape allows the construction of a dimensionless duration curve (Figure 5.9), which represents the average duration characteristics of all the streams in a region. The ordinate of the dimensionless duration curve is not discharge in cfs but the ratio of discharge to bankfull discharge. Each ordinate value is the average value of this ratio for the gaging stations considered. The data define a single curve representing all stations. The ratio of unity stands for bankfull discharge.

On this duration curve it can be seen that in Idaho bankfull discharge is equaled or exceeded 4.5 percent of the time, or 16 days a year. In the Colorado Front Range bankfull is equaled or exceeded 1 percent of the time, or about 3.6 days a year.

The flow occurring on the average 0.1 percent of the time, or 0.37 day per year (1 day in about 3 years) is about double the bankfull discharge in both Idaho and Colorado.

Average annual discharge is, on the average, equaled or exceeded 25 percent of the time. In Idaho this discharge would be 0.23 bankfull and in the Colorado Front Range 0.13 bankfull discharge. The Idaho streams sustain low flow, never flowing at less than 0.05 bankfull, whereas the Colorado streams flow less than that amount 50 percent of the days in a year.

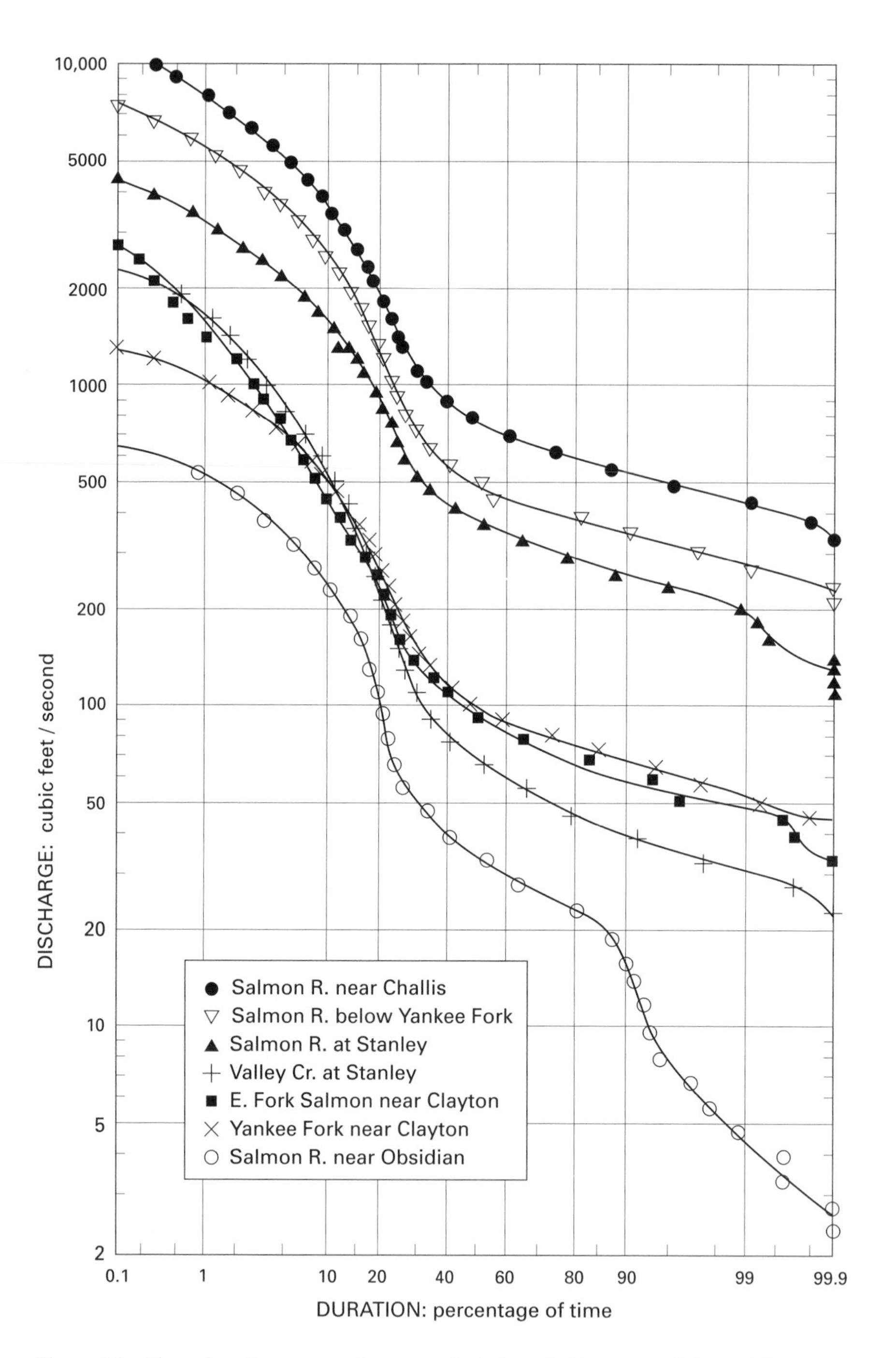


*Figure 5.8* Flow duration curves for several stations in the upper Salmon River, Idaho. (After Emmett 1975.)

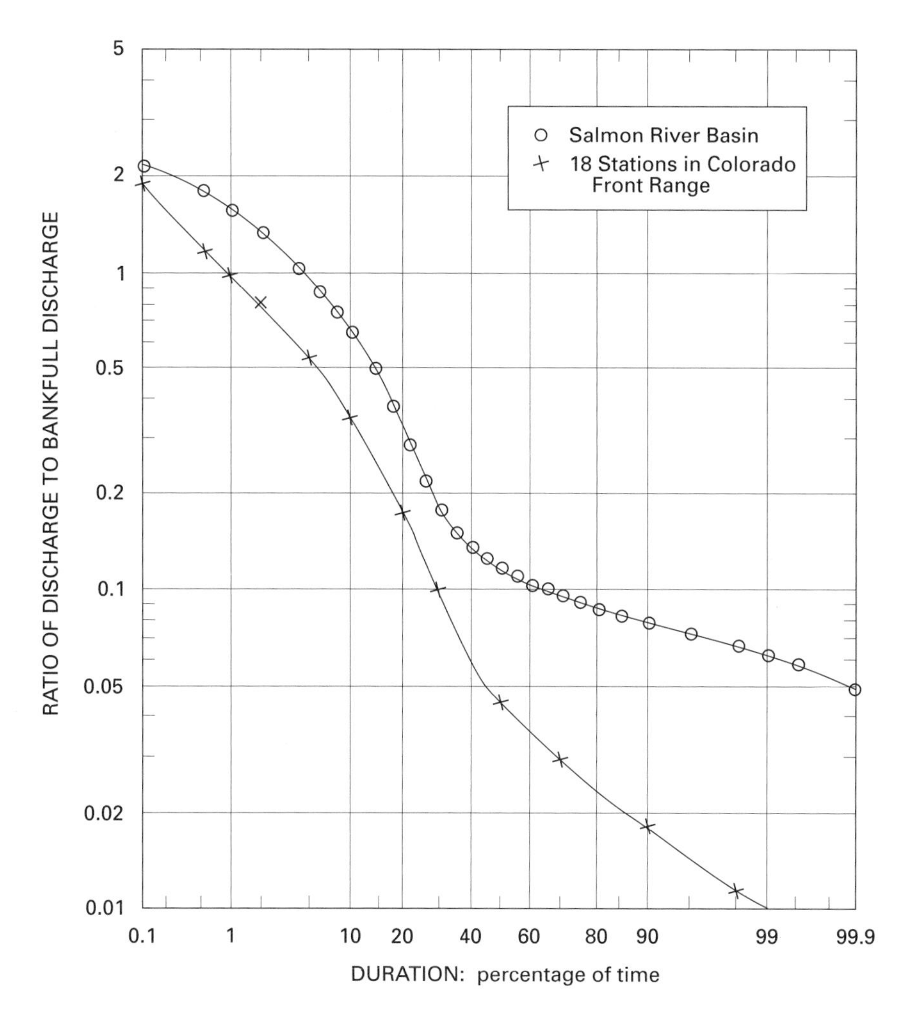


*Figure 5.9* Dimensionless flow duration curves for the upper Salmon River area, Idaho, and for 18 stations in the Front Range, Colorado.

## Flow Volume

For most engineering purposes the primary interest in river flow is how much water will probably pass some point of diversion or power plant, or how much would be stored in a reservoir at that point. For these purposes expression of the flow in acre-feet or some other volumetric unit is most appropriate. If the goal is to understand geographic patterns, a unit should be used that eliminates drainage area as a variable. Inches of runoff per year or cfs per square mile satisfy this requirement. An inch of runoff means the volume of water covering the drainage basin in question to a depth of 1 inch. The volume of 1 inch on 1 square mile is 53.3 acre-feet. The rate of discharge of 1 inch per hour of runoff from 1 acre is 1 cfs.

To show a distribution of annual runoff from unit area, inches of runoff is the unit most commonly used. The distribution of annual runoff in inches for the United States is published in the form of a map in the *National Atlas of the United States,* prepared by the U.S. Geological Survey in 1970. A paucity of runoff in the semiarid areas, punctuated with islands of high runoff in the mountains, dominates the western United States, except for the coastal zone receiving the winter storms of Pacific origin. The eastern half of the country has a much more uniform pattern, with a belt of high runoff extending from New England down the Appalachian Mountain chain.

Streamflow is what is left over after precipitation has supplied the demands of evaporation from vegetation, soil, and water bodies. Leftovers or differences tend to vary greatly with time; that is to say, the variability of the residues is greater than the variations in the original quantities. Suppose the rainfall in a certain year is 40 inches, which would be typical for a location such as Washington, D.C. Evaporation and transpiration might take 20 inches during the year, leaving 20 inches to be carried off by streamflow. Suppose that in the following year the precipitation is 30 inches, 25 percent less than in the previous year. If evaporation and transpiration are the same, which is quite possible, streamflow would be only 10 inches, or 50 percent of that which occurred in the year previous. Thus a 25 percent change in precipitation becomes a 50 percent change in runoff—a demonstration of the sensitivity of streamflow to changes in rainfall. Variability of runoff tends to be greater with more arid conditions. Climatic variability is a characteristic inherent in the subhumid zones of the world.

There is a secular or irregular variation in the mean values of most of the climatic parameters: precipitation, cloudiness, temperature, to name a few. The overall picture of temporal runoff variation in the United States in this century may be described as follows. There was a general decrease in streamflow during the first three decades. The period of lowest streamflow of record occurred in the decade 1930–1940. This trend was reversed during the next decade, but in parts of the West, 1977 was the driest year.

These observed trends in runoff have been similar in direction to records of other climatologic parameters. During the period of record through most of the Northern Hemisphere, a tendency toward warmer and drier conditions has existed, at least since about 1870. By about 1870 many glaciers had advanced farther downvalley than they had for some centuries previously.

Temperatures at long-term stations have also shown a trend toward increasing values. These changes toward drier and warmer conditions have been more marked and consistent at high latitudes than at low, and have been generally more marked in the Northern Hemisphere than in the Southern. The differences between hemispheres are, however, not clearly established because there are relatively few observation stations in the Southern Hemisphere.

It can be said unequivocally that the observed changes in streamflow are not sufficiently cyclic to allow any meaningful forecast of future streamflow by extrapolation of apparently repetitive behavior of flow.

CHAPTER SIX

# Rivers of the World

## Total Runoff from the Continents

As part of his work on the International Hydrologic Decade, Raymond L. Nace extended available lists that depicted the relation of drainage area to mean annual flow of rivers of the world. His tabulation, expressed in English units, is presented in Table 6.1. The rivers in this table are those that flow into the ocean; they are not tributary to other larger rivers.

If the discharge figures are cumulated from largest to smallest, the total runoff of the world's continents can be approximated, because the cumulative curve becomes nearly asymptotic to the quantity $24.7 \times 10^6$ cfs (700,000 cms).

## Size Distribution of Basin Characteristics

When a large array of physical values, such as this list of discharges and drainage basins in the world, is available, the size distribution is of interest because it may indicate hydrologic relations and possibly causative factors. To examine these possibilities, the average annual discharge values were arranged in order of magnitude producing an array of 211 items, the largest flow being that of the Amazon with a mean of 6.2 million cfs.

The size distribution gives a different insight into hydrologic relations than a correlation diagram does. For example, it was shown that for much of eastern United States the average annual discharge is directly porportional to the drainage area, with an average of 1 cfs per square mile of basin area. This follows from the fact that in that region average precipitation is of roughly the same order of magnitude, and topography is rather subdued.

*Table 6.1* Average flow of rivers of the world, in order of rank (Nace 1970)

| Rank order of discharge | River | Country | Drainage area (sq mi × $10^{-3}$) | Length of river (mi) | Average annual discharge (cfs × $10^{-3}$) | Rank order of length | Rank order of drainage area |
|---|---|---|---|---|---|---|---|
| 1 | Amazon | Brazil | 2,300 | 4,100 | 6,200 | 3 | 1 |
| 2 | Congo | Congo | 1,430 | 2,850 | 1,460 | 8 | 2 |
| 3 | Yangtze | China | 750 | 3,570 | 1,170 | 5 | 9 |
| 4 | Orinoco | Venezuela | 380 | 1,700 | 812 | 20 | 16 |
| 5 | La Plata | Uruguay | 1,200 | 2,420 | 779 | 14 | 4 |
| 6 | Brahmaputra | Bangladesh | 360 | 1,800 | 699 | 17 | 19 |
| 7 | Yenisei | USSR | 1,010 | 3,690 | 699 | 4 | 6 |
| 8 | Ganges | India | 410 | 1,560 | 657 | 24 | 13 |
| 9 | Mississippi | U.S. | 1,240 | 4,160 | 650 | 2 | 3 |
| 10 | Lena | USSR | 960 | 2,680 | 568 | 10 | 7 |
| 11 | Mekong | Indochina | 307 | 2,640 | 551 | 11 | 23 |
| 12 | Irrawaddy | Burma | 166 | 1,250 | 480 | 26 | — |
| 13 | Ob | USSR | 946 | 3,460 | 438 | 6 | 8 |
| 14 | Tocantins | Brazil | 350 | 1,640 | 388 | 21 | 20 |
| 15 | Amur | USSR | 716 | 2,700 | 385 | 9 | 10 |
| 16 | Mackenzie | Canada | 527 | 263 | 381 | 12 | 17 |
| 17 | Saint Lawrence | Canada | 396 | — | 348 | — | 15 |
| 18 | Columbia | U.S. | 258 | 1,200 | 281 | 28 | — |
| 19 | Magdelena | Columbia | 92.7 | 994 | 265 | — | — |
| 20 | Yukon | U.S. | 328 | 1,980 | 259 | 15 | 21 |
| 21 | Zambesi | Mozambique | 500 | 1,600 | 250 | 23 | 11 |
| 22 | Niger | Nigeria | 429 | 2,500 | 247 | 13 | 12 |
| 23 | Indus | Pakistan | 374 | 1,900 | 236 | 16 | 18 |
| 24 | Danube | Romania | 315 | 1,770 | 222 | 19 | 22 |
| 25 | Pechora | USSR | 124 | — | 141 | — | — |
| 26 | Kolyma | USSR | 257 | — | 138 | — | — |
| 27 | Songkoi (Red) | Vietnam | 46.3 | — | 138 | — | — |
| 28 | Sankai (Si) | China | 45.6 | 1,250 | 127 | 27 | — |
| 29 | Godawari | India | 115 | — | 127 | — | — |
| 30 | Fraser | Canada | 84.8 | 851 | 125 | — | — |
| 31 | North Dvina | USSR | 138 | 800 | 123 | — | — |
| 32 | São Francisco | Brazil | 261 | 1,800 | 99.2 | 18 | — |
| 33 | Kujjuaq | Canada | 51.5 | — | 90.0 | — | — |
| 34 | Pyasina | USSR | 74.1 | — | 89.7 | — | — |
| 35 | Neva | USSR | 109 | — | 89.3 | — | — |
| 36 | Nile | Egypt | 1,150 | 4,200 | 87.9 | 1 | 5 |
| 37 | Nelson | Canada | 279 | 1,600 | 83.7 | 22 | — |
| 38 | Rhine | Netherlands | 56.0 | 820 | 78.4 | — | — |
| 39 | Krishora | India | 117 | — | 68.8 | — | — |
| 40 | Kuskokwim | U.S. | 47.9 | — | 68.8 | — | — |

*Table 6.1* (continued)

| Rank order of discharge | River | Country | Drainage area (sq mi × $10^{-3}$) | Length of river (mi) | Average annual discharge (cfs × $10^{-3}$) | Rank order of length | Rank order of drainage area |
|---|---|---|---|---|---|---|---|
| 41 | Saguenay | Canada | 34.8 | 470 | 64.2 | — | — |
| 42 | Indigirka | USSR | 139 | — | 63.9 | — | — |
| 43 | Mobile | U.S. | 44.4 | 400 | 63.1 | — | — |
| 44 | Hwang-Ho (Yellow) | China | 260 | 3,000 | 62.5 | 7 | — |
| 45 | Copper | U.S. | 24.4 | — | 61.1 | — | — |
| 46 | Skeena | Canada | 21.2 | — | 61.1 | — | — |
| 47 | Dnepr | USSR | 195 | 1,420 | 60.0 | 25 | — |
| 48 | Stikine | U.S. | 19.0 | — | 60.0 | — | — |
| 49 | LaGrande | Canada | 37.4 | — | 59.7 | — | — |
| 50 | Anadyr | USSR | 77.2 | — | 59.3 | — | — |
| 51 | Rhone | France | 37.8 | 500 | 58.9 | — | — |
| 52 | Usumcinta | Guatemala | 15.4 | — | 58.2 | — | — |
| 53 | Churchill | Canada | 30.8 | — | 55.8 | — | — |
| 54 | Po | Italy | 27.0 | 400 | 54.7 | — | — |
| 55 | Salween | Burma | 108 | — | 52.9 | — | — |
| 56 | Shatt-al Arab | Iraq | 209 | — | 50.8 | — | — |
| 57 | Nushagak | U.S. | 12.4 | — | 49.8 | — | — |
| 58 | Albany | Canada | 51.7 | — | 49.4 | — | — |
| 59 | Moose | Canada | 41.9 | — | 48.7 | — | — |
| 60 | Susitna | U.S. | 20.0 | — | 46.6 | — | — |
| 61 | Churchill | Canada | 109 | — | 42.4 | — | — |
| 62 | Susquehanna | U.S. | 28.9 | — | 40.2 | — | — |
| 63 | Nottaway | Canada | 25.1 | — | 40.2 | — | — |
| 64 | Saint John | Canada | 21.4 | — | 39.9 | — | — |
| 65 | Kouila-Niari | Congo | 25.5 | — | 38.8 | — | — |
| 66 | Vistula | Poland | 76.1 | 655 | 37.9 | — | — |
| 67 | Rupert | Canada | 16.7 | — | 35.6 | — | — |
| 68 | Yana | USSR | 91.9 | — | 35.3 | — | — |
| 69 | Nass | Canada | 7.99 | — | 33.8 | — | — |
| 70 | Negro | Argentina | 37.0 | — | 33.5 | — | — |
| 71 | Maniconagan | Canada | 17.6 | — | 33.4 | — | — |
| 72 | Don | USSR | 163 | — | 33.0 | — | — |
| 73 | George | Canada | 17.3 | — | 32.5 | — | — |
| 74 | Eastman | Canada | 18.3 | — | 31.5 | — | — |
| 75 | Mezen | USSR | 30.1 | — | 31.3 | — | — |
| 76 | Pur | USSR | 36.7 | — | 30.8 | — | — |
| 77 | Loire | France | 42.1 | — | 30.0 | — | — |
| 78 | Thelon | Canada | 55.0 | — | 29.7 | — | — |
| 79 | BioBio | Chile | 28.0 | — | 28.0 | — | — |

*Table 6.1* (continued)

| Rank order of discharge | River | Country | Drainage area (sq mi × $10^{-3}$) | Length of river (mi) | Average annual discharge (cfs × $10^{-3}$) | Rank order of length | Rank order of drainage area |
|---|---|---|---|---|---|---|---|
| 80 | Karun | Iran | 22.4 | — | 27.2 | — | — |
| 81 | Valdivia | Chile | 27.0 | — | 27.0 | — | — |
| 82 | Appalachicola | U.S. | 20.0 | — | 26.7 | — | — |
| 83 | Kamchatka | USSR | 17.6 | — | 26.2 | — | — |
| 84 | Saint Maurice | Canada | 18.4 | — | 25.7 | — | — |
| 85 | Arnaud | Canada | 19.1 | — | 24.8 | — | — |
| 86 | Santa Cruz | Argentina | 5.98 | — | 24.7 | — | — |
| 87 | Murray-Darling | Australia | 409 | — | 24.7 | — | — |
| 88 | Chao Phraya | Thailand | 40.0 | — | 24.6 | — | — |
| 89 | Neman | USSR | 38.0 | — | 24.2 | — | — |
| 90 | Elbe | Germany | 52.1 | — | 24.0 | — | — |
| 91 | Glomma | Norway | 16.0 | — | 24.0 | — | — |
| 92 | Garonne | France | 22.0 | — | 23.9 | — | — |
| 93 | South Dvina | USSR | 33.9 | — | 23.9 | — | — |
| 94 | Olenek | USSR | 49.0 | — | 23.8 | — | — |
| 95 | Grand Rivière Baleine | Canada | 16.3 | — | 23.5 | — | — |
| 96 | Sacramento | U.S. | 27.0 | — | 23.0 | — | — |
| 97 | Ebro | Spain | 33.2 | — | 22.0 | — | — |
| 98 | Aux Feuilles | Canada | 16.6 | — | 21.9 | — | — |
| 99 | Rivière Baleine | Canada | 12.1 | — | 21.9 | — | — |
| 100 | Hudson | U.S. | 13.4 | — | 21.3 | — | — |
| 101 | Hayes | Canada | 41.7 | — | 20.8 | — | — |
| 102 | Connecticut | U.S. | 11.2 | — | 19.3 | — | — |
| 103 | Kazan | Canada | 27.6 | — | 19.2 | — | — |
| 104 | Delaware | U.S. | 11.4 | — | 18.8 | — | — |
| 105 | Back | Canada | 41.4 | — | 18.5 | — | — |
| 106 | Onega | USSR | 22.0 | — | 17.8 | — | — |
| 107 | Colville | U.S. | 19.3 | — | 17.8 | — | — |
| 108 | Klamath | U.S. | 12.1 | — | 17.1 | — | — |
| 109 | Harricanaw | Canada | 11.3 | — | 17.1 | — | — |
| 110 | Severn | Canada | 38.9 | — | 17.0 | — | — |
| 111 | Tone | Japan | 4.63 | — | 16.8 | — | — |
| 112 | Penobscot | U.S. | 9.53 | — | 16.7 | — | — |
| 113 | Petit Mecatina | Canada | 7.45 | — | 16.4 | — | — |
| 114 | Oder | Poland | 42.2 | — | 16.1 | — | — |
| 115 | Santee | U.S. | 15.7 | — | 15.4 | — | — |
| 116 | Amgun | USSR | 15.8 | — | 15.3 | — | — |
| 117 | Pee Dee | U.S. | 6.3 | — | 15.2 | — | — |
| 118 | Attawapiskat | Canada | 19.4 | — | 15.0 | — | — |

*Table 6.1* (continued)

| Rank order of discharge | River | Country | Drainage area (sq mi × 10⁻³) | Length of river (mi) | Average annual discharge (cfs × 10⁻³) | Rank order of length | Rank order of drainage area |
|---|---|---|---|---|---|---|---|
| 119 | Winisk | Canada | 26.0 | — | 14.9 | — | — |
| 120 | Kuban | USSR | 17.7 | — | 14.8 | — | — |
| 121 | Moisie | Canada | 7.41 | — | 14.7 | — | — |
| 122 | Natashkwan | Canada | 6.49 | — | 14.7 | — | — |
| 123 | Narva | USSR | 21.6 | — | 14.6 | — | — |
| 124 | Ishikari | Japan | 4.9 | — | 14.6 | — | — |
| 125 | Rioni | USSR | 4.25 | — | 14.5 | — | — |
| 126 | Potomac | U.S. | 13.7 | — | 14.0 | — | — |
| 127 | Aux Outardes | Canada | 7.30 | — | 13.9 | — | — |
| 128 | Mae Klong | Thailand | 9.85 | — | 13.8 | — | — |
| 129 | Duero | Spain | 26.6 | — | 13.2 | — | — |
| 130 | Drini | Albania | 4.79 | — | 13.2 | — | — |
| 131 | Konkoure | Guinea | 6.56 | — | 12.7 | — | — |
| 132 | Altamaha | U.S. | 14.2 | — | 12.1 | — | — |
| 133 | Savannah | U.S. | 10.6 | — | 12.0 | — | — |
| 134 | Mitchell | Australia | 26.8 | — | 11.4 | — | — |
| 135 | Mino | Spain | 6.25 | — | 11.2 | — | — |
| 136 | Wenlock | Australia | 6.79 | — | 10.8 | — | — |
| 137 | Weser | Germany | 14.6 | — | 10.8 | — | — |
| 138 | James | U.S. | 10.0 | — | 10.7 | — | — |
| 139 | Gordon | Tasmania | 2.19 | — | 10.3 | — | — |
| 140 | Burdekin | Australia | 50.2 | — | 10.3 | — | — |
| 141 | Orange | South Africa | 50.2 | — | 10.3 | — | — |
| 142 | Kennebec | U.S. | 5.75 | — | 9.88 | — | — |
| 143 | Dnestr | USSR | 27.8 | — | 9.71 | — | — |
| 144 | Roper | Australia | 31.4 | — | 9.64 | — | — |
| 145 | Coleman | Australia | 9.15 | — | 9.64 | — | — |
| 146 | Kem | USSR | 10.7 | — | 9.64 | — | — |
| 147 | Pascagoula | U.S. | 6.60 | — | 9.57 | — | — |
| 148 | Roanoke | U.S. | 9.65 | — | 9.57 | — | — |
| 149 | Cape Fear | U.S. | 9.15 | — | 9.46 | — | — |
| 150 | Ijssell | Netherlands | 61.8 | — | 9.35 | — | — |
| 151 | Archer | Australia | 7.64 | — | 9.28 | — | — |
| 152 | Pearl | U.S. | 6.64 | — | 8.90 | — | — |
| 153 | Tajo | Spain | 29.4 | — | 8.86 | — | — |
| 154 | Sabine | U.S. | 9.34 | — | 8.68 | — | — |
| 155 | Saint Johns | U.S. | 8.72 | — | 8.40 | — | — |
| 156 | Tiber | Italy | 6.37 | — | 8.26 | — | — |
| 157 | North Kennedy Normandy | Australia | 7.64 | — | 9.28 | — | — |

*Table 6.1* (continued)

| Rank order of discharge | River | Country | Drainage area (sq mi × $10^{-3}$) | Length of river (mi) | Average annual discharge (cfs × $10^{-3}$) | Rank order of length | Rank order of drainage area |
|---|---|---|---|---|---|---|---|
| 58 | Daly | Australia | 14.5 | — | 8.26 | — | — |
| 59 | Ceyhan | Turkey | 8.49 | — | 8.15 | — | — |
| 60 | Adige | Italy | 3.77 | — | 7.97 | — | — |
| 61 | Merrimack | U.S. | 4.79 | — | 7.94 | — | — |
| 62 | Orange | South Africa | 115 | — | 7.52 | — | — |
| 63 | Uribante | Venezuela | 3.86 | — | 7.45 | — | — |
| 64 | Umpqua | U.S. | 3.68 | — | 7.45 | — | — |
| 65 | Pieman | Australia | 0.981 | — | 7.42 | — | — |
| 66 | Brazos | U.S. | 44.0 | — | 7.38 | — | — |
| 67 | Trinity | U.S. | 17.2 | — | 7.27 | — | — |
| 68 | Yodo | Japan | 2.75 | — | 7.13 | — | — |
| 69 | Tuloma | USSR | 6.76 | — | 7.02 | — | — |
| 70 | Eel | U.S. | 3.11 | — | 6.99 | — | — |
| 71 | Fitzroy | Australia | 33.4 | — | 6.88 | — | — |
| 72 | Victoria | Australia | 13.5 | — | 6.88 | — | — |
| 73 | Vjosa | Albania | 2.49 | — | 6.88 | — | — |
| 74 | Seyhan | Turkey | 7.88 | — | 6.57 | — | — |
| 75 | Androscoggin | U.S. | 3.47 | — | 6.42 | — | — |
| 76 | Neches | U.S. | 7.95 | — | 6.39 | — | — |
| 77 | Suwanee | U.S. | 7.10 | — | 6.35 | — | — |
| 78 | Kilzil Irmak | Turkey | 30.2 | — | 6.35 | — | — |
| 79 | Fitzroy | Australia | 52.9 | — | 6.28 | — | — |
| 80 | Guadalquivir | Spain | 22.2 | — | 6.25 | — | — |
| 81 | Escambia | U.S. | 3.82 | — | 6.07 | — | — |
| 82 | Rogue | U.S. | 3.94 | — | 6.00 | — | — |
| 83 | Neuse | U.S. | 5.60 | — | 5.89 | — | — |
| 84 | Ord | Australia | 17.8 | — | 5.89 | — | — |
| 85 | Sebou | Morocco | 4.17 | — | 5.58 | — | — |
| 86 | Clarrence | Australia | 6.45 | — | 5.51 | — | — |
| 87 | Escaut | Belgium | 7.45 | — | 5.47 | — | — |
| 88 | Coruh | Turkey | 7.68 | — | 5.47 | — | — |
| 89 | Tana Gaussa | Kenya | 3.50 | — | 5.22 | — | — |
| 90 | Santa | Peru | 2.08 | — | 5.12 | — | — |
| 92 | Avacha | USSR | 1.83 | — | 4.91 | — | — |
| 93 | Kemalpasha Simav | Turkey | 8.65 | — | 4.69 | — | — |
| 94 | Pamlico | U.S. | 4.40 | — | 4.69 | — | — |
| 95 | Colorado | Argentina | 4.94 | — | 4.62 | — | — |
| 96 | Chowan | U.S. | 4.94 | — | 4.62 | — | — |
| 97 | Sakarya | Turkey | 22.5 | — | 4.59 | — | — |

*Table 6.1* (continued)

| Rank order of discharge | River | Country | Drainage area (sq mi × $10^{-3}$) | Length of river (mi) | Average annual discharge (cfs × $10^{-3}$) | Rank order of length | Rank order of drainage area |
|---|---|---|---|---|---|---|---|
| 198 | San Joaquin | U.S. | 14.0 | — | 4.45 | — | — |
| 199 | Gilbert Staaten | Australia | 28.0 | — | 4.34 | — | — |
| 200 | Nagara | Japan | .786 | — | 4.31 | — | — |
| 201 | Chehalis | U.S. | 1.29 | — | 4.20 | — | — |
| 202 | Tumbes | Peru | 3.86 | — | 4.13 | — | — |
| 203 | Seman | Albania | 2.04 | — | 3.99 | — | — |
| 204 | Ponoy | USSR | 3.94 | — | 3.95 | — | — |
| 205 | Ogeechee | U.S. | 4.63 | — | 3.88 | — | — |
| 206 | Chira | Peru | 1.93 | — | 3.85 | — | — |
| 207 | Yuzhny-Bug | USSR | 24.6 | — | 3.81 | — | — |
| 208 | Saco | U.S. | 1.73 | — | 3.67 | — | — |
| 209 | Arno | Italy | 3.15 | — | 3.64 | — | — |
| 210 | Busuk Mendereo | Turkey | 9.64 | — | 3.56 | — | — |
| 211 | Oum-Er Rebia | Morocco | 11.6 | — | 3.56 | — | — |

In the western states, however, areal variation of precipitation is very high and strongly influenced by mountains. This kind of variation also characterizes the different continents, so one would not expect a high correlation between drainage area and mean annual discharge for the major rivers of the world. Indeed, this is the case when the data from Table 6.1 are plotted.

But ranking the data in order of magnitude permits a different view of the array. The distribution of the values of average discharge is plotted in Figure 6.1 on log probability paper. Through most of the array the data plot as a nearly straight line with one important exception—the Amazon, which has an unusually high discharge. This plot means that the logarithms of the discharge values are distributed in a normal or Gaussian manner. Such a distribution of hydrologic quantities is seen in many instances.

Figure 6.1 shows that 50 percent of the rivers of the world have an annual discharge of 17,000 cfs or greater, and half have a smaller discharge.

A Gaussian or normal distribution is the ordinary curve of probability. It implies that, in some manner, random chance is involved in governing the distribution of values in the array. Some physical justification comes

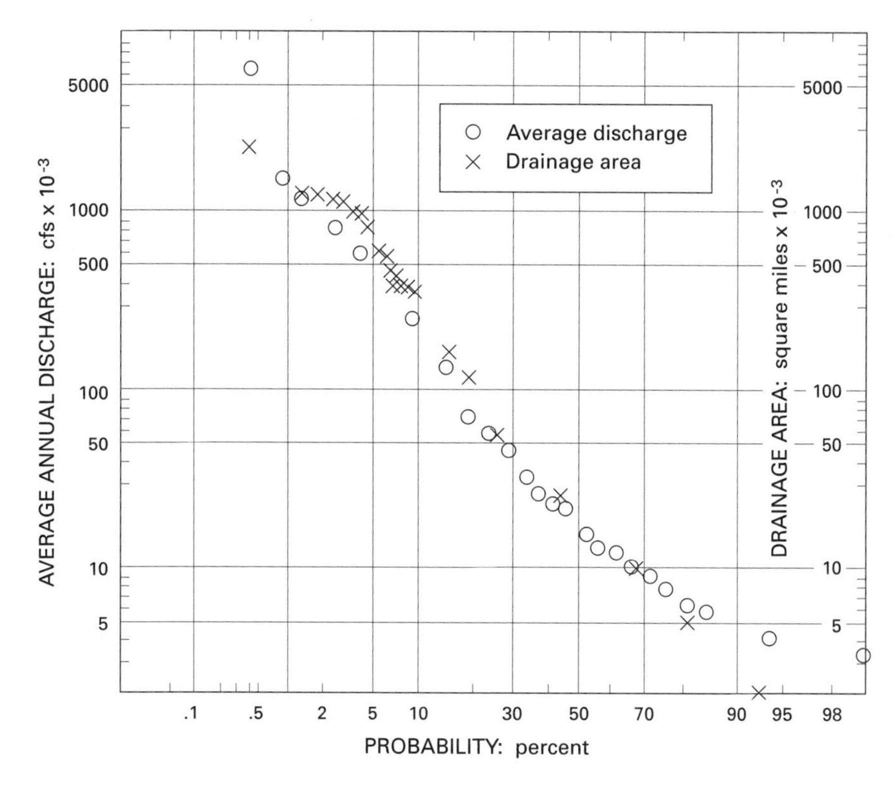


*Figure 6.1* The size distribution of values of average discharge and drainage areas for rivers of the world. The percent probability equals 100 times the rank order divided by one more than the number of items.

from the size distribution of drainage areas of the same 211 river basins in the worldwide tabulation. The distribution of drainage areas was similarly arranged in order of magnitude and the probability calculated for each basin size. As can be seen in Figure 6.1, there is near-coincidence of the data for basin size and for annual discharge. That is, the logarithms of both annual discharge and drainage area are normally distributed. The distribution of basin size in this large sampling leads to a comparable distribution of average annual discharge values.

This normal distribution of drainage areas has an important geomorphic implication. Rivers draining a continent compete with one another. At their headwaters one basin may capture area from another if its rate of land lowering is greater than that of the adjacent basin. This tendency to compete means that chance or random events can play an important role in determining the size of a basin and thus its average discharge.

The work of Robert E. Horton on the configuration of the network of

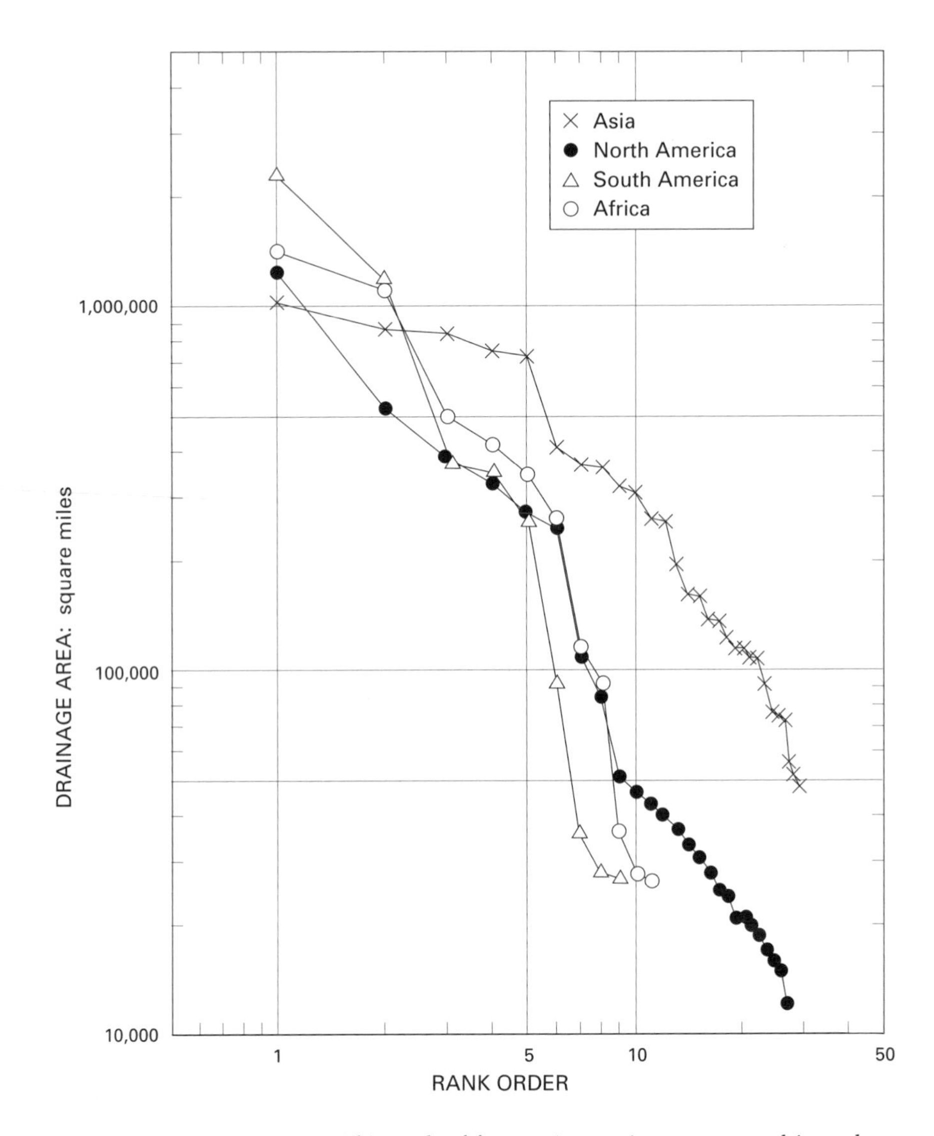


*Figure 6.2* Drainage areas within each of four major continents, arranged in rank order of drainage basin size.

river channels led later workers to show that the channel net is random in its topology. The data presented here show that the array of basin size in the world also is distributed in a manner that can be expressed as a probability function. This finding reinforces a principal argument of this book and will be further elaborated in the comparison of randomly constructed basin areas with those occurring in nature, discussed in Chapter 12.

## Basin Features within the Continents

The continents of the planet are of vastly different size and therefore have very different numbers of drainage basins. The geology of the continents also influences their location, size, and discharges. An example of the effect of geography and geology is the difference between the two largest basins in China. The Yangtze has a basin area of 750,000 square miles and a discharge of 1.17 million cfs, or 1.56 cfs per square mile. It drains the Tibetan Plateau north of the Himalayan Range and is fed by monsoon precipitation. The basin adjoining it to the north, the Yellow River, has a basin area of 260,000 square miles and a discharge of 62,000 cfs, or 0.24 cfs per square mile. This basin drains a dry and cold steppe and flows through part of the great loess deposits of that arid region. Thus geography and climate substantially influence river character.

The geologic history of the Indian continental mass pressing northward against southern Asia pushed the channels of three major rivers close together. The channels of the Yangtze, Mekong, and Salween are at one place within 50 miles of one another in the Ning Ting Shan, a mountain range near the border of Burma and China.

In South America, the vast expanse of land at elevation less than 1,000 feet above sea level is drained by one very large river, the Amazon. Because of the tropical climate, its runoff is also great—2.7 cfs per square mile. This is five times that of the Mississippi, which produces 0.52 cfs per square mile. Geology and geologic history are major determinants of geomorphology.

As we saw in Figure 6.1, when all continents are included the annual discharge and drainage area values are normally or randomly distributed. The summation of all the rivers in such a variety of geologic and climatic conditions tends to average out the differences and expose the overall influence of random effects. But when individual continents are considered, these differences become overriding. Figure 6.2 plots the drainage area of each basin versus its rank order of size within the

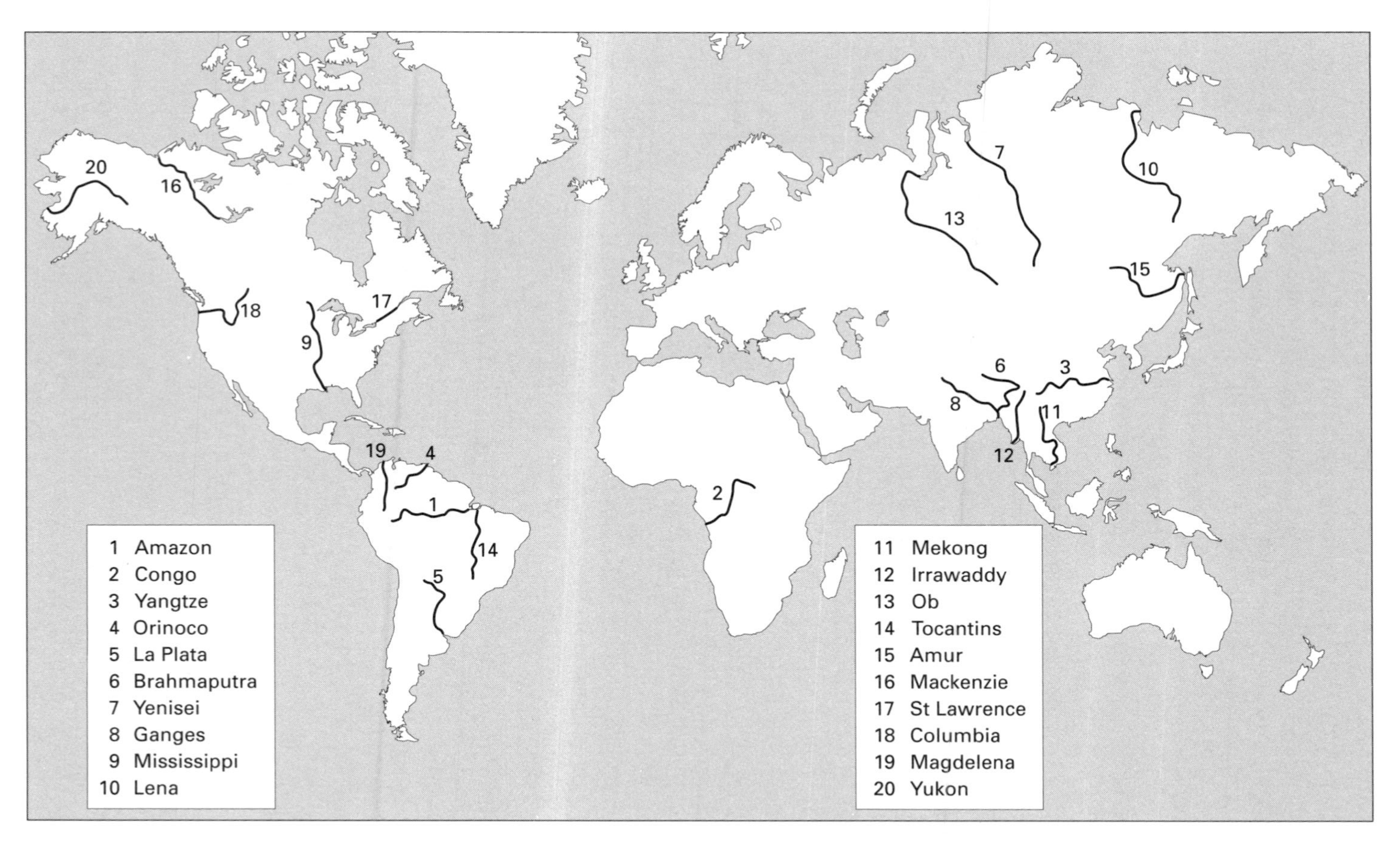


*Figure 6.3* World map showing twenty of the major rivers of the world, listed in rank order of drainage basin size.

continent. Asia, far larger in size than South America, has a maximum basin size of 1.01 million square miles (the Yenisei), whereas the Amazon has an area of 2.3 million square miles. The Mississippi, with a basin area of 1.24 million square miles, is intermediate in basin size.

The five largest basins in Asia are of nearly the same order of magnitude, 1.1 million to 0.72 million square miles. The five largest in South America range from 2.4 million to 0.26 million square miles, and in North America from 1.24 million to 0.28 million. Figure 6.3 shows the location of the 20 largest rivers.

Table 6.1 includes 211 rivers. In that list the number of rivers in Asia is much greater in any other continent: North America has 27, Africa 16, and South America 15.

In summary, when the rivers of all continents are considered, their major features—size and discharge—show the effect of randomness or probability. That fact supports a thesis of this book, that random chance plays an important role in river features and the forms assumed all tend toward the most probable.

CHAPTER SEVEN

# Flow Variability and Floods

## Changes with Time

The analysis of time fluctuation of river flow is based on the concept that a record of river flow represents a sample taken from an indefinitely large population. The variance characteristics of the sample are assumed to be a measure of the variance of the population. On this basis we can compute the probability that the mean flow in a series of future years will equal, exceed, or be less than the mean of the record period. This probability is central to the design of storage reservoirs.

The study of river discharge as a function of time is complicated by the expansion of reservoir storage, river diversion, and other engineering works that render the streamflow data a record of releases from storage, not a measure of natural river discharge. In some engineering studies it is necessary to estimate these effects, so occasionally a careful record has been reconstituted from early gaging data, correlations, and climatic information. Two of these reconstructed records are (1) the annual flow of the Colorado River at Lees Ferry, Arizona, a 61-year array prepared by the Bureau of Reclamation, representing the recorded flow corrected for upstream depletion; (2) the unimpaired inflow to the delta of the Sacramento River, California, 1921–1983, a 63-year array compiled by the Division of Planning, Department of Water Resources, State of California. I have analyzed the Colorado River data in some detail (Leopold 1959), so I will use here the inflow to the Sacramento River delta as an example.

Annual values of streamflow often have a normal or Gaussian distribution. When arranged in order of magnitude and plotted on arithmetic-probability paper, they fall in essentially a straight line with discharge as the ordinate and probability as the abscissa. The plotting position on the abscissa is the probability of a given flow being equaled or exceeded: $p = m/(n + 1)$, where $p$ is probability, $m$ is rank, and $n$ the number of

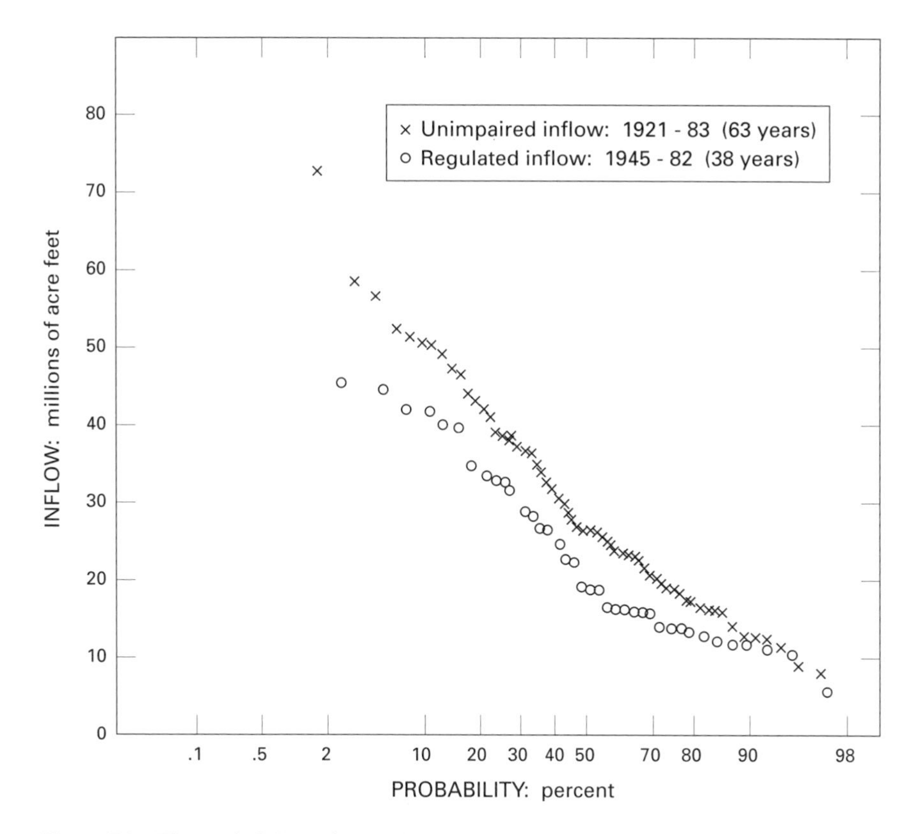


*Figure 7.1* The probability of annual discharge of the unimpaired inflow to the delta of the Sacramento River, California.

years of record. This probability is the reciprocal of the recurrence interval. Such a plot of the unimpaired inflow to the delta of the Sacramento River is given in Figure 7.1.

The arithmetic mean of the record is 29.8 million acre-feet. Deviations from the mean of each year's flow were computed. Having squared the deviations, we can compute the standard deviation $\sigma$:

$$\sigma = \sqrt{\frac{\Sigma x^2}{n - 1}},$$

where $x$ is the deviation from the mean and $n$ the number of items in the array. The standard deviation of this record is 14.1 million acre-feet. The probable deviation is 0.674 times the standard deviation.

These terms may be interpreted as follows. In a normally distributed sample, there is a 50 percent chance (probability 0.50) that a given event or quantity will be within one probable deviation on either side of the sample mean, and a 66 percent chance (probability 0.66) that it will be within one standard deviation on either side of the mean.

Not all hydrologic data are normally distributed. Even though the individual values are not, the logarithms of those values will usually be normally distributed. Volumes of runoff such as annual values tend to be nearly normally distributed, but flood flows tend to have their logarithms normally distributed.

## Volume of Flood Water

When rain falls upon the earth, an astounding amount of water is involved, even in a light shower. A two-day rain is commonplace in the central United States and western Europe, and in that period an inch of rainfall is very common. An inch of rain falling on the state of Ohio would be 2.19 million acre-feet or 712 million cubic meters. A comparable rain of 25 millimeters falling over Belgium would be 760 million cubic meters.

Such a rainstorm is not unusual and would not necessarily cause a flood. Ordinarily runoff constitutes less than half the water falling in a given storm. The greater part either is retained in the soil, from which it is drawn by plants and thus returned to the atmosphere, or drains gradually to groundwater storage and then, in part, to river channels.

The volume of water that falls as rain is large, but so is the volume of channels of the river system. W. B. Langbein computed the following for a 300-square-mile basin above Saint Paul, Indiana. A flow of 10 cfs per square mile at every point in the channel system would be within the bankfull stage at most points. The volume in channels would then be about 6,000 acre-feet, almost 2 billion gallons, or about 0.4 inch over the drainage basin. Under the same flow conditions a nearby basin with a drainage area 10 times the size, or 3,000 square miles near Spencer, Indiana, would have in its channel system 120,000 acre-feet, 39 billion gallons. This is 20 times as much as in the basin above Saint Paul. Thus the volume of the channel system increases rapidly downstream.

Within the banks of stream channels there is a great amount of space capable of holding water. But the space is not utilized efficiently, because storm rainfall is seldom distributed geographically in a uniform manner. The channel system may be overtaxed in one place, while in another the streams flow at less than full capacity. Despite the capacity to handle large total volumes of water without overflows, floods still occur.

The river channel is constructed by the river. On most days of the year only the bottom part of the channel carries water. Several days a year the channel is three-quarters full, and once or twice a year on average the river flows bankfull.

Flows so large that they cannot be contained in the channel must spread out over the adjacent floodplain. The floodplain is where nearly all flood damage occurs, because there humans have chosen to grow crops or build buildings. We have encroached on areas that the river must at times cover with water.

## Flood Discharges

A flood may be defined as the occurrence of a flow of such magnitude that it overtops the natural or artificial banks in a reach of river channel. Where a floodplain exists, a flood is any flow that spreads out over the floodplain. Eons before the Neanderthals began flaking flint, the overflow of floodplains was occurring much as it does today. Flood damage, therefore, is a consequence of human use of the floodplain. On the average, the water that is discharged at rates in excess of channel capacity constitutes about 5 percent of the total annual discharge of the basin. In other words, compared with the volume of water flowing at within-bank or nonflooding rates, the volume of flood water is not large.

Extreme floods have been noted in historical records or in the mythology of nearly all countries. There are some very long and extremely well documented records of flood heights, not only on the Nile, where the record is famous, but on several rivers in Europe and Scandinavia. The record of measured flood discharges, in contrast to flood heights, is much more limited. In the United States river gaging in a modern sense did not even begin until 1895. In 1992 there were 7,590 operative gaging stations. The record would, at first glance, make it appear that extreme floods are increasing in magnitude, but this is an artifact of measurement history. The records are becoming more numerous and longer, so the number of recorded floods of large magnitude increases with time.

Record flood discharges from drainage basins of various sizes in the United States compiled in 1955 are presented in Figure 7.2, along with an envelope curve existing in 1890. The figure shows how more and longer records have included much larger floods for each basin than were on record half a century earlier.

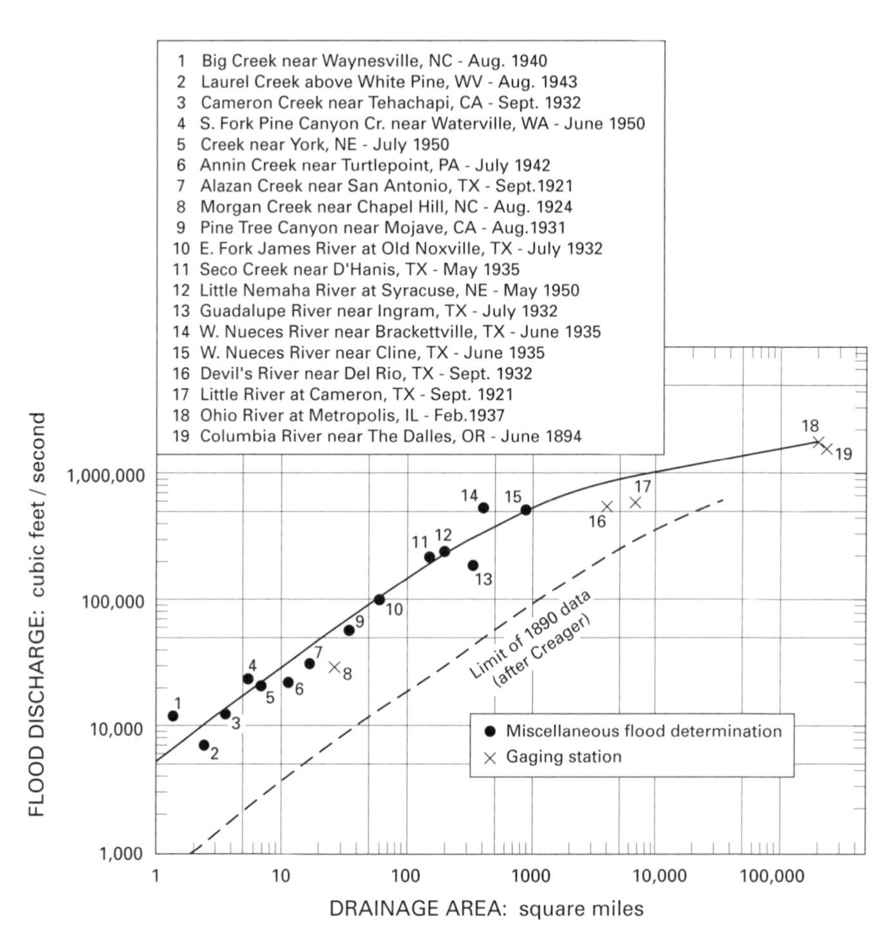


*Figure 7.2* Extreme flood discharges measured for drainage basins of various size in the United States. For comparison, the envelope curve of known flood experience as of 1890 is shown. (After Hoyt and Langbein 1955.)

## The Concept of Flood Frequency

The relation of the magnitude of a variable to its frequency of occurrence is a frequency curve. As discussed previously, the curve is an estimate of the size distribution of a population prepared from a sample of the population. In flood-frequency studies, the population is the spectrum of flood occurrences at a given point in a river system over a long period, but the sample selected to represent that population is the measured record of flood experience. That record may encompass only a few years, but sometimes is as long as several decades.

There are two principal ways in which a flood-frequency array is constructed. In the annual flood series, the highest momentary peak in each water-year of record (October 1 to September 30 in the United States) is tabulated. The alternative, a partial-duration series, is an array made up of all individual momentary peak flows above some lower limit, including all the peaks above that limit even if several occurred in a single year and none occurred in other years. The items in the flood series are arranged in order of magnitude or in rank order. In the usual practice, the flood data are the values of the annual flood, that is, the highest momentary discharge each year of record. These values are plotted against the recurrence interval *(RI)* in years:

$$RI = \frac{n+1}{m}$$

where $n$ is the number of items in the array, and $m$ is the rank number. (In the earlier example the probability of occurrence was used; probability is the reciprocal of the recurrence interval.)

As an example, in the flood record of Seneca Creek, Maryland (Figure 7.3), the sixth-highest flood in the 61-year record occurred in 1980. Its recurrence interval is then $62/6 = 10$ years approximately. A flood this size has the probability $1/10$ of occurring in any one year, that is, a probability of 10 percent. There is a chance of 1 in 10 that such a discharge will occur this year or any other year.

Each discharge is plotted against its recurrence interval or probability of occurrence. Any of several types of plotting paper may be used. Some arrays of hydrologic data, of which flood occurrence is one, usually have a nearly Gaussian distribution of the logarithms of the quantities, in which case the data would plot as a straight line on log-probability paper. The paper chosen should be the type on which the plotted data most closely approximate a straight line. In Figure 7.3 the plot is on the paper generally used by the U.S. Geological Survey—what is called Pearson Type III distribution paper. Whereas probability paper is based on a distribution of data that is symmetrical, flood data tend to have one tail of the distribution curve somewhat elongated. The Pearson distribution is of this nature.

I have chosen Seneca Creek at Dawsonville, Maryland, to illustrate flood-frequency plotting not only because it is used several other places in this book to illustrate river characteristics, but also because it exemplifies some common aspects of the flood record for a single station.

In the first 38 years of record, the three highest discharge values deviated from the straight line established by the bulk of the data and ap-

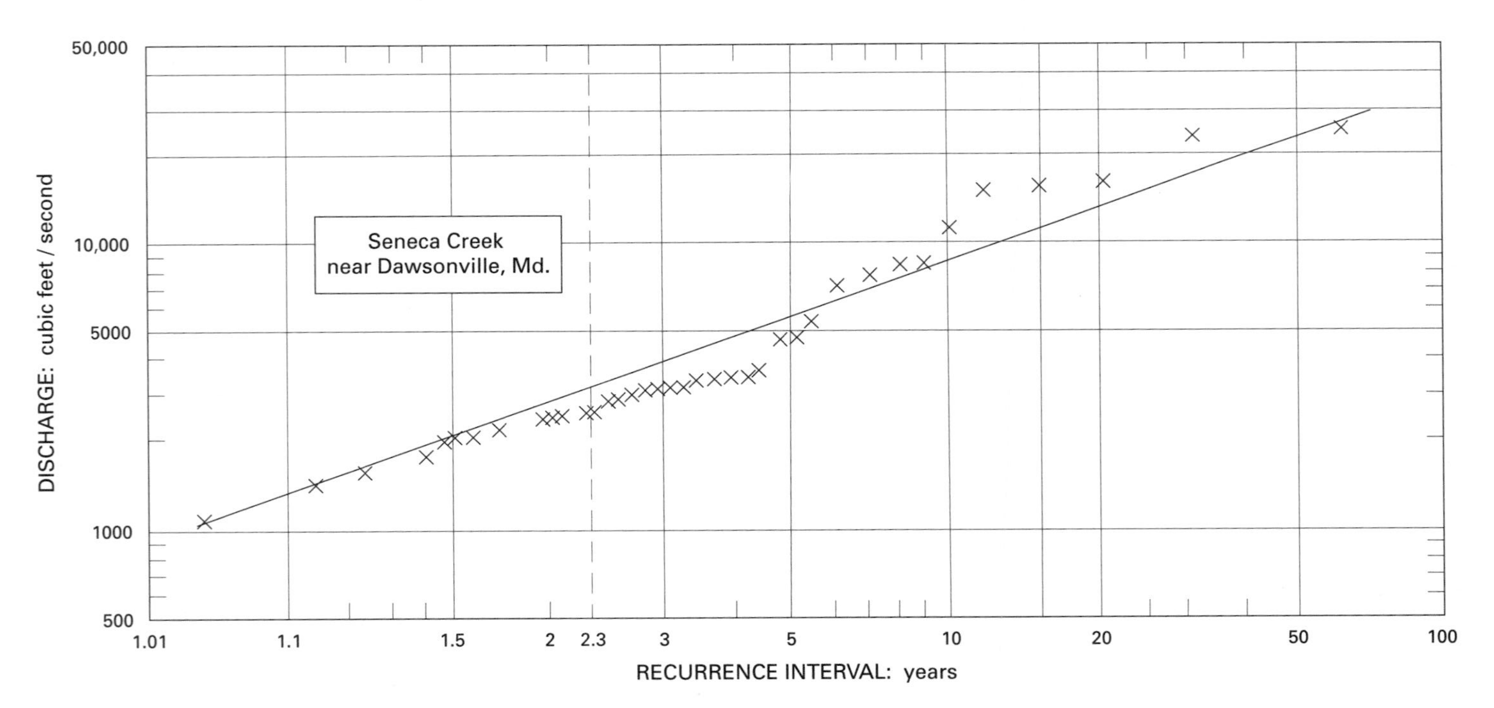


*Figure 7.3* Flood-frequency plot of the 61-year record at Seneca Creek near Dawsonville, Maryland.

peared excessively large. The expected highest discharge in that period was about 7,000 cfs, but 15,000 cfs was actually experienced. When the record had been taken for 60 years, it became evident that those high discharges were not exceptional.

## Regional Flood Data

The record of any particular river may not be representative of the general region, inasmuch as records are short and human activity varies geographically. To represent a whole region, some sort of averaging is needed to smooth out the variation among individual station records. Accordingly, the U.S. Geological Survey combined the flood records within each broad geographic area to derive regional flood-frequency curves. The results have been published for the entire United States, except arid regions where the regional method did not apply, and are available in the U.S. Geological Survey Water-Supply Papers 1671–1688 inclusive. With each volume devoted to a major river system or region, the regional frequency curve applicable to any location can be constructed with little effort.

The regional curves are prepared as two types of graphs, the first being the relation of mean annual flood to drainage area and the second the relation of recurrence interval to mean annual flood. The annual flood is defined as the highest momentary peak each year. The arithmetic average of the annual floods is called the mean annual flood. This value plots on a flood-frequency curve at a recurrence interval of 2.3 years. For that reason a special dashed vertical line is printed on flood-frequency plotting paper at an abscissa value of 2.3 years.

It is necessary to emphasize one aspect of the recurrence interval, which is the average interval of time within which a flood of given magnitude will be equaled or exceeded but once. It is also a statement of probability. A flood with a recurrence interval of 10 years is one that has a 10 percent chance of recurring in any one year. Such a flood might occur next year. A flood so large that it would not be expected more than once in a hundred years has a 1 percent probability of occurring next year. But on the average the 10-year flood may be expected to occur once in a 10-year period. The logic stems from the fact that floods occurring during a period of time constitute a sample of a large population in time. If in a period of 61 years, as in the Seneca Creek example, the largest flood recorded was of a certain size, it is probable that the next 61 years will also contain a flood of that magnitude. The recurrence interval is not a forecast, it is merely a statement of average expectancy.

## Causes of Floods

Until the early part of the present century there was a widespread belief that forests could prevent floods. Hydrologic research illuminated the basic fact that great floods occur when a drainage basin is incapable of infiltrating additional water, and all precipitation or snowmelt must run off the surface. Such a basin condition may occur after prolonged rainfall that gradually saturates the soil surface. Under such conditions additional water provided by rainfall or snowmelt leads to surface runoff and there is a mixed effect of vegetation. Forests and other vegetation then markedly restrain erosion and slow the movement of sediment.

In 1993 great floods devastated not only the central United States, but in the same months caused tremendous damage in India, Nepal, and Bangladesh. In those countries flooding and landslides killed 2,100 people and left 6 million homeless when unusually heavy monsoon rains occurred in Bengal, Punjab, and Assam. Because of the flooding, 6.6 million people were located in emergency camps. In Nepal alone, 50,000 families were left homeless.

During July 1993, rivers in much of the upper Mississippi River Basin flooded when severe thunderstorms continued for weeks over land that was already wet from the usual spring runoff. Many stations over a wide area reported several inches in a day, from the Dakotas east through the north central states.

On July 22–23 a large area in southern Iowa and northern Missouri received more than 6 inches of rain in one and a half days. I estimate that, in those two states alone, the storm dumped 8.58 million acre-feet of rain over 48,000 square miles. This amount of water would be twice the usable reservoir space in Shasta Lake, California, or three times the usable capacity of Elephant Butte in New Mexico, or one-third the capacity of Lake Mead in Arizona.

It was the persistence of heavy local rains that was unusual. Flooding was particularly extensive along the lower Missouri, Des Moines, and middle Mississippi rivers. Many discharge records were broken as higher flows occurred than had previously been experienced. The Mississippi River near Cairo had a discharge three times that which flood-control plans are meant to contain.

The record-breaking 1993 floods probably represent an aspect of the climatic shift that began about 1950 in the United States, and that caused the increased severity and frequency of storms in Britain and northwest Europe discussed by Hubert Lamb. One of the characteristics of the climatic regime since 1950 is an increase in variability. For example, after 7 years of drought California experienced in 1993 one of the wettest winter-spring seasons on record.

The floods of 1993 resulted from unusual weather conditions and not from human activities. Flood damage, however, was materially altered in pattern and amount by levees, dams, and other control works. Protection via engineering construction causes floodplains to be developed for urban and agricultural use. As long as the protection works contain the flood waters, the human influence is beneficial. But when those works fail, damage is increased because of encroachment onto the floodplains.

When great floods and large river basins are considered, human works have limited effect. But for smaller drainage areas, the alteration of the land surface is very important. Urbanization is the major contributor. Roads, street gutters, roof downspouts, paved parking areas, and storm drains all move water downhill faster than would occur on natural slopes. The speeding of runoff means that a given volume of water must be discharged in a shorter period of time. This requires the peak rate to be increased.

Figure 7.4 illustrates the principle. The area under a hydrograph is the volume of water that runs off from a given area. The time elapsed between the center of rainfall and the center of runoff is called the lag time. As might logically be assumed, the lag depends on the speed with which rainfall is moved downchannel. If this speed is increased, lag time decreases. All the aspects of suburbia tend to move rainwater quickly downstream. The net effect is to increase the peak flow in channels that carry urban runoff. Note in Figure 7.4 that decreased lag time increases peak discharge.

## Need for Information on Flood Expectancy

As in most aspects of life, there is always a possibility that extreme events will disrupt daily routine. It is for this reason that insurance coverage is an important part of modern life. But from the standpoint of the engineering that supports civilization, the probability of interruption sets the condition for financial as well as organizational preparedness.

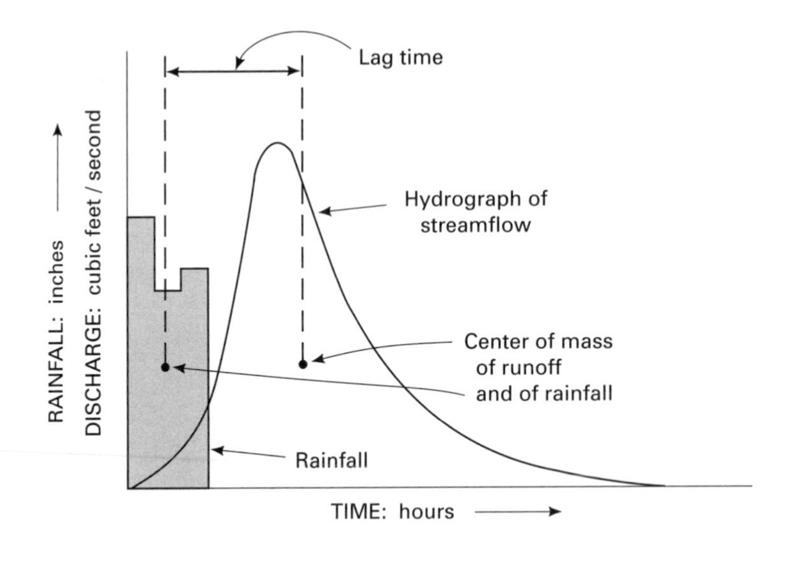


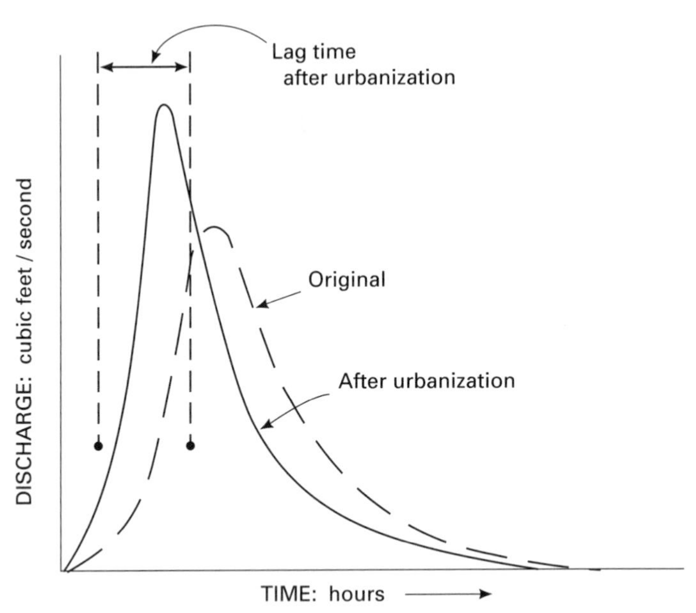


*Figure 7.4* Hypothetical unit hydrographs relating runoff to rainfall, with definition of significant parameters, showing the effect of decreased lag time on flood peak.

For example, there are hundreds of thousands of culverts and small bridges over which roads and highways span channels of various size. If the capacity of such culverts were too small and many of them washed out, the expense would be very great. But if overdesign were the rule, money for construction would be needlessly spent. After years of study it was decided as a matter of policy that culverts constructed with federal money would be designed to carry without failure a flood discharge with a recurrence interval of 50 years. Such a decision places an enormous burden on hydrologists. They must develop simple but dependable methods to estimate flood discharge in ungaged areas in diverse climates and topography and for various basin sizes. Bridge design, storm sewers, city street drainage, culvert size, and flood insurance rules are a few other applications of flood-frequency calculations.

A different kind of social cost results from the fact that people use flood-prone areas. Despite their knowledge of the importance of flood estimates, individuals continue to build on floodplains, encroaching on parts of rivers needed to carry rather ordinary discharges. The floodplain level is equaled or exceeded about once a year. So despite the vast sums spent on controlling floods, flood damage continues to rise year after year and is measured in billions of dollars. Per-capita flood damage increased two and a half times in the United States between 1916 and 1985.

The great floods of 1993 emphasize the importance of reexamining the policies of the United States that include flood control. It is not widely understood that these policies, enunciated by the Congress and materially elaborated by the many concerned federal agencies, are not merely directed at flood control. They include a wide variety of interconnected aims, desires, programs, and fiscal considerations, and involve agriculture as well as engineering. The policies cover a complicated mixture of land development and disaster prevention. This large subject was covered in my 1954 book with Thomas Maddock, *The Flood Control Controversy*. The analysis in that study is, for the most part, still applicable today (see especially pp. 130–154). A brief discussion of human occupancy of flood-prone land is in Dunne and Leopold 1978, chap. 11.

In general, flood control in the United States should place much greater emphasis on restriction of development of floodplains, flood-proofing of individual sites or local areas, an insurance program in which premiums are proportional to risk, and planned utilization of floodplains for peak flow reduction. Reliance on engineering protection by levees and dams should be a much less prominent part of flood-damage mitigation.

## Methods of Computing Flood Discharge

Gaging stations have been operated for short or long periods at more than 20,000 sites in the United States, although at any one time far fewer are in operation. In 1992 the number of operating gages was 7,590. At these sites the discharge for any chosen recurrence interval can be determined by the type of frequency analysis just described. But we have previously estimated that, including first-order channels, there are approximately 3.25 million miles of river channels in the United States, so the number of gaged sites is a minuscule part of the locations where flood discharges might potentially be needed. Hydrologists have spent much effort developing methods of estimating discharge values for ungaged locations. These have concentrated on using combinations of precipitation, topography, vegetal cover, soils, geology, and sediment materials as indicators. Some of these procedures are remarkably good predictors, but all have restrictions on applicability and require information that is not readily available.

The recently developed procedures based on simple measurements of the river channel are now available for nearly all parts of the country, for ephemeral as well as for perennial streams, and for various climatic zones. Nearly all depend on field measurement of bankfull width or width of the active channel. The equations permit computation of discharge for recurrence intervals of 2 to 50 years. The method is based on the logical notion that the channel is carved and maintained by the discharges it experiences and thus reflects those events. Such studies have been published for the western United States, differentiated on the basis of alpine, northern plains, southern plains, and Rocky Mountain areas, and for California, the Missouri Basin, Kansas, Idaho, Colorado, Utah, the Piedmont region, Wyoming, Nevada, and New Mexico. Typical is the report by Hedman and Osterkamp (1982) on the channel geometry of streams in the western United States, based on analysis of 151 gaging stations and extensive field testing.

## Examples of Human Influences

In recent years an important source of discharge variability and flood experience has emerged: anthropogenic causes, especially urbanization and flood control engineering. Gaging station records no longer reflect climatic variability, for the natural changes are obscured and usually overpowered by river regulation, land-use effects on runoff, and control by reservoirs. Charles Belt has demonstrated that flood heights, not discharge values, in the confined channel of the Mississippi have been raised as much as 10 feet by the construction of levees. These structures prohibit the river from utilizing its floodplain when discharge exceeds the capacity of the original or normal channel.

Discharges recorded on most rivers in the United States do not reflect natural values of hydrologic conditions. Seasonal and annual runoff measurements are materially influenced by man's choices for influencing irrigation and navigational needs, water supply, and energy requirements. In reality, only the bench-mark gaging station can now be trusted to reflect natural variations in streamflow.

These man-made effects are of exceptional importance, because there is little organized effort to monitor land-use changes through time. Therefore, separation of natural and human effects is virtually impossible.

Two examples illustrate the point. One of the most detailed studies of river channel change with time and the effect of urbanization is the 19-year span of observations on Watts Branch near Rockville, Maryland. As the urban area expanded during the period 1954–1965, the growth in the number of houses increased the frequency of overbank flooding. After the field study was completed in 1972, another 15 years of record accumulated before the station was discontinued in 1987.

Urbanization began on a large scale in about 1950. In the period 1958–1972, the mean annual flood was 781 cfs. In the period 1973–1987 the value was 959 cfs, an increase of 23 percent.

The bankfull discharge at the natural section of the study area was 220 cfs. The number of times bankfull is equaled or exceeded in most rivers is once every 0.9 year (about once a year). This is the recurrence interval in the partial-duration flood series and is equal to a 1.5-year recurrence interval in the annual flood series.

On Watts Branch the numbers of times discharge was greater than bankfull were as follows:

1958–1967: 49, or 4.9 per year
1968–1977: 74, or 7.4 per year
1978–1987: 73, or 7.3 per year

Before urbanization became important in 1958–1964, the number of overbank flows was 10 in 7 years, or 1.4 per year. The effect of change in land use is especially large on a small basin such as Watts Branch, which has an area of 3.7 square miles.

The change can also be seen in the records of larger basins. Seneca Creek, about 8 miles north of Watts Branch, drains an area of 100 square

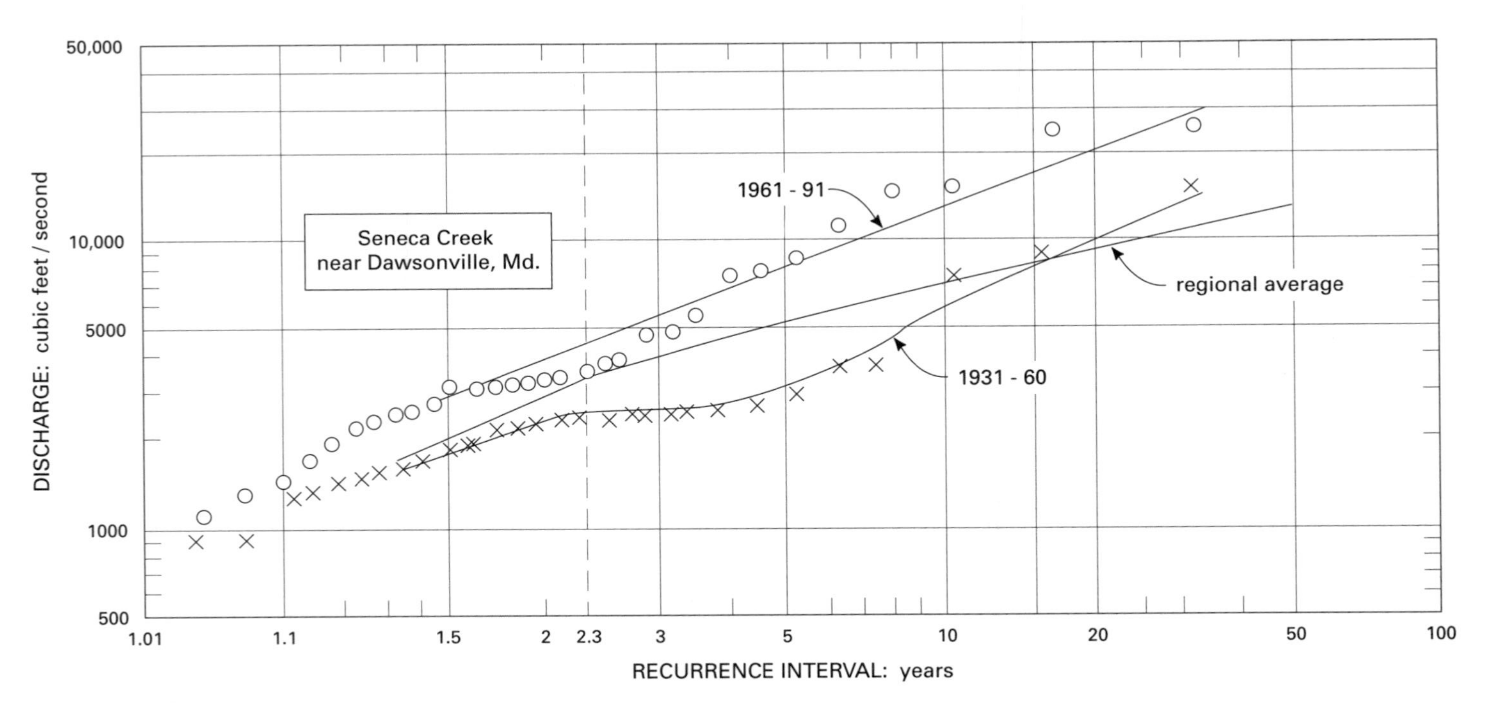


*Figure 7.5* Flood-frequency curves for Seneca Creek, Maryland, during the periods 1931–1960 and 1961–1991, showing the significant difference in the two periods. Also included is the regional flood-frequency curve for the area that includes Seneca Creek.

miles and has also been influenced by the growing urbanization. The frequency curve for Seneca Creek, using the 61 years of record, suggests that important changes in flood experience occurred during the period. To examine this surmise, we divided the record into two parts, 1931–1960 and 1961–1991. The two recurrence interval curves were plotted on the same graph (Figure 7.5), along with the regional flood-frequency curve. The regional graph represents all the gages in the region, expressed for a drainage area of 100 square miles.

It is obvious that the second half of the record experienced higher discharges at all recurrence intervals. The mean annual flood increased from 2,300 cfs to 3,490 cfs. The flood of 10-year recurrence interval increased from 7,300 cfs to 16,000 cfs. The regional curve was based on data prior to 1962 and therefore more closely approximates the earlier record. The mean annual flood from the regional data was 3,400 cfs.

Seneca Creek has a bankfull capacity of 2,000 cfs. The average annual flood in the period 1931–1960 was 2,973 cfs. and in the period 1961–1991 was 6,014 cfs, an increase of 100 percent. In the earlier period Seneca Creek experienced discharge exceeding bankfull 35 times, or 1.2 times per year. But in the later period the experience occurred 66 times, or 2.2 times per year, an increase of 83 percent.

These examples show that variation in discharge characteristics must be examined carefully to estimate what portion humans have contributed to the observed change.

CHAPTER EIGHT

# Relationships between Channel and Discharge

## Channel Capacity and Effective Discharge

A river constructs and maintains a channel of such form that all discharges less than a moderate size are contained. Larger discharges exceed the channel capacity and overflow the floodplain. Those that exceed the capacity are floods that cause damage. Channel capacity has a direct importance to humans and is also significant to river morphology and process. Therefore, the factors determining the size of channel and its physical form have both practical and intellectual significance.

The floodplain is defined as the flat area adjacent to the river channel, constructed by the present river in the present climate and frequently subject to overflow.

Each part of the definition is important to an understanding of the nature and characteristics of the floodplain. Especially important is the phrase "constructed by the present river in the present climate." For if the river channel degrades or deepens, what was formerly the floodplain is abandoned and becomes a terrace, an abandoned floodplain.

Why is the channel created and maintained by the normal flow of water relatively small? The processes involved are erosion and deposition, and their effectiveness increases with increased discharge. But also operative is the frequency of events of various magnitude. The concept of frequency effectiveness, set forth by Wolman and Miller, elucidates this matter. They pointed out that the work done by a river in erosion and deposition—in simple terms, effectiveness—increases with discharge. But low discharges that are ineffective and do little work are common. The higher the discharge, the more uncommon it is. High discharges, when they do occur, are very effective in the erosion-deposition process. It follows that some intermediate discharge, neither high nor low, is both sufficiently frequent and sufficiently effective to be most important in forming and maintaining the channel.

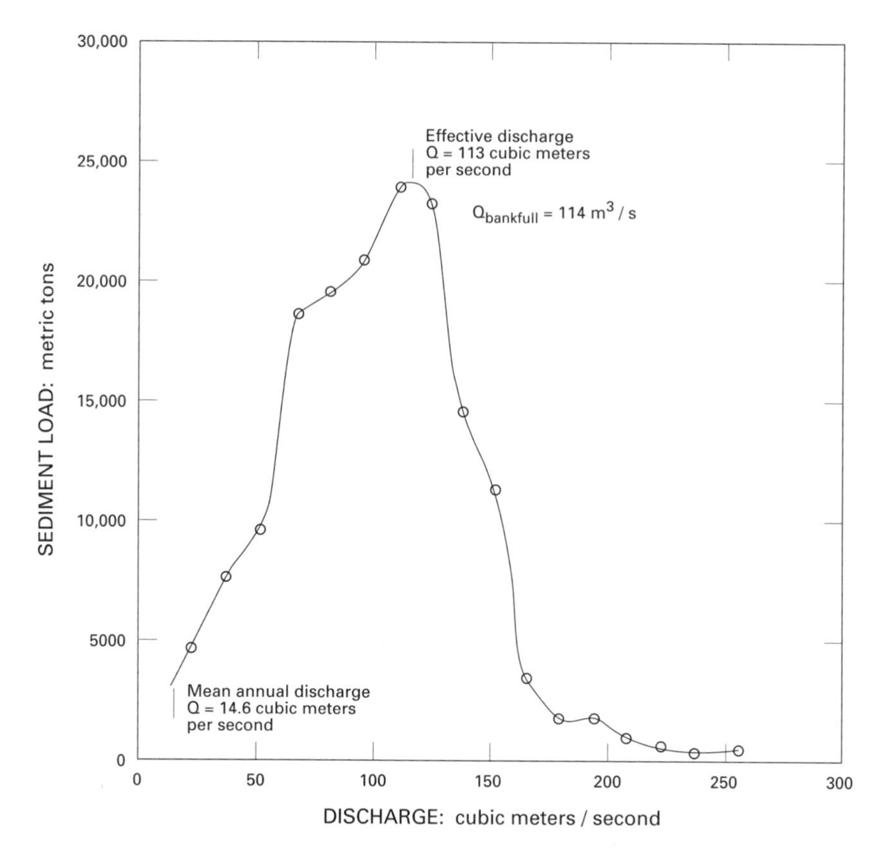


*Figure 8.1* The annual sediment load transported by various discharges of the Little Snake River near Dixon, Wyoming. The effective discharge is defined as the discharge that carries the largest amount of sediment over a long period of time. The bankfull condition here coincides closely with the effective discharge. (After Andrews 1980.)

In 1964 there were practically no data on the movement of coarse material, or bedload. So we used data on suspended sediment to show that over a period of time the bankfull discharge carried more suspended sediment than either higher or lower discharges. Recently a practical sampling device for measuring bedload became available. The earlier hypothesis was proved correct by accumulation and analysis of adequate sediment data obtained over a wide range of discharge values.

The work of E. D. Andrews is especially important in demonstrating the verity of the concept. He called the discharge that carried the largest amount of sediment the effective discharge. This term had been used previously, but in a qualitative and hypothetical sense with only a vague reference to the relation of effectiveness to bankfull flow. In that same

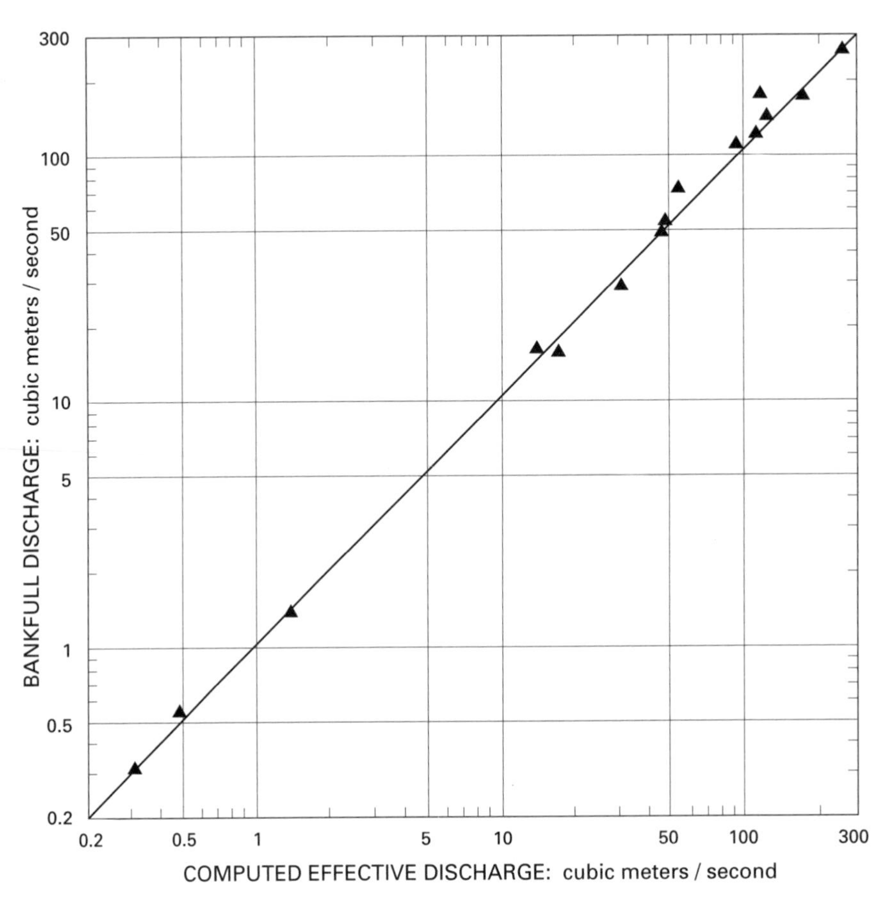


*Figure 8.2* The relation of computed effective discharge to bankfull discharge for stations in the Yampa River Basin, Colorado and Wyoming. (After Andrews 1980.)

sense the words "channel-forming discharge" had been used. Andrews put the concept on a quantitative basis, as shown by his analysis of the relation of discharge to total load carried over time by the Little Snake River (Figure 8.1). The near-coincidence of bankfull discharge to the flow carrying the largest load illustrates the concepts and is borne out by the relation of effective discharge and bankfull, as illustrated in Figure 8.2.

Recall that most channels are curved, not straight. Over time a curve tends to migrate by erosion of one bank balanced by deposition on the opposite bank. Actual field observation confirms that the erosion rate, the sediment transport rate, and the bar building by deposition are most active when the discharge is near bankfull. The highest discharges carry the most sediment during their passage; but they are so infrequent that

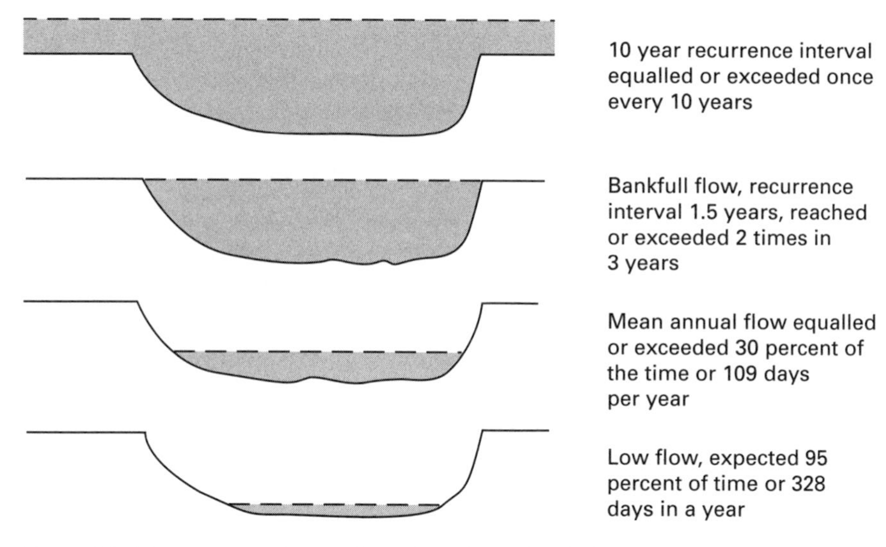


*Figure 8.3* The amount of water in a river channel and the frequency with which such an amount occurs.

over time they do not accomplish as much work as the more frequent events.

## The Channel and Flows of Different Frequencies

We have said that the mean annual flood has a recurrence interval of 2.3 years. A discharge of this magnitude is slightly larger than bankfull and flows over the floodplain. The relationship of flows of this and other magnitudes is shown in simplified form in Figure 8.3.

The mean annual flow of a river is equaled or exceeded 25 to 30 percent of the time, or about 91 to 109 days a year, so about 265 days a year the discharge is less than the average value. In other words, the average discharge is a rather large flow.

The average discharge fills the channel about one-third full; that is, the depth is one-third bankfull when the average discharge occurs. Bankfull discharge occurs 1 or 2 days each year and has a recurrence interval that averages 1.5 years. Less frequent discharges exceed the channel capacity and overflow the flood plain.

The preceding remarks express typical relations. Details differ somewhat among geographic regions and from one climate to another. Therefore any quantitative use requires data from local gaging stations.

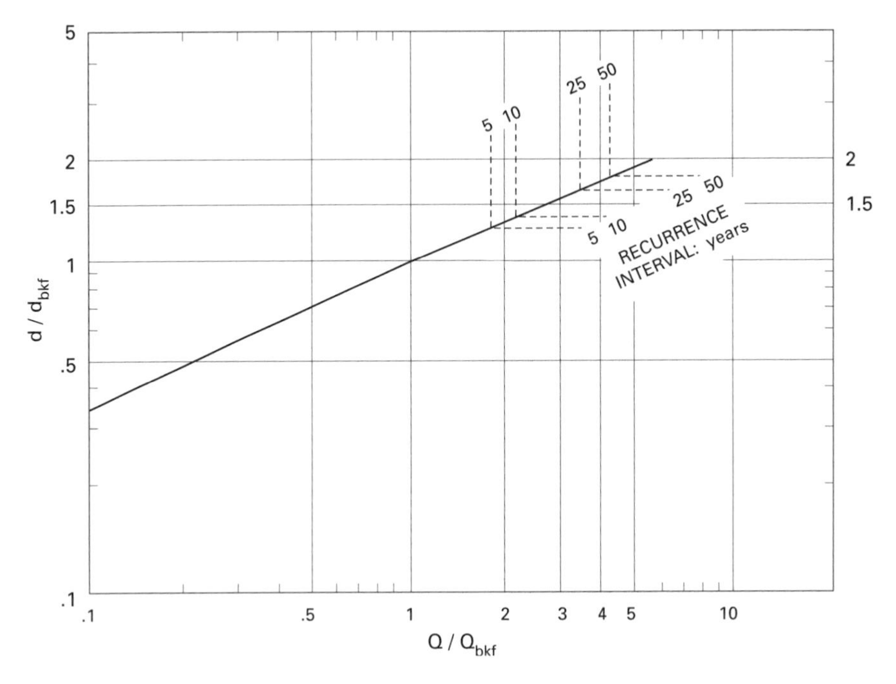


*Figure 8.4* A dimensionless rating curve for stations in the eastern United States.

As explained in the discussion of gaging station records, the discharge rating curve is a plot of the gage height as a function of discharge. The rating curves of different stations are similar but not identical, owing to the differing shapes of channel cross sections and the differing vertical placement of the stage gage scale. As in many aspects of hydrology, comparison is facilitated by expressing quantities in dimensionless form as ratios. Rating curves of different stations become quite similar if expressed in dimensionless coordinates. The ordinate of the dimensionless rating curve, instead of being in feet, is the ratio of observed depth to bankfull depth. The abscissa is the ratio of observed discharge to bankfull discharge. As shown in Figure 8.4, at bankfull stage the mean line through individual points must pass through the coordinates $d/d_{\text{bankfull}} = 1$, $Q/Q_{\text{bankfull}} = 1$.

The relations among depths and discharges at average annual flow and at bankfull are shown for several regions at the end of this chapter.

Dimensionless rating curves differ somewhat among stations, for they are influenced by climate, geology, and size of drainage basin. Yet Figure 8.4 for the eastern United States is nearly identical to the comparable rating curve for the Salmon River, Idaho. For present purposes, our example exhibits the principles and conveys a generalized picture of the relation of discharge values to the average river channel.

Figure 8.4 shows the values of ordinate and abscissa for floods of different recurrence intervals. These values differ somewhat for different regions, but those depicted are applicable to the eastern United States.

One of the interesting results of the analysis leading to Figure 8.4 is the expected average depth of the 50-year flood. The data show that for rivers of moderate size that have a floodplain, the 50-year flood is expected to overflow the floodplain to a depth equal to 80 percent of the bankfull depth of the river.

## Criteria for Bankfull Stage in the Field

It is a nice generalization to draw a cross section of an average channel with an ideal shape and flat floodplain. In the field the situation is often far less clear. Every location along a channel reach is slightly different in shape, vegetation, location and form of bars, and nature of bank materials. Therefore it is important to have a set of criteria for identifying the bankfull level.

The first step is to ascertain the expected level of the floodplain, the bankfull level, for a river of given size in the region. To this end the curves in Figure 8.5 have been developed from a large number of field observations in various parts of the United States. The lines are drawn through data points for basins in (1) high-rainfall areas such as Pennsylvania, with average annual precipitation of 45 inches, (2) Mediterranean climates of winter rainfall such as the San Francisco area, and (3) mountain areas of the western states such as Idaho, Colorado, and Wyoming. The graphs show the mean channel width and mean depth at bankfull as functions of drainage area. The latter is particularly useful because if the drainage area is known, the mean height above the streambed can be read where the bankfull level is expected. This gives a first, important clue regarding where to look for indicators of bankfull. Terraces or remnants of abandoned floodplains are common, and these may be easily mistaken for currently building point bars or for the floodplain.

For example, to identify the bankfull level in a basin of 20 square miles area in a mountainous part of Utah for which no curve is shown, one would look at curves for the Green River, Wyoming, and the Salmon River, Idaho. Figure 8.5 shows that the bankfull depth might be expected at 1.5 to 1.9 feet above the bed. This knowledge would constrain the

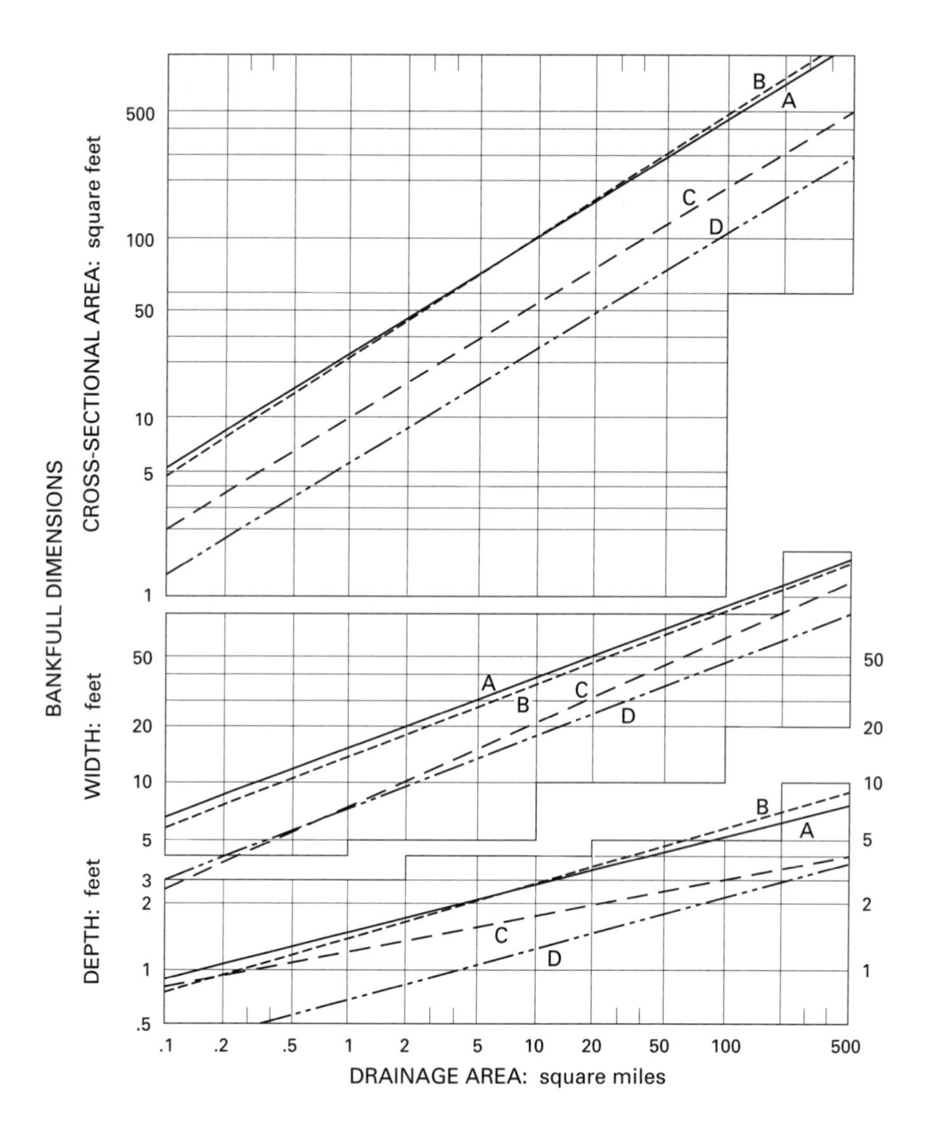


*Figure 8.5* Average values of bankfull channel dimensions as a function of drainage area for four regions. (After Dunne and Leopold 1978.)

range of indicators sought in the field inspection. The principal indicators in order of usefulness are as follows.

1. The point bar is the sloping surface that extends into the channel from the convex bank of a curve, as shown in Figures 1.3 and 1.4. The top of the point bar is at the level of the floodplain because floodplains generally result from the extension of point bars as a channel moves laterally by erosion and deposition through time.

2. The bankfull level is usually marked by a change in vegetation, such as the change from bare gravel bar to forbs, herbs, or grass. Shrubs and willow clumps are sometimes useful but can be misleading. Willows may occur below bankfull stage, but alders are above bankfull. In Idaho the lichens on rocks changes species and thus color at bankfull level. In ephemeral channels the bankfull stage is marked by changes of plant species.

3. There is usually a topographic break at bankfull. The stream bank may change from a sloping bar to a vertical bank. It may change from a vertical bank to a horizontal plane on top of the floodplain. The change in topography may be as subtle as a change in slope of the bank.

4. Bankfull is often registered by a change in the size distribution of materials at the surface, from fine gravel to cobbles, from sand to gravel or even fine gravel material. It can change from fine to coarse or coarse to fine, but a change is common.

5. Even more subtle are changes in the debris deposited between rocks, such as the amount of leaves, seeds, needles, or organic debris. Such indicators are confirmation rather than primary evidence. Flood-deposited debris alone should not be trusted.

## Procedure for Designating Bankfull in the Field

Experienced geomorphologists will often be able to specify the bankfull elevation by mere inspection, but even they are sometimes in error. It is best not to depend on primary indicators at just one location or just one cross section. Even reliable evidence may lie at somewhat inconsistent elevations, so the best procedure is to use as many local indicators as can be found in a short reach of channel.

The following is the procedure I prefer. Choose a reach of length equal to about 20 widths and carry to it flagging material of two colors and a handful of nails. One color is used to flag bankfull indicators and the other to flag terrace remnants. When you find an indicator of the type described above, put a nail through a piece of flagging and push the nail

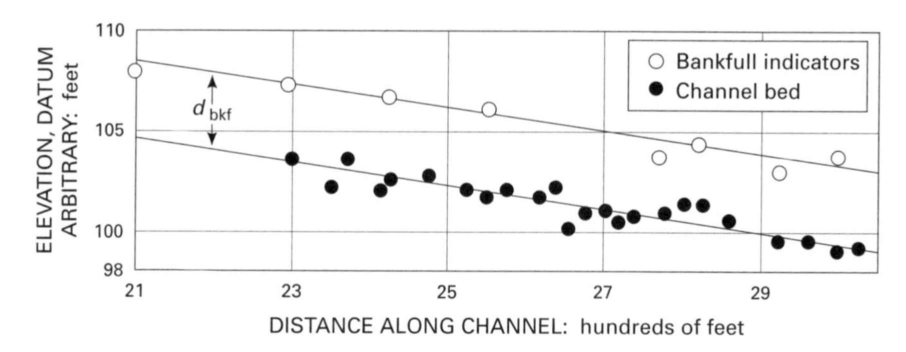


*Figure 8.6* Profile of channel bed and indicators of bankfull, Watts Branch near Rockville, Maryland. Note that the bankfull profile is parallel to the mean streambed profile.

into the ground. When all the indicators are flagged, tape along the channel setting stakes at distances of, say, 20 feet, marking each stake with its distance. Then survey the elevations of the water surface and bankfull indicators at all flagged locations. Plot the data in your field notebook while still at the scene, to ensure that there is some consistency in the profiles of terrace, bankfull locations, and water surface. Figure 8.6 is an example of such a profile.

If the procedure is carried out at a gaging station, choose as the datum some precise and describable object, preferably the bronze plate established as the bench mark when the station was constructed. Survey the elevation of some definite gage height such as the top of the lowest gage plate, usually 3.32 feet above the zero of the gage. When you plot the profile, the elevation where the bankfull line crosses the gage cross section can be read as a gage height. The bankfull discharge can be determined by entering that gage height in the discharge rating curve.

## Frequency of Bankfull Discharge

It has been mentioned that bankfull discharge has an approximately constant recurrence interval, 1.5 years in the annual flood series. This is a remarkable fact that deserves discussion.

Terrace heights above the present streambed vary, although the three terraces prominent in many western U.S. valleys usually lie within a definite range of heights. Typical relations were illustrated in Figure 1.9, in which the three terraces stand above the present stream at approximately 20 to 30 feet, 7 to 10 feet, and 2 to 4 feet. The range is broad enough to preclude any semblance of a constant recurrence interval flood. Terrace heights are reached by floods of no constant recurrence interval, and this makes the bankfull level unique.

The computation of the recurrence interval at which bankfull stage is reached requires two independent types of evidence. The first is a record of discharge at the site sufficiently long so that a flow-frequency analysis is possible. The second is a field survey of the type just described, which includes flagging or noting the field indicators and performing a levelling survey of them. The level of the bankfull stage can then be entered into the discharge rating curve, and the bankfull discharge is obtained.

Once the field-determined bankfull discharge is known, its recurrence interval can be determined from the flood-frequency curve of the gaging station. This procedure has been carried out for a reasonable number of stations. Because many of these analyses were performed by persons who did not themselves make the field surveys, an unknown degree of possible error or variance exists. Nevertheless, all investigations conclude that the bankfull discharge has a recurrence interval in the range of 1.0 to 2.5 years, and the value 1.5 appears to be a reasonable average.

Surveys made by me and my colleagues in the upper Green River Basin, Wyoming, and stations in Colorado, Idaho, and Pennsylvania are compiled in Figure 8.7. The 1.5-year recurrence interval discharge was read from the frequency curve. That value and the field-determined bankfull discharge are compared in the figure. Of the 42 points, 3 are strongly aberrant. But the agreement of the preponderance of the data shows that the 1.5-year recurrence interval gives a reasonable estimate of bankfull discharge.

## Bankfull Discharge and Drainage Area

The field surveys to determine bankfull are more tedious than the office work of developing a flood-frequency graph. Field surveys require travel and equipment. Therefore, many more computations of bankfull discharge have been made using the 1.5-year recurrence interval than are available from fieldwork.

In some basins the relation of bankfull discharge to drainage area is quite definite, with only modest scatter of points about the mean line. Such a plot was shown in Figure 5.6 for the Salmon River Basin, from field surveys made by W. W. Emmett.

Where such reasonable relations exist, values of bankfull discharge can be derived from frequency studies for regional comparisons. An example

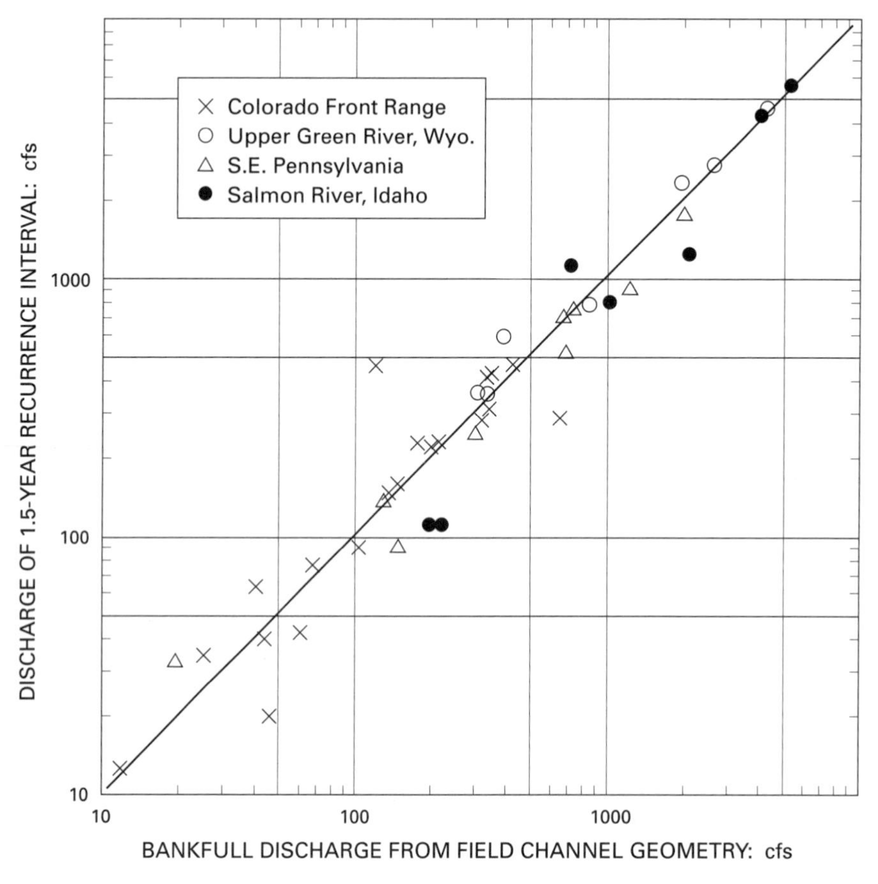


*Figure 8.7* Comparison of field-determined bankfull discharge with discharge of 1.5-year recurrence interval for 4 regions in the United States.

was shown in Figure 5.7, in which bankfull discharge was related to drainage area for several regions. Let me emphasize that these mean lines have been drawn through points with considerable scatter in the plot.

The different slope of the lines is presumably indicative of the importance of widespread rainstorms as a causal factor in the incidence of large floods. Wyoming and Idaho streams are fed primarily by snowmelt, and heavy rainfalls are local when they occur. As a result, bankfull discharge for large basins is small compared with both eastern and far western locations. For basins of 100 square miles, Pennsylvania streams have bankfull discharge three times larger than that of Wyoming. For the same drainage area, both North Carolina and coastal California have discharges five times larger than those of Wyoming.

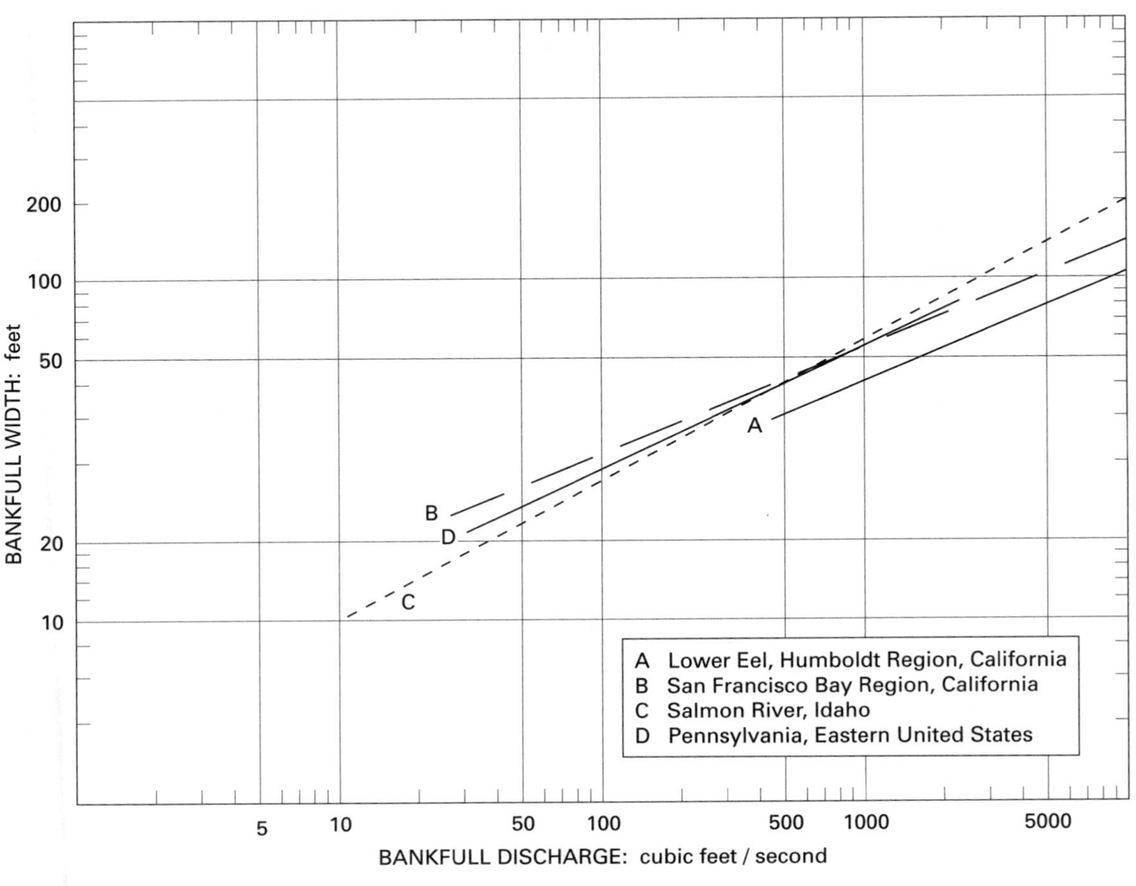


*Figure 8.8* Bankfull width as a function of bankfull discharge for several regions in the United States.

## Bankfull Width—A Conservative Parameter

It is worthy of note that the channel width at bankfull is closely related to bankfull discharge over a wide range of channel sizes and regional climates. If we ignore the individual station points and present only the mean curves, the plot in Figure 8.8 shows data for four widely separated regions. The bankfull width is similar for the same discharge at these locations.

The determination of bankfull discharge requires special fieldwork or the availability of well-developed regional curves. In contrast, average annual discharge is published for every gaging station, and the parameters of width, depth, and velocity for any in-channel discharge can readily be ascertained from the current-meter data. Therefore, the relation of channel width at average discharge and average annual discharge is available for a much larger number of river basins. An example for 11 basins in California is shown in Figure 8.9.

A similar plot including only the mean lines, not the individual gaging stations, is shown in Figure 8.10 for 11 other river basins, the mean discharge ranging from 0.05 cfs in ephemeral channels of New Mexico to 6.5 million cfs for the Amazon. Through 8 orders of magnitude, channel width increases as the square root of discharge in the downstream direction. For any discharge value the width-to-depth ratio is large in channels of types $B_4$, $B_5$, $C_4$, and $C_5$ in the Rosgen system (as in the Republican River). The ratio is small in channels of fine-grained material, such as $B_6$ and $C_6$ (as in the Tombigbee). The relation is sufficiently definite that an approximation of mean discharge can be obtained by a simple measure of width if stream type is considered. Even if we look at extremes of width, at 1,000 cfs mean flow the narrowest channel is about 105 feet and the widest 230, but in most basins channel width is in the range 140 to 180 feet.

So consistent is the relation of channel width to discharge that many empirical studies have developed correlations that permit an estimate of mean channel discharge from a simple measurement of bankfull width. Moreover, the measurement of width in the field is not subject to large error even if the elevation of the bankfull condition is not correct. The width is not much different whether the elevation chosen is at bankfull or somewhat below bankfull. Depth, on the other hand, is very sensitive to an exact choice of bankfull elevation.

The correlation of average discharge with width is not the only utility of the width measurement. Discharge at various overbank frequencies is also correlated with width, though with a somewhat larger standard

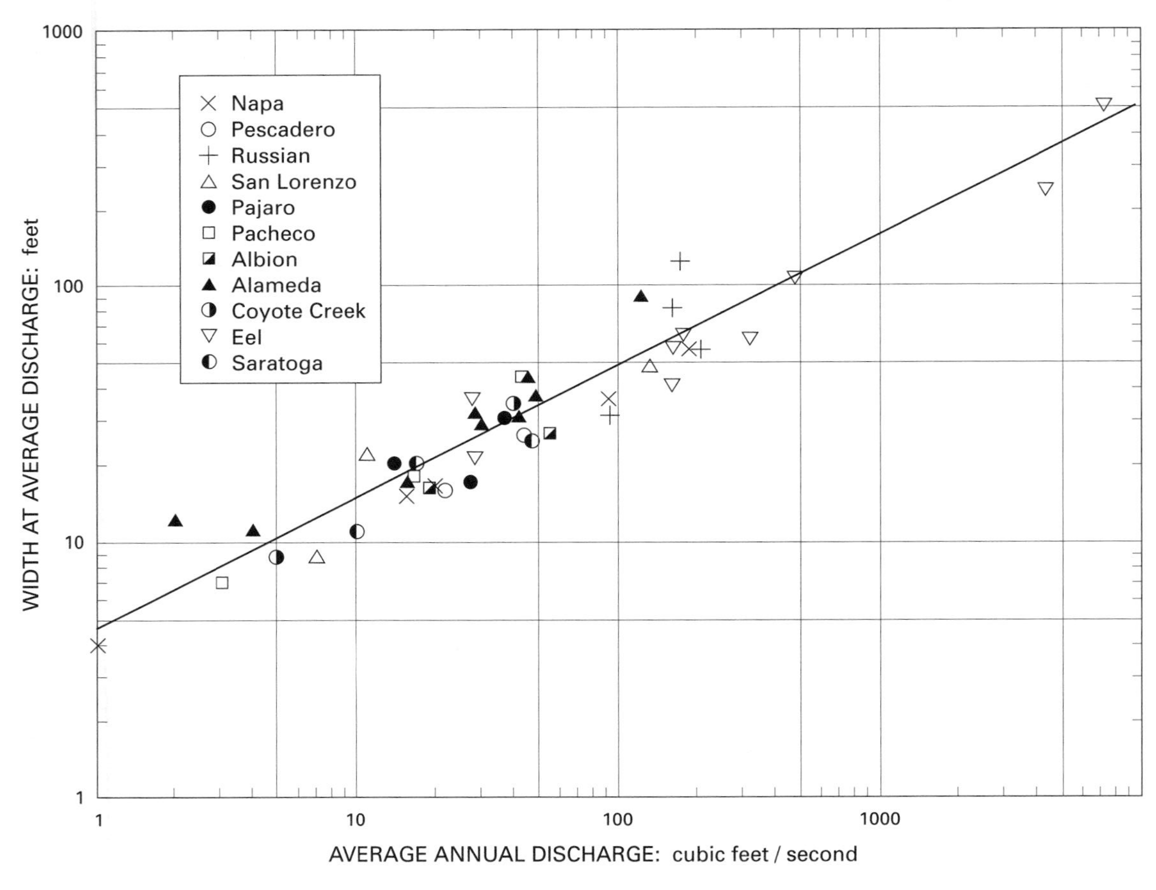


*Figure 8.9* Channel width at average annual discharge plotted against average annual discharge for 45 gaging stations in 11 basins in California.

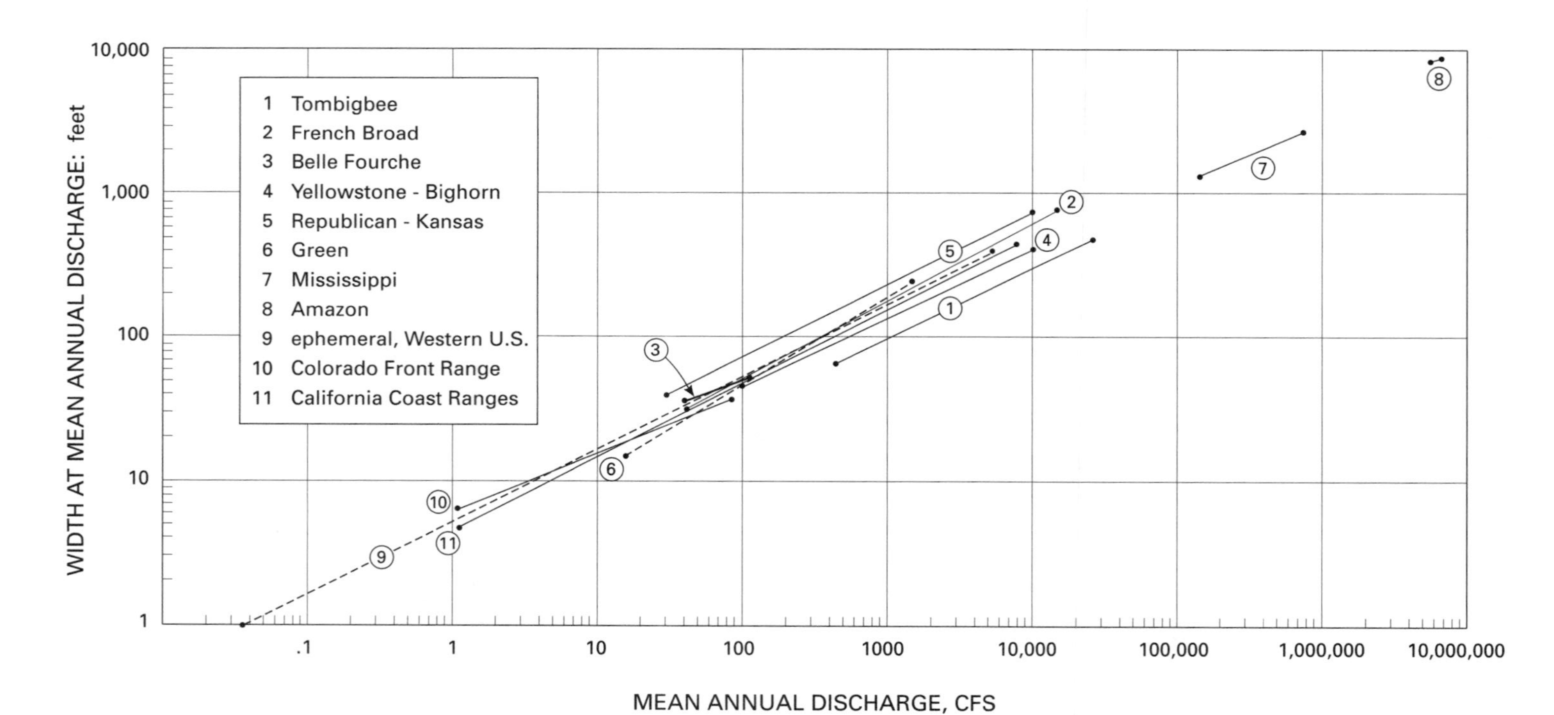


*Figure 8.10* Channel width at mean annual discharge in relation to mean annual discharge for very small to very large basins.

error. Typical examples of such correlations permit an estimate of the discharges of recurrence intervals of 2, 5, 10, and 25 years. Hedman, Moore, and Livingston (1972) estimated the discharge of the 10-year flood at 53 stations in Colorado. The data gave the equation using bankfull width, $W$: $Q_{10} = 3.64\ W^{1.604}$. The comparable equation based on 135 gaging stations in Idaho is $Q_{10} = 1.73\ W^{1.66}$.

Regressions based on width alone gave somewhat more consistent correlations than when other parameters were added and led to considerably better results than equations using basin parameters such as drainage area, elevation, and precipitation.

Such equations have been developed for a number of regions including Utah, California, Kansas, Missouri River Basin, Colorado, Wyoming, Nevada, and New Mexico.

## A Channel Geometry Survey

The bankfull discharge is very important in river science. It is the discharge that forms and maintains the channel. It is the discharge that, over a long period, carries most of the sediment. Very little bedload is moved by the river at discharges less than bankfull. It is the discharge above which flood damage begins.

For these reasons the determination of bankfull has been prominent in litigation concerning river preservation and instream flow. I am therefore including recommended procedures for field determination of bankfull. Later I will give a series of useful average values of river discharge and the appropriate morphologic relations.

Carrying out the steps described below is in itself a learning process of great utility. The procedure takes about one day in the field and one day in the office for each gaging station. Its purpose is twofold: to organize basic data in a form that can be used for many practical needs, and to gain direct experience in the most common hydrologic procedures.

### *Section 1*

1. Go to the nearest U.S. Geological Survey office or to a library that carries government documents, and look at the volumes called *Water Resources Data, Colorado* (or other state name). There is a separate volume for each year. The new volumes have a map of the gaging

stations. Record the names of those in and near your area.

Consult not only the most recent year but every volume of the last 10 years to determine the stations that have been discontinued. Photocopy the page of the publication for each gaging station.

2. Contact the central state office of the Geological Survey and ask them to copy for you the Form 9-207 for each gaging station on your list. This form summarizes current-meter discharge measurements at the station. Ask also for a flow duration curve of daily discharges for each station where these curves have been compiled.
3. Plot the location of each gage from the latitude and longitude given in the published record. Determine how to drive to each location.
4. Prepare to visit each station. Your equipment list should include survey level and rod, 100-foot cloth tape, a dozen ½-inch rebars 2 to 3 feet long, maul, colored plastic ribbon, field notebook, engineer's scale, pencils, topographic quadrangle (topog quad) maps of the area of each station, the page of xerox data for each gage, 10-to-the-inch cartesian plotting paper, some tenpenny nails and string for unanticipated purposes, and wading boots.

*Section 2*

5. Go to the gage. Inspect the valley for terraces. Draw a sketch of the valley cross section showing terraces, and draw a plan view of the river reach showing the gage, river curves, bars, pools and riffles—to scale, but estimated by eye.
6. Look for the bench mark near the gage, often a brass plate set in concrete. If not found, set a rebar nearby as a bench mark and show it in detail on your map. Read the gage height (water level on the outside staff gage plate). Record the gage height and time of reading. Place the rod on top of the gage plate, or where a good reading can be obtained, and determine the elevation of the zero reading of the gage plate, that is, the gage datum relative to your bench mark.
7. Consult the graph of bankfull depth versus drainage area to get an idea of where to look for bankfull indicators. Walk the reach, and flag with colored ribbon each place you can find a reasonable bankfull indicator. Flag also elevations where low terrace seems clear.
8. Beginning at the bench mark, survey a long profile with shots on the water surface, the center line of the channel, on the flags set for bankfull, and flags on low terrace if present. Plot the long profiles in your notebook immediately and see if the profile of water surface,

channel bed, bankfull, and terrace are about parallel, as they should be. Insert on the plot the staff gage scale. Where the bankfull profile crosses the gage plate, read the gage height of the bankfull stage.

9. Find a nearby location that you consider representative of the channel reach, preferably in a straight portion at the lower end of a pool or head of a riffle.
10. Establish the two ends of the cross section with rebars. Stretch a tape, zero end on left bank. Survey the cross section and immediately plot it in your notebook before moving the instrument. Show on the sketch map the location of the cross section and the elevations of the iron pins marking the section endpoints.
11. Find the bankfull elevation where the profile crosses the cross section and draw the bankfull level of the water surface on the cross section.
12. Write a short description of the bank material and its stratigraphy.
13. Make a pebble count of 100 rocks as close to the center of the stream as possible and at the lower end of a pool (upper end of a riffle). Compute the cumulative numbers and immediately plot data representing the size distribution in your notebook. The size fractions are in millimeters: <4, 4, 5.6, 8, 11, 16, 22, 32, 45, 64, 90, 128, 180, 256. If sand or silt, call it <4
14. Now that the slope, width-depth ratio, and size of bed material are known, classify the reach in the Rosgen classification.
15. Read over this checklist before leaving the site.

*Section 3 (office work following fieldwork)*

1. Replot the cross section on a convenient scale and show on the ordinate axis both the elevation scale from the survey and the gage height scale for the gaging station. Counting squares, construct a new graph of cross-sectional area as a function of elevation.
2. From the Forms 9-207 plot the hydraulic geometry curves of width, depth, and velocity versus discharge. It is useful to copy onto one sheet the hydraulic geometry curves, the rating curve (gage height versus discharge), and cross-sectional area—all versus discharge.
3. For each gaging station tabulate the annual flood for each year of record, arrange in rank order, and compute the recurrence interval. Read the discharge value for the 1.5-year flood from the flood-frequency curve. Tabulate these values with the drainage area for each station. Plot bankfull $Q$ (1.5-year recurrence interval) against drainage area. Compare the graphs with those of other regions.

*Table 8.1* Summary of river data for Napa River Basin, California

| Station | Redwood Creek, near Napa |
|---|---|
| Years of record | 15 |
| Drainage area (square miles) | 9.79 |
| USGS station number | 11 45 82 |
| $Q_{average}$ (cfs) | 10.8 |
| $Q_{1.5}$ (cfs) | 1,160 |
| Values at $Q_{average}$ | |
| Width (feet) | 14 |
| Depth (feet) | .75 |
| Velocity (feet per second) | 1.0 |
| Exponents in the hydraulic geometry | |
| $b$ (exponent of width) | 0.30 |
| $f$ (exponent of depth) | 0.30 |
| $m$ (exponent of velocity) | 0.40 |
| Values at $Q_{1.5}$ (approximately bankfull) | |
| Width (feet) | 62 |
| Depth (feet) | 3.0 |
| Velocity (feet per second) | 6.2 |
| Average annual precipitation | 22.7 inches |
| Slope | |
| Map | 0.010 |
| Field | 0.0096 |
| $Q_{1.5}/Q_{average}$ | 107 |
| $Q_{1.5}$/drainage area | 118 |
| Bed material | |
| $D_{50}$ (millimeters) | 90 |
| $D_{84}$ (millimeters) | 140 |
| $Q_5$ (cfs) | 1,230 |
| $Q_{10}$ (cfs) | 1,450 |
| $Q_{25}$ (cfs) | 1,600 |
| $Q_5/Q_{1.5}$ | 1.06 |
| $Q_{10}/Q_{1.5}$ | 1.25 |
| $Q_{25}/Q_{1.5}$ | 1.38 |

4. Now begin to fill out the summary sheet for each gaging station (see Table 8.1). The average annual flow is given in the published record along with drainage area, years of record, and values for extreme events. From the hydraulic geometry curves read off the width, depth, and velocity for $Q_{average}$ and for $Q_{1.5}$. Measure the slope from the profile. Read bed material sizes from the pebble count graph. Obtain average precipitation from a good isohyetal map, as in many of the volumes of the series, *Magnitude and Frequency of Floods in the United States*, USGS Water Supply papers 1671 to 1688.
5. From the flagged elevations as estimates of bankfull, the gage height and elevation were determined. These entered on the hydraulic geometry curves give the field estimate of bankfull discharge. Among the ways of estimating bankfull discharge, the above is the best—field elevation used with measured values in the hydraulic geometry.
6. Estimate discharge from bankfull width. Read bankfull width from the graph of regional relations of width versus discharge.
7. Go to the published flood-frequency report for the basin in which you work and use the nomogram or equations to plot the flood-frequency graph from the regional data. This result can be compared with the result from the individual station.
8. Read the cross-sectional area at bankfull. Multiply the area by calculated velocity to get $Q$. This calculated velocity should be compared with the real velocity from the hydraulic geometry.

When all the field and office steps are completed, it has been found very useful to enter the measurements into a standard or specific table, a sample of which for one gaging station is shown in Table 8.1.

## Some Useful Average Values

In many hydrologic problems it is convenient to know the order of magnitude of a quantity you are about to compute. Also, it is useful to carry in your field notebook average values of quantities so that when you look at a river, you will know the order of magnitude of various quantities that might apply to that river. The following data are typical of those I carry in my field book.

### *Relations to Drainage Area*

Drainage area is the parameter most easily determined. On a topographic map the outline of drainage area upstream of a chosen point can be readily drawn. Trace the outline on a piece of tracing paper. Then lay it on a sheet of cross-section paper. Count the number of squares in the area and convert to square miles or acres as needed.

Given the drainage area and considering the region, estimate bankfull width, depth, and cross-sectional area from Figure 8.5. Estimate bankfull discharge from Figure 5.7. Compare with the following data on mean annual discharge and bankfull discharge for other basins.

*Mean Annual Discharge*

| | |
|---|---|
| Most of eastern United States | 1 cfs per square mile |
| Western Cascade Range | 5 cfs per square mile |
| California Coast Range | |
| Mountainous | 3 cfs per square mile |
| Inland | 0.2 cfs per square mile |

Compare mean annual discharge values $Q_{ave}$ with values of bankfull discharge $Q_{bkf}$:

Bankfull discharge in relation to mean annual discharge:

For 21 stations in Coast Range area of California, $Q_{ave} = 0.034\ Q_{bkf}$

For 13 stations in eastern United States, $Q_{ave} = 0.12\ Q_{bkf}$

For 20 stations in Front Range of Colorado, $Q_{ave} = 0.14\ Q_{bkf}$

Flood discharge in relation to bankfull discharge, the latter approximated by 1.5-year recurrence interval (subscript numbers refer to recurrence interval of the discharge):

For 42 stations in Coast Range of California, $Q_5 = 4.5\ Q_{bkf}$

$Q_2 = 1.9\ Q_{bkf}$

This depends in part on average precipitation:

For Eel River Basin, 9 stations, wet climate

$Q_5 = 2.4\ Q_{bkf}$

$Q_2 = 1.3\ Q_{bkf}$

For Alameda River Basin, 9 stations, dry area

$Q_5 = 7.1\ Q_{bkf}$

$Q_2 = 1.9\ Q_{bkf}$

Percentage of channel depth filled at average annual discharge,

8 stations in Wyoming

$d$ at $Q_{ave} = .43\ d$ at bankfull

Many stations in eastern United States

$d$ at $Q_{ave} = .30\ d$ at bankfull

Frequency of occurrence of discharge equal to mean annual discharge.
Average for most regions in United States
flow is $\geq Q_{ave}$ 25 percent of time
Ratio of instantaneous peak discharge to average daily discharge for same day
30 examples, Maryland—Delaware
$Q$ peak $= 2.9\ Q$ average for day
30 examples, Yellowstone River Basin, Wyoming
$Q$ peak $= 1.5\ Q$ average for day

Ratio of momentary discharges of given recurrence interval to bankfull discharge:

| Location | $Q_5/Q_{1.5}$ | $Q_{10}/Q_{1.5}$ | $Q_{25}/Q_{1.5}$ | $Q_{50}/Q_{1.5}$ |
|---|---|---|---|---|
| Salmon River, Idaho (9 stations) | 1.6 | 1.9 | 2.2 | 2.5 |
| Seneca Creek, Maryland (60 years) | 2.7 | 4.2 | 7.5 | 12 |
| Eastern United States (13 stations) | 1.8 | 2.1 | 3.3 | 4.2 |
| Eel River Basin, California (12 stations) | 2.4 | | | |
| Coast Range, California (42 stations) | 4.5 | | | |

Ratio of depth of flood flow of given frequency to depth when bankfull:

| Recurrence interval (yr) | Eastern U.S. | Idaho | Western Cascade, Wash. | Puget Lowland, Wash. | Average |
|---|---|---|---|---|---|
| 1.5 | 1.0 | 1.0 | 1.0 | 1.0 | 1.0 |
| 5.0 | 1.3 | 1.2 | 1.3 | 1.3 | 1.2 |
| 10.0 | 1.4 | 1.3 | 1.5 | 1.4 | 1.4 |
| 50.0 | 1.8 | 1.4 | 2.0 | 1.7 | 1.7 |

CHAPTER NINE

# A Field Example: Watts Branch

## Flow Data and Morphology

When you realize that nearly 20,000 stream gaging sites have been measured in the United States, the mere number of figures published each year seems formidable.

From continuous chart records or digital punch tapes, the mean flow for each day is usually computed in 15-minute intervals. The published record includes not only the mean flow each day but momentary peaks, the rating curve data (relation of stage to discharge), location, drainage area, and maximum observed flow.

If, however, you consider the many other geomorphic and morphologic characteristics that might be described for any reach of river, the flow record actually appears minimal. Not every gaging station location need be subjected to a geomorphic inventory, but some useful characteristics of rivers are so little known that even a modest number of observations would be helpful.

Bankfull stage—the discharge at which overflow begins—is one such characteristic. No uniform procedure is used by government agencies to determine this factor, even when it is stated in published records. The definition of bankfull stage is not consistent from one location to another or among government agencies. For example, for large rivers the U.S. Weather Service lists the "flood damage stage" for many river locations within the purview of its flood-forecasting service. But damage stage is dependent not only on the amount of overflow but on the location of damageable property nearby. The definition is not only inconsistent between sites but can change through time at any single site because of development of the floodplain.

In a study of 71 river gages, the recurrence interval of "flood damage stage" varied from 1 year to 8 years with the mode at 2 to 3 years. As shown previously, the morphologic definition of bankfull is a discharge of recurrence interval 1.5 years. So, "flood damage stage" is definitely overbank.

## Drainage Basin and Channel

A fairly complete description of one example demonstrates not only how many geomorphic and morphologic characteristics are observable, but also the nature of a typical small river. Over some years I have observed in detail the Watts Branch, a tributary of the Potomac River in Maryland. The location described is that portion of the stream near the gaging station designated as Watts Branch at Rockville, Maryland, drainage area 3.7 square miles. I selected this location as an observation area in 1952 and had the gaging station installed in 1956. I did not devote much time to the location on a continuous basis, but observed many phenomena sporadically or repetitively over a span of 20 years. I did some work at the site about every month each year. The modest amount of time spread over the years has paid dividends both in facts and in improved understanding. I highly recommend this continuity of observation: a small amount of effort each time but spread over a long period. I would hope that the types of phenomena observed and measured suggest other parameters to observe, even if large blocks of time are not available.

The relief of the basin of Watts Branch is 215 feet and more or less circular. In 1956 the drainage area was mostly in agricultural land and pasture. Of the 3.7 square miles, 0.73 square mile was woods (20 percent) and 0.20 built up or residential (5 percent). About 426 houses were in the basin at that time, but urbanization has taken over appreciably since then. The stream characteristics discussed here refer to the channel in its 1956 condition.

The reach of the river that has been the object of principal study is in a pasture extending some 2,000 feet downstream from the gaging station. This portion of the stream has a few well-developed meander bends, but over its length its sinuosity (thalweg length divided by valley length) is 1.45. Typical views of this reach are shown in Figure 9.1.

## Channel Cross Section

At the start of the work in 1952, I selected the most uniform and straight reach of the natural channel for discharge measurement. The channel cross section appears in Figure 9.2.

The figure shows that the channel is roughly rectangular, with the right

bank somewhat sloping. The top width is 22.0 feet and the depth is 3.3 feet. Using the peak flow records of a 7-year period, 1958–1964, before urbanization of the basin became rampant, I computed the number of times per year given depths or stages were equaled or exceeded. These are indicated on the cross section. In this stream at this location, the peak flow reaching bankfull stage occurs on the average between once a year and twice a year. The peak flow equaled or exceeded once a year is 0.7 foot above bankfull.

As we know, peak flows refer to momentary peaks. Because peak flows usually last only a few hours, the average flow for a given day is smaller than the peak. It is the average for the day that is used to compute the flow duration curve, which is a cumulative curve of the percentage of time a given flow parameter is equaled or exceeded. From such a curve for the daily flows in Watts Branch the data have been translated into gage height in the channel and are shown on the right side of Figure 9.2.

Direct observation of the discharge when overflow began in Watts Branch gave a gage height of 3.3 feet as the bankfull stage at the natural section. This established the bankfull discharge as 240 cfs. A plot of the flood frequency graph for the 30 years of record, 1958–1987, shows the mean annual flood (recurrence interval 2.3 years) to be 630 cfs and the 1.5-year recurrence interval discharge as about 380 cfs. Thus, in this river the actual bankfull discharge is smaller than the discharge of the 1.5-year recurrence interval. The recurrence interval for 240 cfs is read as 1.15 years.

Gage height is synonymous with stage and refers to the reading on a vertical scale or gage plate, which in the United States is graduated in feet and hundredths of feet. The placement of the zero of the gage plate is arbitrary. Depth of the flow, as defined in Chapter 1, is the height of a rectangle of the measured cross-sectional area and width. It is the quotient of cross-sectional area divided by width.

In this particular stream the mean annual discharge is 4.2 cfs, which corresponds to a gage height of 0.48 foot. The natural channel therefore fills to 15 percent of its bankfull depth. The average annual flow is exceeded in this channel 22 percent of the time. In other words, 78

---

*Figure 9.1* Watts Branch at Rockville, Maryland, in the reach studied over a period of 20 years. Above, looking upstream at a typical straight section. The staff gage in the right foreground marks the location of one of the natural cross sections where flow measurements have been taken. Below, looking upstream at the flood flow over the point bar of a meander at the bankfull stage. The exposed bank in upper center is the edge of a low terrace that stands about a foot above the present floodplain and the currently building point bars.

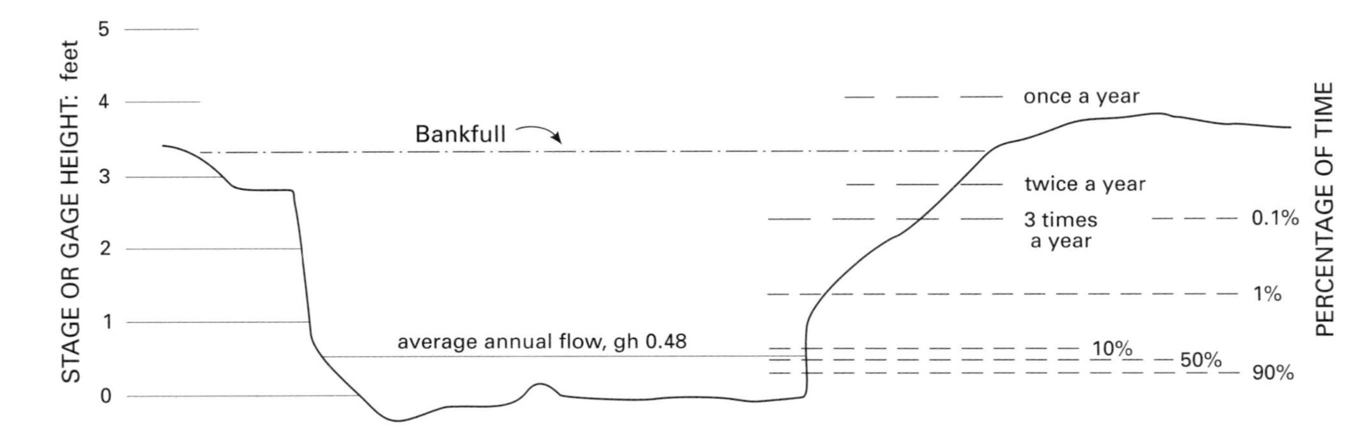


*Figure 9.2* Cross section of the natural channel of Watts Branch. The scale near the right bank indicates the depth of water in the channel equaled or exceeded one or several times each year by flood peaks. The scale on the far right shows the percentage of time that the depth of water in the channel is equaled or exceeded.

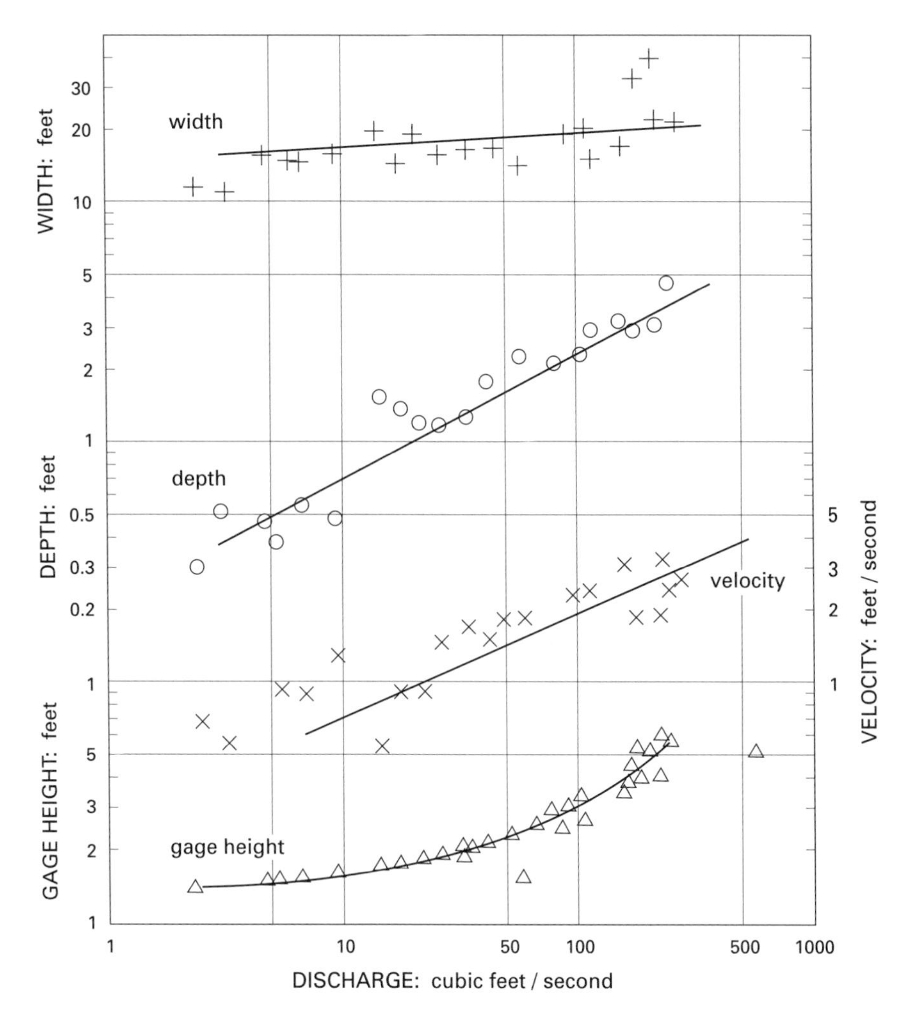


*Figure 9.3* The hydraulic geometry of the natural channel of Watts Branch.

percent of the days in a year, the flow is less than mean annual discharge. The figure for most rivers in the humid-temperate half of the United States is comparable: 75 percent of the time the flow is lower than mean annual.

## Channel Geometry

The flow characteristics of the natural reach of channel are presented in Figure 9.3: velocity, gage height, mean depth, and width as functions of discharge. In natural channels the relation of gage height to velocity, area, and depth do not plot as straight lines on log-log paper, but it is usual that these parameters plot straight lines against discharge.

Although the discharge rating curve, the plot of gage height against discharge, does not plot as a straight line on log-log paper, a major portion of the curve is more or less straight and generally has a slope of 0.34. Indeed, that is exactly the slope of the rating curve for Watts Branch between discharge values of 30 and 200 cfs. This characteristic slope is often very useful when dealing with ungaged basins and limited data.

In Watts Branch the field-determined bankfull discharge does not have a recurrence interval of 1.5 years as expected. Therefore, other indicators of bankfull should be sought to test the field indicators. Here the discharge rating curve is helpful. In Figure 9.3 note the measurements deviating from the line at discharge between 160 and 240 cfs. Notice also the large values of water surface width at 180 and 220 cfs. These tend to confirm the bankfull discharge of 240 cfs from field observation, a value supported by the hydrographer's entry on the Form 9-207. At measurement 105 on September 14, 1966, the hydrographer measured 401 cfs in the water over the floodplain and outside the channel. Of a total discharge of 658 cfs, the portion in the channel was 257 cfs.

## The Floodplain and Point Bar

The actual floodplain of Watts Branch is quite narrow. A wide valley floor stands slightly above the presently building floodplain.

This situation is typical of many parts of the eastern United States, where there has been a minor degradation of small streams. I conjecture that the cause was the extensive planting of cotton and tobacco in the eighteenth and early nineteenth centuries. The result is a pair of remnants standing slightly above the floodplain. On Watts Branch this terrace is about 1 foot high. In places it is not a scarp but a slope so gentle that identification as an abandoned floodplain is not obvious. The present floodplain, being only slightly lower than the much wider valley flat, could be overlooked by the casual observer. The valley flat could be called a floodplain, when in fact it is a low terrace—low enough to be often flooded, just as a floodplain would be. The general pattern of the river, its relation to the valley flat, and the streambed profile can be seen in Figure 9.4.

Because the floodplain should be parallel to the streambed in profile, a line is fitted through the observation points representing point bars and other bankfull indicators. If this line is lower in elevation than another family of points representing observed levels of a berm, this higher level is presumed to be an abandoned floodplain or terrace.

To obtain the average bankfull stage at a gaging station nearby, the

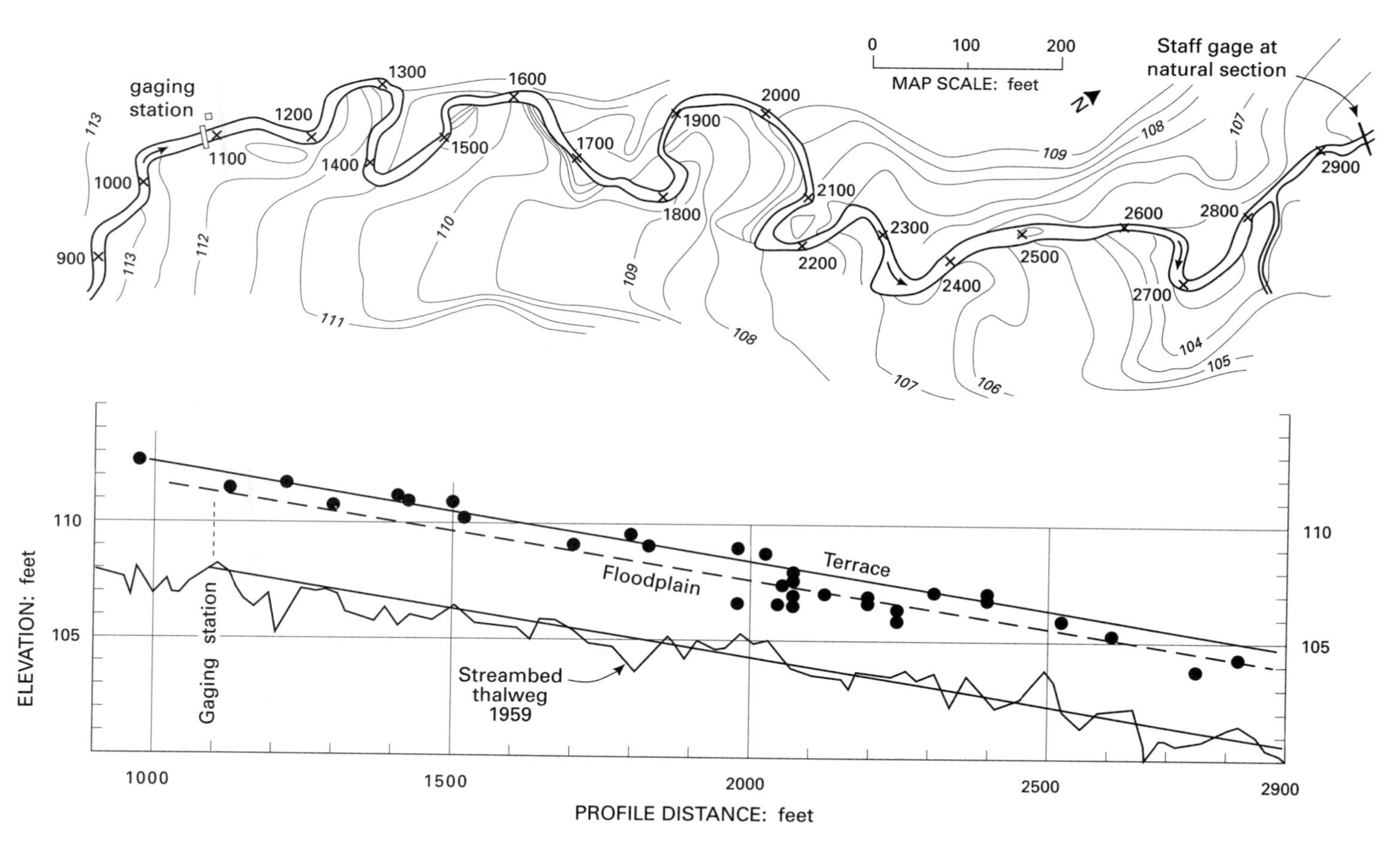


*Figure 9.4* Plan and profile of the study reach of Watts Branch located 500 feet downstream of Highway 28.

profile line representing the floodplain is extended to the location of the gaging station, as shown in Figure 9.4. The elevation of the point of intersection, expressed in terms of the staff gage of the station, will be the gage height of the bankfull stage. This gage height, entered in the rating curve of the station, gives the bankfull discharge. As can be seen, the elevation of the intersection is 111.4 feet and the zero of the staff gage of the gaging station is 107.7 relative to the datum of the profile survey. The gage reading at bankfull stage is then 3.7, which from the rating curve gives a bankfull discharge of 200 cfs. For comparison with the gaging station location, the cross section of the natural channel was an an independent estimate of bankfull discharge, 240 cfs. The agreement is satisfactory considering the local variability of stream bank height.

The streambed of a river is undulating, consisting of alternating deep and shallow sections, or pools and riffles. In the reach studied there are 13 sequences of pool and riffle. The average length of such a pair is 164 feet. The bankfull width is 21 feet, so the average length of a pool-riffle sequence is 7.4 channel widths. This value is in the range typical of most rivers. If we measure the wavelength of individual meander bends along the valley direction, rather than along the channel bed, the average is 127 feet or 6.1 channel widths.

In 1953 I established a series of channel cross sections, the ends of which were monumented with iron posts driven below the ground surface so that they could be relocated by measurement but would not be disturbed by surface use. These cross sections were resurveyed nearly every year from 1953 to 1970. Some had to be abandoned owing to lateral movement of the channel and new cross sections were installed. As of 1970, annual resurvey data were available for 17 years for 8 cross sections, and a minimum of 8 years of record for the others.

The building of point bars and the retreat of the concave bank opposite are well shown in the data for a sample cross section (Figure 9.5). It is evident that during these years the channel moved laterally a distance equal to more than one channel width. Concurrent deposition on the point bar kept the channel width about constant. Various sections measured over the years showed that the net volume of deposition was about equal to erosion. At the point of maximum curvature, erosion slightly exceeded deposition. Just downstream from this point deposition exceeded erosion.

The sections also show that no appreciable deposition occurred on the floodplain surface away from the sloping nose of the point bar. That is, overbank deposition was either absent or too small to measure.

My observations of the material deposited in successive years showed that most of the fine-grained materials that generally characterize the

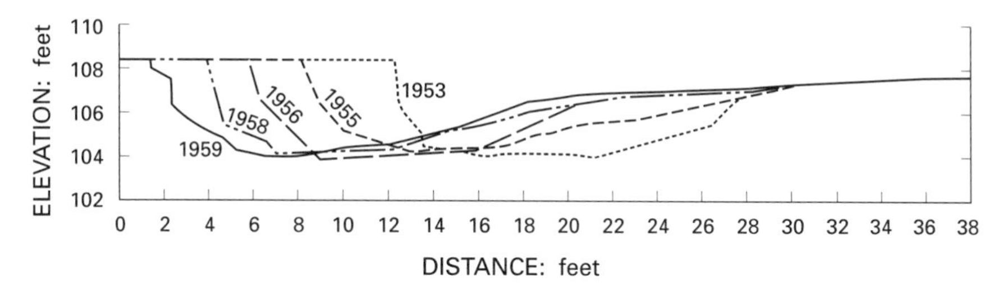


*Figure 9.5* Successive cross sections of Watts Branch from 1953 to 1959 showing the retreat of the left bank from the monument and the extension of the point bar on the right bank. (After Leopold, Wolman, and Miller 1964.)

upper portion of the floodplain of Watts Branch were deposited as point bars or within the channel, rather than by water spreading over the floodplain across the valley in what would be called overbank flow.

## Relation of Valley Alluvium to Bedrock

The geomorphologist concerned with rivers usually has fewer opportunities to observe the relation of a river to the bedrock underlying the valley than does the stratigraphic geologist. A geomorphologist may, on occasion, observe a section opened temporarily to lay bridge foundations or a sewer line. In 1960–1961 a sewer line was laid in the valley of Watts Branch. Even though the construction seriously interfered with studies of the channel, it provided a fortuitous opportunity to observe the relation of valley alluvium to underlying bedrock. A very uneconomical engineering design called for placement of the invert 9 to 12 feet below the valley surface, and thus 6 feet into solid rock, for more than a mile. The average construction engineer unfortunately is unaware of river morphology and of the information that can easily be obtained by inquiry as well as by test boring. I do not know in this instance if the contractor went bankrupt or if the taxpayer shouldered an unnecessarily large cost, but the construction of a simple sewer line took 9 months to proceed about 1 mile, because every foot involved blasting into hard rock.

During those months the exposed stratigraphy was mapped as the trench construction proceeded. The trench was dug in the valley flat over most of the mile. The stratigraphy in Figure 9.6 represents a longitudinal profile down the valley. Although the trench alignment crossed the valley flat and over spurs of colluvium or bedrock, these sections (not representative of the valley floor) have been eliminated in the figure.

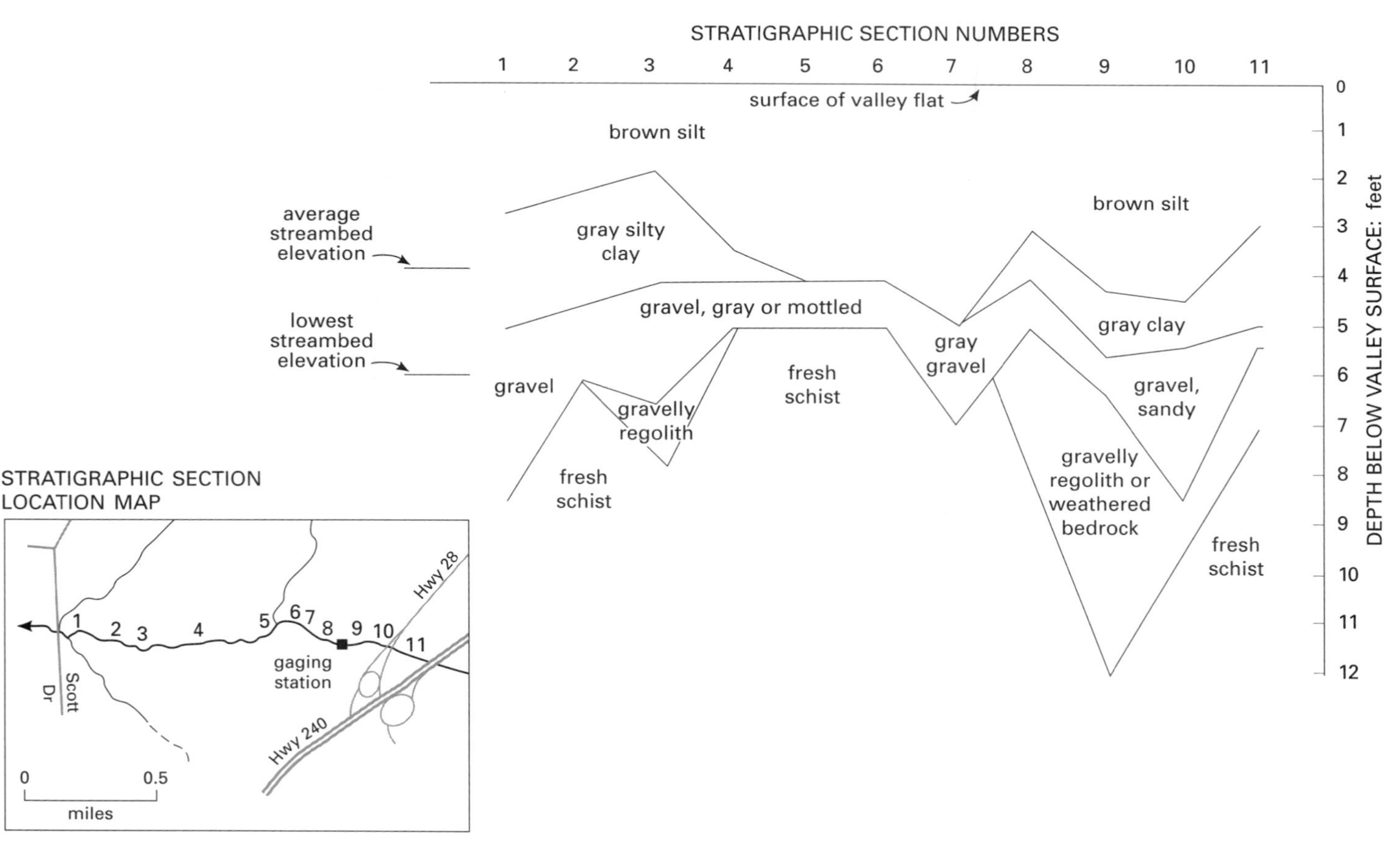


*Figure 9.6* Stratigraphy down the valley of Watts Branch, recorded during the construction of a sewer line. The average and deepest elevations of the stream channel bed are shown at the left. The inset indicates the locations of the sections studied.

There are three principal stratigraphic horizons. The upper 3 feet of the valley flat, or low terrace, is brown silt without obvious bedding; even though it contains some sandy lenses, it is free of gravel. Near the low water in the river, 3 to 4 feet below the surface, the materials are gray, often mottled with limonite staining, and may be clayey, silty, or sandy. The gray color appears within a few years after initial deposition and appears to be the result of reducing processes in materials constantly saturated. Gravel lies at a depth of 4 to 5 feet, coinciding in elevation with the material lying just below the average streambed level. The gravel is usually subrounded, contains much quartz from veins in the schist bedrock, and is gray, purple, or mottled yellowish in color. It is apparently transported by the river.

Bedrock (Wissahickon schist) directly underlies the gravel in most sections and is sometimes fresh, but in places there is a thin zone of rotten or decomposed rock. In several sections the trench exposed an unsorted regolith of angular rocks, sand, and silt, stained with limonite and hematite, often purple with what appeared to be a manganese stain, and without bedding. It has neither the slight rounding nor the stratification associated with the overlying basal gravel of the streambed. Nor is there any pattern of an earlier valley system connecting these pockets of unsorted materials.

The general impression gained from the exposed trench is of a nearly flat valley floor of bedrock underlying the present streambed, only 2 feet or so beneath the bed. The principal stratigraphic units in the valley fill are similar to what the stream is presently depositing on point bars.

The basal gravel is generally pea or cobble size, but occasionally coarser. A size analysis of gravel at a depth of 5 feet below the surface shows a maximum size of 256 millimeters and a median size of 50 millimeters. Both the large size of some of the basal gravel and its gray-to-mottled appearance suggest that the coarsest gravel found a foot or two below the present streambed may be relict, unrelated to the present river regimen. At one section a considerable amount of organic material in the form of sticks and leaves was found on the basal gravel. A piece of wood at the base of the basal gravel was dated by carbon 14 as 250 B.P. The basal gravel, which appears more coarse than the present bed material and possibly relates to an earlier period, was laid down in modern time. This modern date is in keeping with a pollen sample in the brown silt and the gray clayey silt obtained in the stream bank near the gage plate of the natural section. The report on pollen identification stated:

In the gray silt overlying the basal gravel, pollen density was moderately good (ca. 100 grains per square inch slide area); pollen predominantly pine; oak, hemlock, hickory, and walnut present. Herbaceous and weed groups represented by many genera of which ragweed was the most abundant. Corn probably present.

In the upper brown and tan silt and sand, pollen density very low (ca. 20 grains per sq. in. slide area). Tree pollen almost entirely pine. Herbaceous genera poorly represented; ragweed the dominant weed. No corn pollen observed.

Age of sediments: The presence of hardwoods, even in small numbers, plus the high frequency of ragweed and the probable corn grains suggest that the sample corresponds to upper pollen zone C-3. Therefore . . . the date will probably read less than 800 years.

The similarity of present point-bar stratigraphy to that of the valley fill and the two types of geochronologic evidence indicate that present processes account for the valley fill. There is the possibility, suggested by the regolith in pockets of the bedrock floor, that there existed in the Pleistocene or Tertiary a valley slightly more deeply cut into the bedrock than the present one, but it could not have been very wide. The flat valley within which Watts Branch flows was developed by lateral movement of the stream, whose characteristics have changed but little for a period probably measured in tens of thousands of years.

## Suspended Sediment and Bed Debris

Over my years of observation of Watts Branch, most of the work was concentrated between 1954 and 1961. My colleagues and I have collected most of the 63 suspended-load samples available. The number is much smaller than would be available at a regularly operating suspended-load measurement station, but as can be seen in Figure 9.7, the spread of points is sufficient to yield a fairly consistent relation of suspended load to discharge.

Many geomorphologists assume that suspended-load sampling is too routine and time-consuming a chore to be attempted and should be left to an operational office with no duties other than hydrologic data collection. Still, it is quite reasonable for a researcher working on rivers to construct, with personally gathered samples, a usable suspended-load rating curve as an adjunct to the research. A series of samples collected at a wide variety of river stages during the passage of a single flood event will yield a fair approximation of the position and slope of the sus-

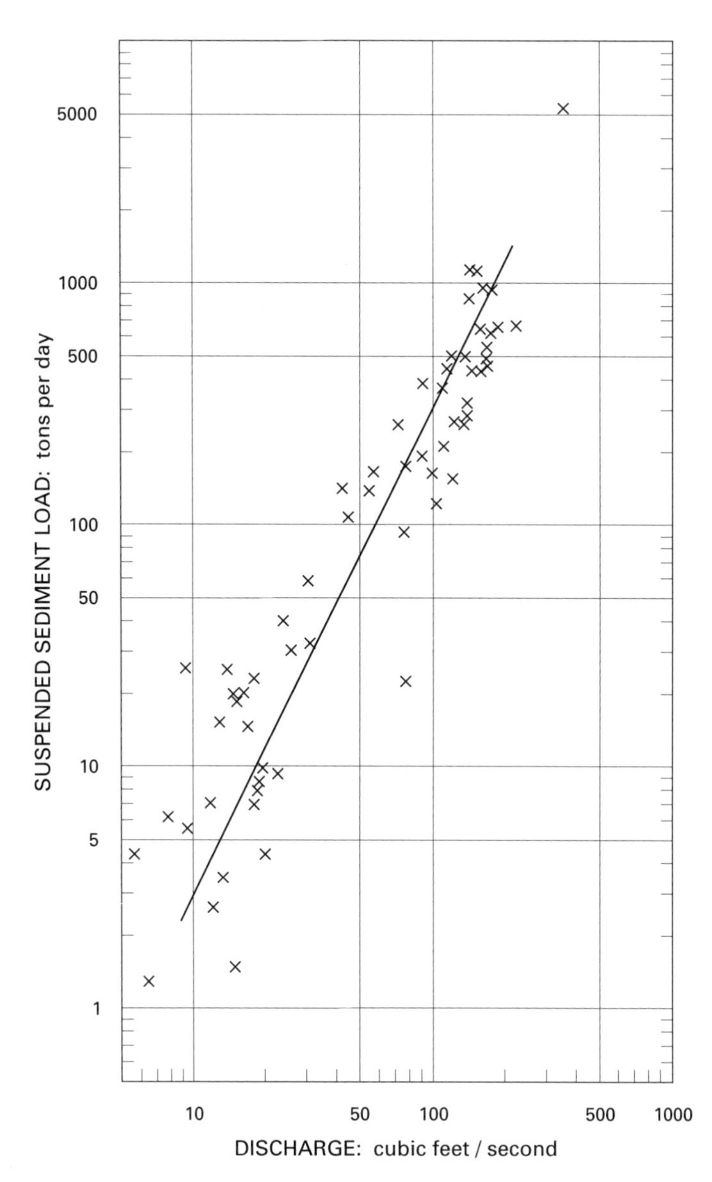


*Figure 9.7* Suspended sediment load as a function of discharge, Watts Branch, 1953–1970.

pended-load rating curve (load versus discharge). Thus, assiduous collection of samples during each of two or three flow events will approximate the sediment rating curve compiled by a data collection organization over a score of years.

The average annual flood, 630 cfs in Watts Branch, carries suspended sediment at a rate of about 10,000 tons per day. Note that no sediment sample was taken at a discharge even approaching the average annual flood, and only one above bankfull discharge. There are no samples at any of the highest peak discharges of record. Despite these gaps, the sampling at more modest flows generated a usable curve.

An average annual suspended-sediment discharge of 230 tons per square mile per year was derived by combining the suspended-load rating curve with the flow duration curve of Watts Branch. As will be seen in Table 11.1 (the tabulation of sediment production for various regions), this estimate for Watts Branch is of the expected order of magnitude for the east central United States.

Nothing is known about the rates of debris transport as bedload, but some observations are available on the movement of individually marked rocks on the bed.

No measurements of bedload transport rate have been obtained. An attempt was made, however, to determine at what relative stage rocks of different size will be moved. This estimation is from initial motion studies in flumes but needs verification by actual measurement in rivers.

At Watts Branch 1,717 rocks varying from 0.5 inch (1.2 centimeters) diameter to 2.5 inches (6.3 centimeters) were painted and placed in a pool location and a riffle location, the two locations identified by different colors of rocks. At nearby Seneca Creek, drainage area 100 square miles, a comparable design was employed using 2,741 painted rocks of similar size. Also, some identifiable angular boulders were placed on the bed. The rock groups were observed over a period of nearly two years. Searches were made periodically for rocks moved out of their original positions.

This experiment to determine the flow needed for initial motion was quite different from the one in New Mexico (described in Chapter 4), the object of which was to measure the effect of rock spacing on movement. The painted rocks on the gravel bars of Watts Branch and Seneca Creek were lying at the surface of the bar. They were picked up, repainted, and returned to the bar surface approximately as they had been deposited by flowing water.

The technique appears easy, but the record keeping is cumbersome and interpretation of results—which are subject to the whim of flood events—often less than satisfying. Independent trials are needed for confirmation

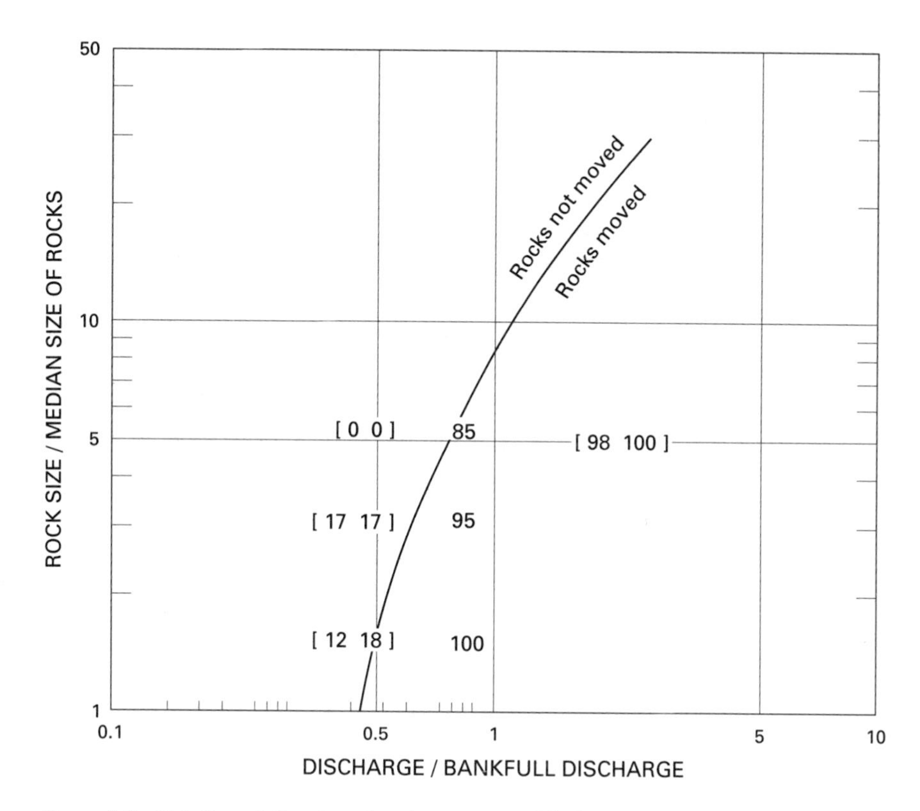


*Figure 9.8* Relation of the size of rock on a gravel-bed river which will be moved by various discharges. Rock size is expressed as the ratio to median size on the bed; discharge is expressed as the ratio to bankfull discharge. The numbers indicate the percentage of rocks in a group that were observed to have moved. Data without brackets are for Seneca Creek at Dawsonville, Maryland. Data in brackets are for Watts Branch at Rockville, Maryland; the left figure is the percentage moved from the pool, and the right figure is the percentage moved from the riffle.

or improvement. To combine the data for two rivers of different sizes, each observation is put in dimensionless form as a plot of two ratios: rock size divided by size of median rock in the streambed, and discharge divided by discharge at bankfull stage (see Figure 9.8). Each observation is shown as an entry on the graph, plotted at the largest discharge value occurring between the dates of observation. For each observation, the numbers represent the percentage of rocks moved from a pool or from a riffle.

A tentative line has been drawn separating no movement from observed movement. It appears that rocks of median size of this gravel-bed stream (1 on the vertical scale) are not moved by flows less than 50 percent of bankfull discharge, which means when the channel is flowing at three-quarters bankfull depth. Rocks about 10 times median size will be moved at about bankfull discharge. Although not indicated by the data used in the graph, I observed that a larger flow is needed to move a given rock size out of a pool than off a riffle.

This early work with painted rocks has been extended and made more sophisticated. E. D. Andrews has shown by direct observation that the $D_{84}$ size on the streambed (84 percent of the rocks are smaller) will be moved by bankfull discharge. His conclusions were corroborated by measurements made by David Rosgen and me in mountain streams in Colorado. These measurements also support the general conclusions reached from the painted rock data of Watts Branch and Seneca Creek.

## Bank Erosion and Sediment Sources

There have been approximately 1,636 suspended-load sampling stations in the United States, but the record at most stations is short. In 1990 the number of stations whose records are available was 662.

In very few cases is it possible to identify exactly the origin of the sediment being measured. Some excellent studies have related sediment yield to drainage area and type of use (see the summary in Table 11.1). Tracing the origin and place of deposition of sediment is a difficult problem and worthy of more work.

In Watts Branch gully head extension is minor. The principal sources of suspended load carried by the river are sheet erosion, seasonal rilling on plowed agricultural land, and bank erosion of channels (which tends, however, to be balanced within the system by deposition on point bars a short distance downstream).

The causes of bank erosion in Watts Branch have been studied by M. G. Wolman and me and the results appear to have applicability to many channels in the region. It is generally supposed that erosion of a riverbank occurs during peak discharge from the shear caused by high-velocity flow against the banks, but in many types of rivers this factor is not important. Rather, bank material is softened, crumbled, granulated, or slumped by other processes which prepare a supply of debris for movement by the high flow. Slumping due to wetting is especially important in semiarid areas, in the steep-walled gullies in valley alluvium. A seepage force develops from water in floodplain materials flowing toward the riverbank after a flood peak has passed. During high flow, water tends to move from the channel into the bank materials, but when the channel water surface declines, this water drains back toward the channel. Flow from the bank materials toward the face of a riverbank creates a pressure toward the channel bank and reduces the ability of the materials to stand as a vertical free face. Slumping of the bank presents the next high flow with an easily available source of sediment. This process is important in the natural channel of the lower Mississippi River. It also tends to weaken a levee as flood water recedes.

Another process of sediment preparation is ice crystal formation, or frost action. Water moves through sediment to the cut bank. If the bank consists of silt and clay, the flow is slow and provides a reservoir of water to the base of any growing ice crystal. The crystals grow during a night when temperature falls below freezing and may attain a length of 2 inches. A granule of bank sediment will be held at the tip of such an ice crystal. On melting, the granule falls to the base of the cut bank, where such material accumulates as a friable mass. Erosion in Watts Branch is concentrated in the winter months and rarely occurs in summer. Maximum flows are most likely from summer thunderstorms, at which time the stream banks are dry and resistant to erosion. There was practically no erosion during the high flow in July 1956, for example.

Flow occurring after accumulation of the ice-prepared debris carries away the sediment at the foot of the bank. Such a flow may leave a high-water mark clearly defined by the still uneroded debris above the last high-water line.

## Effects of Urbanization

At the beginning of the period of observation, 1953, the land in the surveyed zone was used as pasture. During the period 1953–1970, urbanization encroached on portions of the headwaters. The resulting increase in sediment load was due in part to land opened for construction. Home construction has decreased the amount of drainage area devoted to farming and increased the area of paved roads, roofs, and parking lots.

Among the effects of urbanization is an increase in impermeable area, which in turn increases the surface runoff from a storm and thus the volume of runoff. It speeds the delivery of water into stream channels. The greater peak flow from a given storm then causes a flush of sediment, with consequent deposition within the channel system downstream where the gradient declines. In the same general region over time, there has been a tendency for urbanization to cause channel enlargement.

In Watts Branch the channel decreased in cross-sectional area between 1953 and 1970 an average of 1.6 percent per year. For those cross sections measured during the full 20 years, the 1970 channel cross-sectional area was 0.68 of that existing in 1953, a loss of 27 square feet from the original 83. Most of this loss results from narrowing by the plastering of silt on channel banks. The width-to-depth ratio has decreased. The streambed elevation has risen 0.4 foot, but the channel slope has not changed significantly.

There has been deposition overbank, near the channel but outside the usual high-water channel. This feature is not uniform, but occurs only in some locations. The channel has had a strong lateral movement in some places, but in other sections no lateral change of position has occurred. The largest lateral movement on any cross section was 20 feet, or about one channel width.

Urbanization has occurred over a large area near Washington, D.C., and this process is reflected in the growth of housing and infrastructure in the basin of Watts Branch. In the 3.7-square-mile basin drainage area, the number of houses or their equivalent was counted, or was computed from successive editions of the topographic quadrangle map.

The increase is dramatic, as the data show:

| Year | Number of houses |
|---|---|
| 1950 | 140 |
| 1955 | 420 |
| 1965 | 780 |
| 1984 | 2,060 |

This change has had a marked effect on the flow regimen of the creek. The number of momentary peak discharges has increased with urbanization. It is usual for the annual flood to equal or exceed bankfull stage 2 years out of 3. (This is the meaning of a recurrence interval of 1.5 years.) In terms of number of cases per year as judged by the partial-duration series, in a natural basin uninfluenced by humans, bankfull would be equaled or exceeded somewhat more often than once a year (recurrence interval 0.9 year).

In contrast to this normal figure, the first 10 years of operation of the gaging station on Watts Branch recorded 21 cases of discharge exceeding bankfull, 220 cfs. The number of times various discharge values occurred is tabulated below.

| Period | Number of times discharge exceeded— 220 cfs | 350 cfs |
|---|---|---|
| 1958–1967 | 21 | 10 |
| 1968–1977 | 73 | 37 |
| 1978–1987 | 73 | 32 |

Gaging station records show that the number of overbank flows per year increased from 2 to 7 in the three decades of observation.

Another indication of the effect of urbanization is the value of the mean annual flood, recurrence interval 2.3 years. The mean annual flood is the average of the highest peak discharge each year.

| Period | Average annual flood |
|---|---|
| 1958–1973 | 781 cfs |
| 1973–1987 | 959 cfs |

Seneca Creek, a few miles north of Watts Branch, was also affected by urbanization (see Figure 7.5). Its drainage area is 100 square miles.

| Period | Number of times bankfull was exceeded | Mean annual flood |
|---|---|---|
| 1931–1960 | 35 (1.2 times per year) | 2,973 cfs |
| 1961–1990 | 66 (2.2 times per year) | 6,014 cfs |

To summarize, the years of observation of channel and flow conditions demonstrate the value of a modest program taking little time each year but continued for a reasonable number of years. Successive observations are useful only if permanent bench marks are established that can be relocated, and if the cross sections also are permanently monumented.

CHAPTER TEN

# The Hydraulic Geometry

## Hydraulics at a Given Cross Section

One of the surprising characteristics of rivers is that each cross section, on any river, has been shaped and dimensioned over time to accept a range of flows. There is a consistency from one river to another, and from one cross section to another, in the way the hydraulic parameters change from low flow to high flow.

It is easy to visualize that water depth increases as discharge in a river increases. What happens to water velocity or river width is not so intuitively obvious. The hydraulic geometry is a way of describing these changes in quantitative terms. At a river cross section, as discharge changes the following generalities usually hold:

Both depth and velocity increase substantially with increasing discharge, and at about the same rate.
Width increases slightly with discharge.
Channel flow resistance or hydraulic roughness decreases slightly with increasing discharge.
Water surface slope does not change appreciably with discharge when measured over a distance equivalent to several channel widths.
Suspended load increases rapidly with discharge, and at a much higher rate than any other parameter.

There are many relationships in hydrology that can be described as logarithmic functions, or alternatively as power functions. The principal hydraulic parameters are related to discharge as power functions, often with the considerable scatter that is also common in hydrology. The relations to discharge at a given river cross section can be written as

$$w = aQ^b, \quad d = cQ^f, \quad u = kQ^m, \quad L = pQ^j, \quad s = rQ^z$$

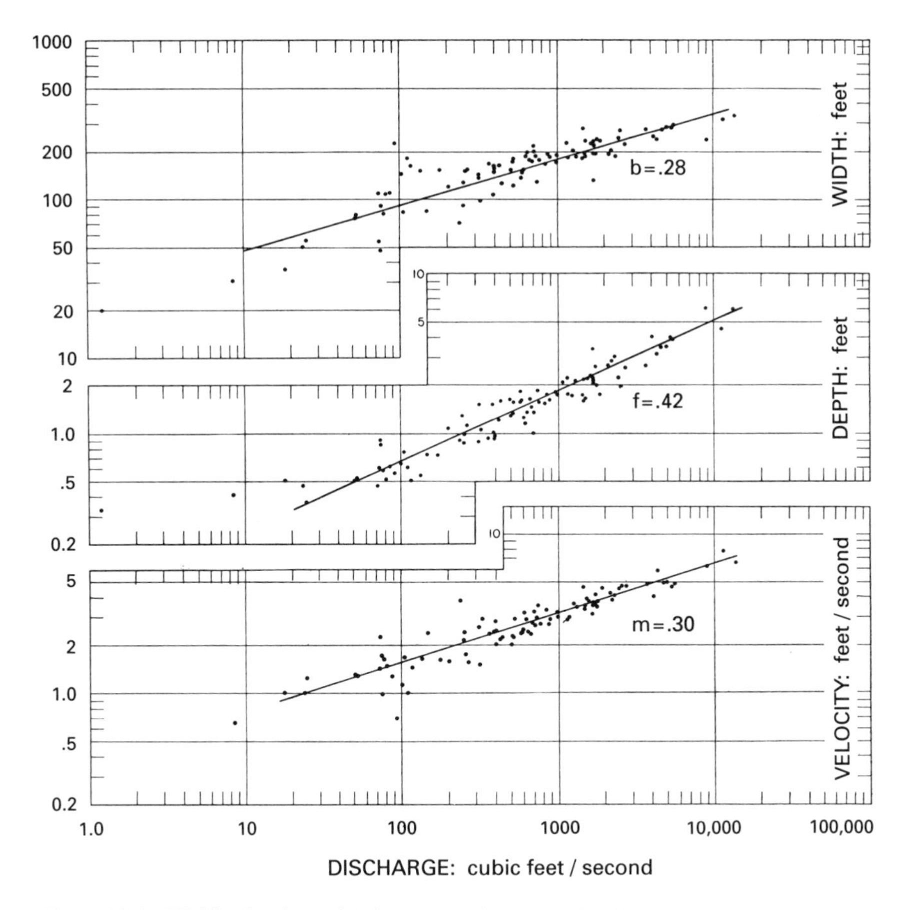


*Figure 10.1* Width, depth, and velocity in relation to discharge. Powder River near Locate, Montana.

where $w$ is width, $Q$ is discharge, $d$ is mean depth, $u$ is mean velocity, $L$ is sediment transport rate, $s$ is slope, the letters $b$, $f$, $m$, $j$, and $z$ are exponents, and $a$, $c$, $k$, $p$, and $r$ are coefficients. These equations applied to changes at a river cross section express how the hydraulic parameters react to discharges with different frequencies of occurrence.

The greatest attention has been devoted to the first three relations, because $Q = wdu$ and thus $b + f + m = 1$. Values of these exponents have been determined by plotting data for gaging stations on many rivers of the world. The field data for plotting come from the individual current-meter measurements, each of which records width, cross-sectional area, gage height, discharge, and mean velocity. In the United States these data are not published but are part of the permanent files of the U.S. Geological Survey, recorded on Forms 9-207. These forms generally are kept in the USGS district office of each state and are available to the public.

A typical plot of the current-meter data at a given river cross section is shown in Figure 10.1 for the Powder River at Locate, Montana. Each plotted point is a measurement. At low flow the hydrographer may move upstream or downstream to pick an appropriate site for current-meter operation. This slight change in measured cross section accounts for some of the scatter of the plotted points. Exponents $b$, $f$, and $m$ in this case have values of 0.28, 0.42, and 0.30, the sum of which is 1.00. The exponent is the slope of the mean line, the tangent measured directly on the plot; for width it is 0.28 unit up for 1.00 unit horizontally. Note that the plot also requires that the products of the three ordinate values must equal the value of the abscissa. In this figure, at 1,000 cfs $w = 185$ feet, $d = 1.8$ feet, and $u = 3.0$ feet/second, or $185 \times 1.8 \times 3.0 = 1{,}000$ cfs.

The exponents $b$, $f$, and $m$ essentially describe both the geometry of the channel and the resistance to erosion associated with the character of bed and banks. For example, a wide, dish-shaped channel would have a rapid rate of increase in width with increasing discharge; a boxlike channel with straight steep sides—such as one might expect to find in cohesive materials—would have a low value for $b$ and a high value for $f$.

## Variation in a Downstream Direction

Discharge increases downstream in a river as tributaries enter. Comparison of various cross sections along the length of a stream requires that the comparison be made for a flow of a given recurrence interval or frequency of flow. The most meaningful discharge for any discussion of channel morphology is the one that forms or maintains the channel. We have seen that this relationship is complex, but that the effective discharge can be approximated by bankfull discharge. In many rivers the bankfull discharge is one that has a recurrence interval of about 1.5 years.

The downstream change of hydraulic parameters can be described in a manner similar to that used to show the changes at a river cross section. Many more data are available on mean annual discharge than on bankfull discharge, so it provides a convenient measure of flow. It is important to recognize, however, that the mean annual discharge is not responsible for the channel form, but simply reflects the result produced by the more effective flows. An example of the change in hydraulic parameters at different stations down a channel system is shown in Figure 10.2 for the

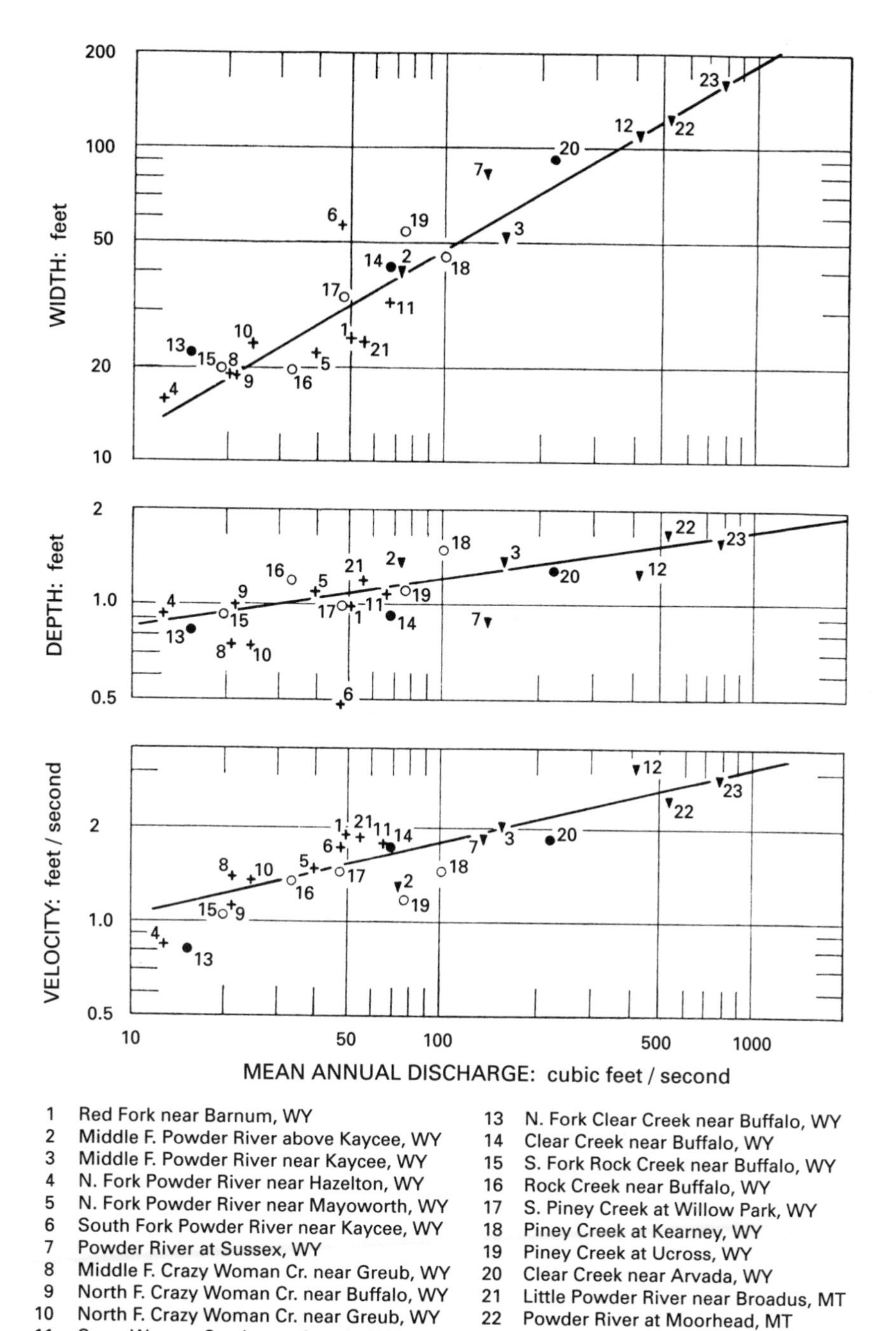


*Figure 10.2* Width, depth, and velocity in relation to mean annual discharge as discharge varies downstream. Powder River and tributaries, Wyoming and Montana.

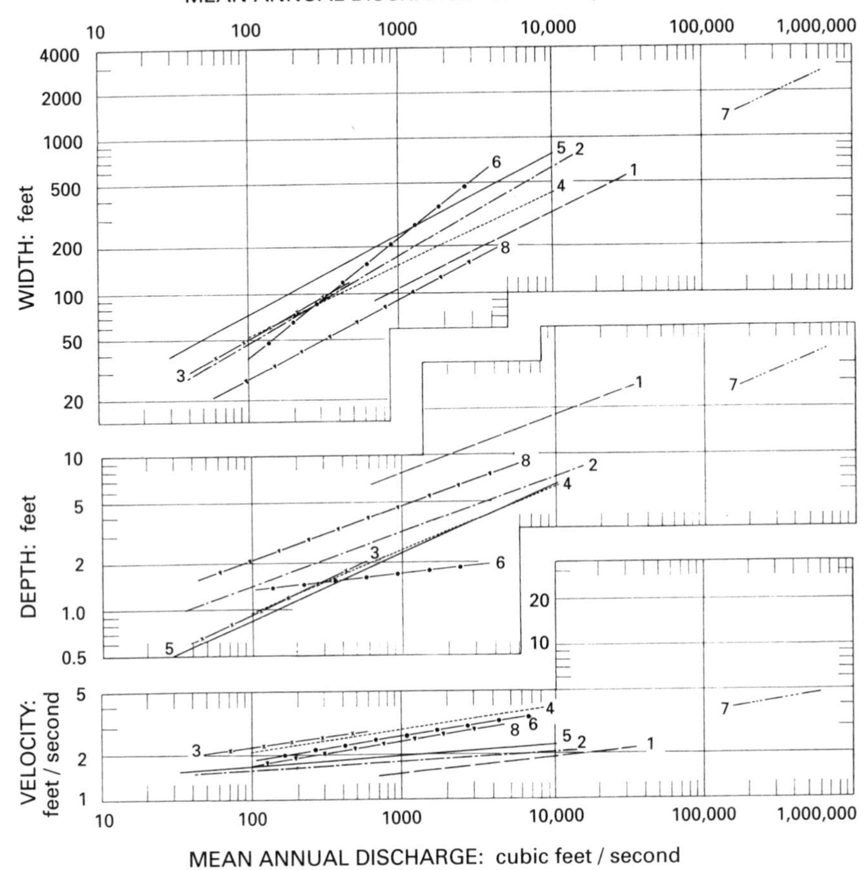


Stream and Location

1 —— Tombigbee, Ala.
2 —·— French Broad, N.C.
3 Belle Fourche, Wyo.
4 Yellowstone - Bighorn, Wyo.
5 —— Republican - Kansas, Kan.
6 Loup, Nebr.
7 Mississippi, main stem
8 Madras irrigation canals, India

*Figure 10.3* Width, depth, and velocity in relation to mean annual discharge as discharge increases downstream in various river systems.

Powder River and tributaries. The discharge chosen for this graph is the mean annual.

Mean annual discharge tends to have similar frequency of occurrence among rivers of different types. This value is equaled or exceeded on the average about 25 percent of the time—that is, about 25 days out of 100. Because mean annual discharge is published for nearly every gaging station in the United States, it is a useful figure.

The width, mean depth, and mean velocity corresponding to flow at mean annual discharge, when plotted against discharge as discharge increases downstream, give curves that constitute further elements of the hydraulic geometry of stream channels. Omitting the points showing values for individual gaging stations, downstream curves for various river systems are shown on a single graph for purposes of comparison in Figure 10.3.

From these curves it can be seen that there is a considerable similarity in the slope of lines among various river systems, even though the intercepts vary. The stable irrigation canals in India (unlined canals that are allowed to scour and fill to achieve an equilibrium form) appear similar to the natural river channels. Most rivers tend to increase downstream—in width, in depth, and in mean velocity respectively—in quite a similar way. These lines may be described by the same equations that were used to describe changes at a station, but with different exponents and coefficients.

The slopes of the lines differ between the at-a-station and the downstream cases. There is a significant difference in the reaction of the hydraulic parameters when discharge increases at a given cross section and when discharge of a given frequency increases downstream. In the latter case, width increases rapidly with discharge, depth has a moderate increase, and velocity increases only slightly or remains constant downstream.

## Interrelation of the At-a-Station and Downstream Conditions

The relation of changes in the hydraulic parameters with change in discharge, at a cross section and downstream, is shown diagramatically by the block diagrams in Figure 10.4. Diagram A shows an upstream section at low flow. (By low flow I mean low discharge.) Diagram C is the same section at high flow. Block diagram B is a downstream section at low flow and D is the same section at high discharge. The lower block diagrams represent the entire river system at low flow and at high flow.

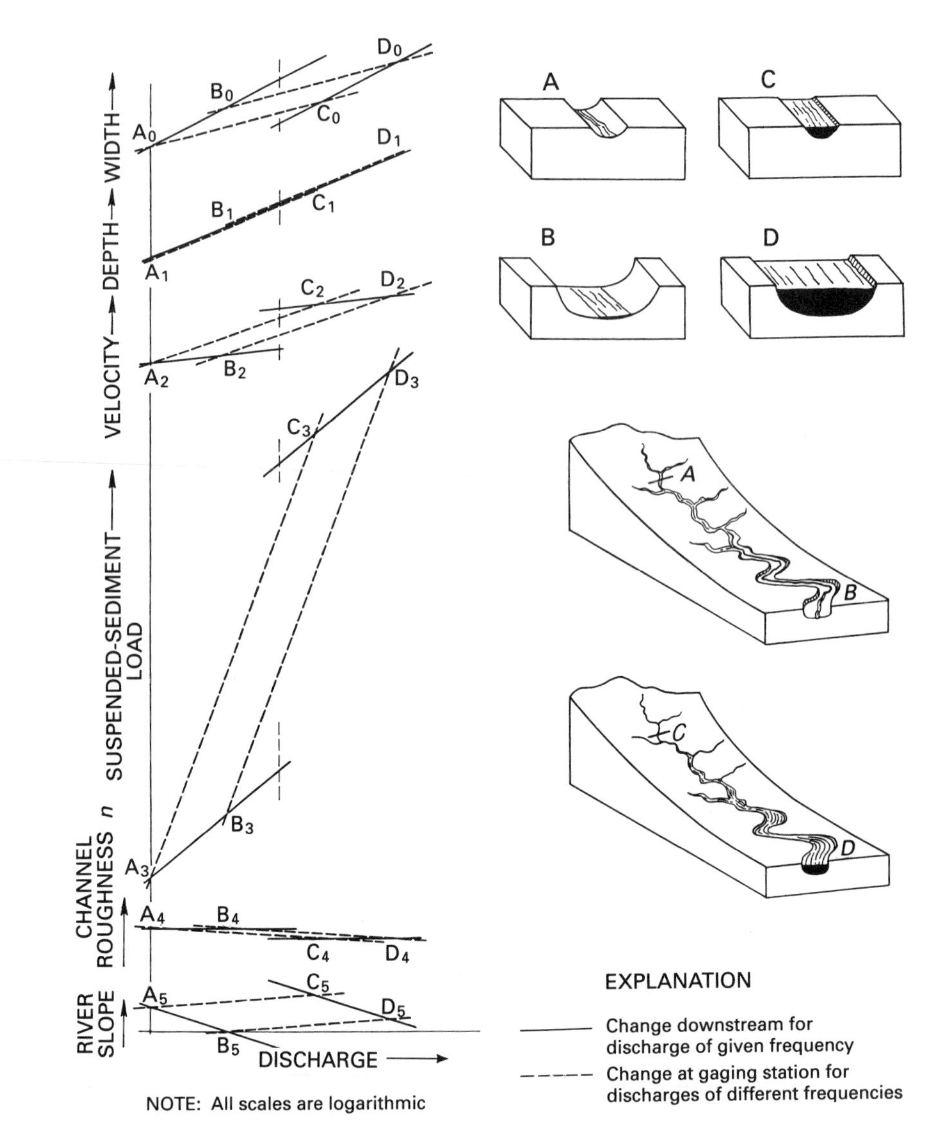


*Figure 10.4* Diagrammatic representation of relation of width, depth, and velocity to discharge at a station and downstream.

The hydraulic geometry shows in a quantitative manner the relations among hydraulic parameters in the channels illustrated in the block diagrams.

There is a pattern of interrelationship of stream characteristics that includes relative frequency of discharge. This hydraulic geometry is basic to understanding the adjustment of channel shape to carry the sediment load supplied to the streams. The relation of hydraulic characteristics at a station to the downstream characteristics can be seen in the graphs at the left of Figure 10.4. The line $A_0$–$C_0$ shows the increase of width with increase of discharge at an individual gaging station near the headwaters of a stream, and $B_0$–$D_0$ shows the same relationship for a downstream station. If the discharge $A$ is of the same frequency (say 50 percent of the time) at the headwater gage as discharge $B$ at the downstream gage, then the line $A_0$–$B_0$ is the increase of width with discharge downstream, corresponding to a frequency of 50 percent. Also if $C$ and $D$ have the same frequency (1 percent of the time), then line $C_0$–$D_0$ is the increase of width downstream at a discharge frequency of 1 percent.

If the slope of the width-discharge curve $A_0$–$B_0$ for a given frequency is 0.5 (exponent $b = 0.5$), a typical value for many rivers, and if the depth curve $A_1$–$B_1$ is parallel, as in Figure 10.4, then the slope of the velocity-discharge curve $A_2$–$B_2$ must be zero ($m = 0$). At that frequency the velocity does not change downstream. The line $A_2$–$C_2$ is the increase of velocity with discharge at the upstream gage, and $B_2$–$D_2$ at the downstream gage. It is clear that at the 1 percent frequency, the velocity is higher at both stations than at the more frequent discharge. Velocity increases rapidly with increase in discharge at a given cross section, but does not change with increasing discharge downstream.

In Figure 10.4 the width-to-depth ratio increases downstream at any frequency. This increase is shown by the fact that $A_0$–$B_0$ slopes more steeply than $A_1$–$B_1$. In nature the relation depicted in Figure 10.4 usually occurs with depth increasing at about the same rate at a station and downstream, and velocity being nearly constant downstream.

The channel characteristics of natural rivers are seen to constitute an interdependent system that can be described by a series of graphs having a simple geometric form, which suggests the term "hydraulic geometry." Channel characteristics of a particular river system can be described in terms of the slopes and intercepts of the lines in the geometric patterns discussed.

Let us now summarize some of the important relations in the geometry, with constant reference to Figure 10.4.

• At a given cross section, channel width does not change much as discharge increases. Width changes very rapidly downstream along the river, and the increase is proportional to the square root of the discharge.
• Depth increases rapidly with increase in discharge at a river cross section. It also increases rapidly downstream. The increase with discharge is about as rapid in the cross section as in the downstream direction. This is shown by the near-coincidence of the lines in the graphs of depth versus discharge.
• Velocity also increases with discharge at a cross section, but in the downstream direction it is nearly constant or increases slightly.
• Sediment load increases at a cross section faster than any other parameter. This is shown by the steep slope of the at-a-station graph of sediment. But the sediment increases downstream at the same rate as the discharge, that is, suspended sediment concentration is about constant downstream.
• Channel roughness or hydraulic resistance decreases slightly both at a cross section and downstream as discharge increases.
• Slope remains fairly constant at all discharges at a cross section, but decreases rapidly downstream.

Even though the numerical values of slopes and intercepts may not provide any visual picture of a river basin, comparison of the values of these factors among rivers has useful aspects. In several publications I have presented values of the exponents $b$, $f$, and $m$ for various river stations and river basins. A far more extensive analysis of values was made by Chris C. Park of Lancaster University, who concluded that presenting single average values of the exponents gives a misleading impression of a regularity that does not exist. Nevertheless, for present purposes I will give some average values to convey to the reader the order of magnitude, and also discuss some of Park's more extensive analysis.

The following are averages of the most common values of exponents:

| Hydraulic exponent | At a station | Downstream |
|---|---|---|
| Width, $b$ | 0.26 | 0.50 |
| Depth, $f$ | .40 | .40 |
| Velocity, $m$ | .34 | .10 |

Park presented two types of diagrams showing the distribution of published values of the exponents at 139 gaging stations and 72 river basins. His histograms of frequency of various values are shown in Figure 10.5, to which I have added a broken line showing the average values quoted above. Park's histograms for channel width are inde-

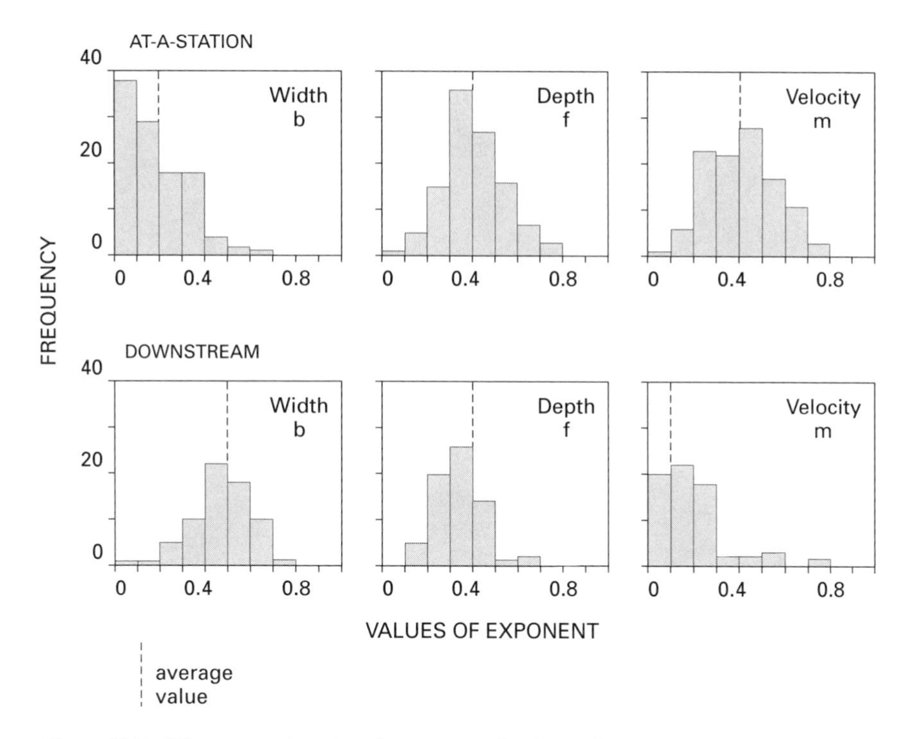


*Figure 10.5* Histogram showing the variety of values of exponents in the hydraulic geometry. The vertical dashed line on the abscissa scale shows the average value of the exponent. (From Park 1977.)

pendently supported by A. D. Knighton. Histograms do not show the simultaneous values of the three exponents, which are of more interest because the three are interrelated. Park's triaxial diagrams for natural channels are shown in Figure 10.6. As he points out, almost any combination has been recorded somewhere.

Although the values of the exponents have considerable variation, one is more consistent than the others. Within any river basin the channel width varies as the square root of the discharge, whether the discharge chosen is mean annual flow or bankfull. This is the value of exponent $b$ in the downstream direction, commonly 0.5. In Figure 8.10 it was shown that this relation exists through eight orders of magnitude. The same square root relation exists also in measurements made by R. I. Ferguson on supraglacial meltwater channels on the surface of the Lower Arolla Glacier in Switzerland. The discharges he measured varied from 0.7 to 3.7 cfs. Similar relations between storm discharge and concurrent widths in large and small semiarid ephemeral washes and gullies were meas-

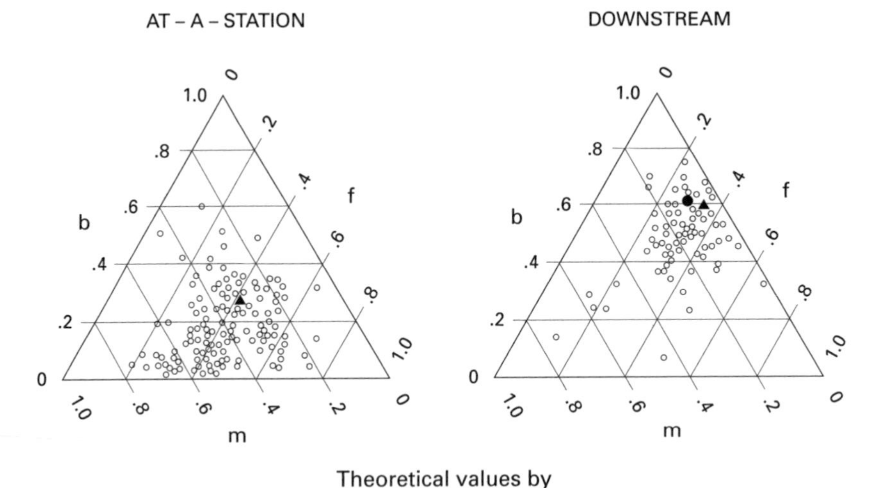


*Figure 10.6* Triaxial diagrams showing the various combinations of exponents in the hydraulic geometry. (From Park 1977.)

ured by John P. Miller and me. Discharges ranged from 0.04 to 5,000 cfs. The largest difference in river width at a given discharge that I have studied in the United States is between a coastal plain river, the Tombigbee, and the Great Plains Republican River, Kansas. The latter is 2.3 times wider than the former. But channels on ice are narrower than on any river of comparable discharge. These differences seem small in the substantial range of discharges in the sample shown in Figure 8.10.

The second important geomorphic aspect of this hydraulic geometry is the influence of chance, that is, the variance introduced by the fact that physical laws do not dictate one and only one combination of the dependent variables. There are eight variables in the relations affecting channels: discharge, width, depth, velocity, slope, roughness, load, and caliber of load. There are not eight equations among them. Therefore, channels can adjust to imposed factors in a variety of ways, which lead to natural variance among examples. The operation of chance or randomness accounts for some of the scatter typically seen in plots of hydrologic data. This has been stated by us and reinforced by the work of Miller and Onesti.

At a given discharge, different rivers exhibit different values of width and depth. Those with fine-grained cohesive banks tend to be deep and narrow, and those with sandy, friable banks tend to be wide and shallow.

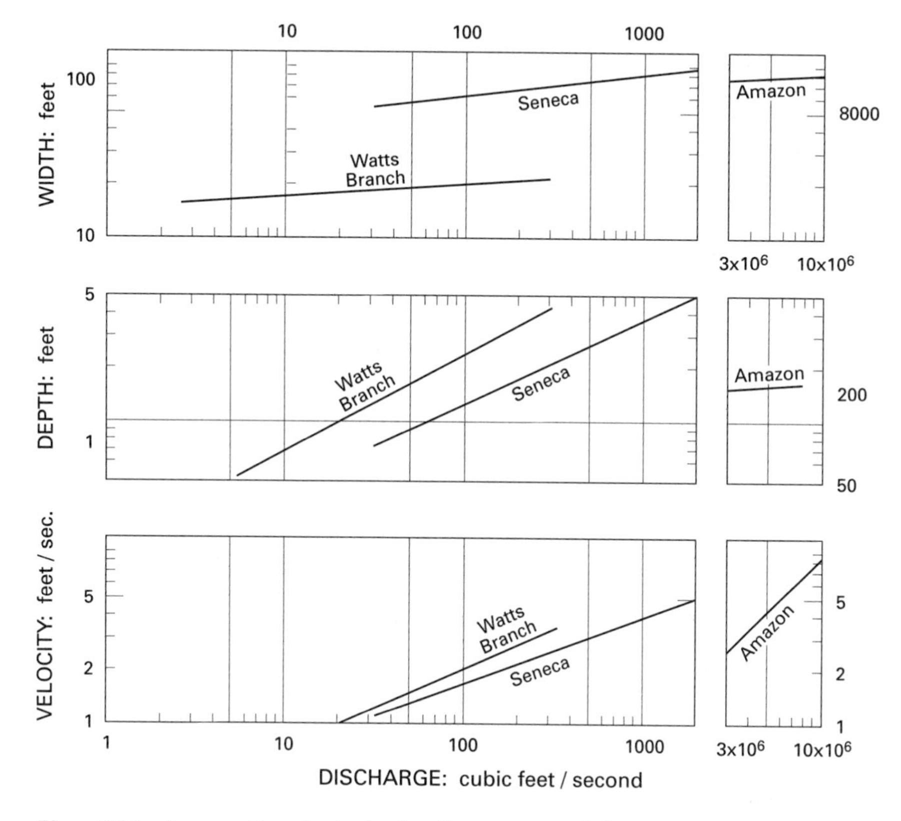


*Figure 10.7* Average lines in the hydraulic geometry of three rivers of different size: Watts Branch drainage area, 3.7 square miles; Seneca Creek drainage area, 100 square miles; Amazon River at Obidos drainage area, 1.9 million square miles. Amazon data from Nordin and Meade, collected 1967–1975.

These differences can be seen in the stream types of the Rosgen classification system, in which the width-to-depth ratio is one of the criteria.

It is interesting to compare the hydraulic curves of small and large rivers. Omitting the individual points, the mean lines of width, depth, and velocity plotted against discharge are shown in Figure 10.7 for a small river, Watts Branch; a medium-size one, Seneca Creek; and the Amazon, the world's largest. The Amazon measurements were made by Carl Nordin and Robert Meade in several expeditions over a period of years. The discharge ranges are 3 to 200 cfs in Watts Branch, 30 to 2,000 cfs in Seneca Creek, and 3 million to 9 million cfs in the Amazon. In all three, width increases little as discharge increases. In fact, the measurements of the Amazon gave a width of about 7,000 feet for all discharges. For the two small rivers, depths varied from 0.5 to about 5 feet. But at a given discharge, the smaller Watts Branch was deeper than the larger Seneca. This comes about because a given discharge is relatively less frequent in a small than in a large river. The depth of the Amazon changes surprisingly little, about 150 to 170 feet.

Velocity, however, is comparable for small and great rivers. The small rivers have a velocity from 0.5 to 5 feet per second, and the largest of all rivers 2 to 7 feet per second. The small rivers accommodated an increase in discharge by increasing both depth and velocity in nearly equal amounts. The Amazon, by contrast, accepted the discharge by increasing velocity only. The bed apprently became increasingly smooth as the discharge increased, presumably by trimming the height of the bed dunes.

## Variations of Slope, Load, and Roughness

The geometry concept can also be applied to other hydraulic parameters—slope, sediment load, and roughness—because they vary with discharge both at a station and downstream. The river data show that the at-a-station slope is conservative and changes but little with discharge, or $s \propto Q^z$ where $z = 0$. Sediment load changes more rapidly than any other parameter, as shown by the value of slope of the sediment rating curve, usually between 2 and 2.5, or $L \propto Q^j$ where $j = 2$ to 2.5. Hydraulic roughness expressed as Manning's $n$ varies as $n \propto Q^y$ where $y = -0.02$, a slight decrease of roughness.

In the downstream direction, there is a large decrease of slope with discharge. The change is quite variable among rivers, but the most common value of $z$ in $s \propto Q^z$ is $-0.75$. Roughness decreases slightly: $y = -0.19$. Suspended sediment concentration appears to be nearly constant or to decrease slightly, for the sediment load increases nearly in direct proportion to the increase in discharge, as shown by the value $j = 0.8$.

## Aspects of Channel Joining

When two channels join to form one, the discharge of the merged channel increases. Its width is not the sum of the widths of the tributaries, but considerably less. Width increases downstream in a river system as the square root of discharge, which may be defined as mean discharge, bankfull discharge, or any other discharge of constant frequency. The widths of the separate channels $w_1$ and $w_2$ may be stated in relation to their respective discharges $Q_1$ and $Q_2$ as

$$w_1 \propto \sqrt{Q_1} \quad \text{and} \quad w_2 \propto \sqrt{Q_2}$$

Because combined discharge $Q_c = Q_1 + Q_2$, and combined width $w_c \propto Q_c$,

$$w_c \propto \sqrt{(Q_1 + Q_2)}, \qquad w_c \propto \sqrt{(w_1^2 + w_2^2)}$$

The combined width is equal to the square root of the sum of the squares of the joining widths.

I took photographs of the Colorado River from a small plane at low elevation. Among them is one of the junction of the Green River and the Colorado River in southeastern Utah (Figure 10.8). On the river, my measurement of the width of the Colorado in Marble Canyon was 220 feet. From the photo, the respective widths of the Green and Colorado above the junction were 150 and 162 feet. The width of the combined rivers was computed as $\sqrt{(150^2 + 162^2)} = 220$ feet, a calculation that agrees with the expected relation.

## Theoretical Derivation of the Hydraulic Geometry Exponents

There has been sufficient interest in the description of channel systems by the equations of the hydraulic geometry that attempts have been made to derive from general principles the observed values of the exponents. The first effort was made by Walter Langbein and me. To satisfy the lack of the necessary equations, we agreed that the minimum production of entropy in an open system is satisfied by the maximum probable distribution of energy in the system. This led to the concept that an exponent in one of the geometry equations is a measure of the variance of that parameter. Because the most probable state is defined as the state in which the sum of the variances is minimum, then the most probable values of the dependent variables will be those in which their variances are minimum, subject to the constraint that they must obey the laws of hydraulics. The computation for a given set of assumptions then leads to a numerical solution for the exponents. This theoretical result is an expression of the most probable state. As has been shown by Park, real rivers show a variety of values of the exponents (Figure 10.6). The theoretical value appears to lie well within the cluster of plotted points shown in the figure.

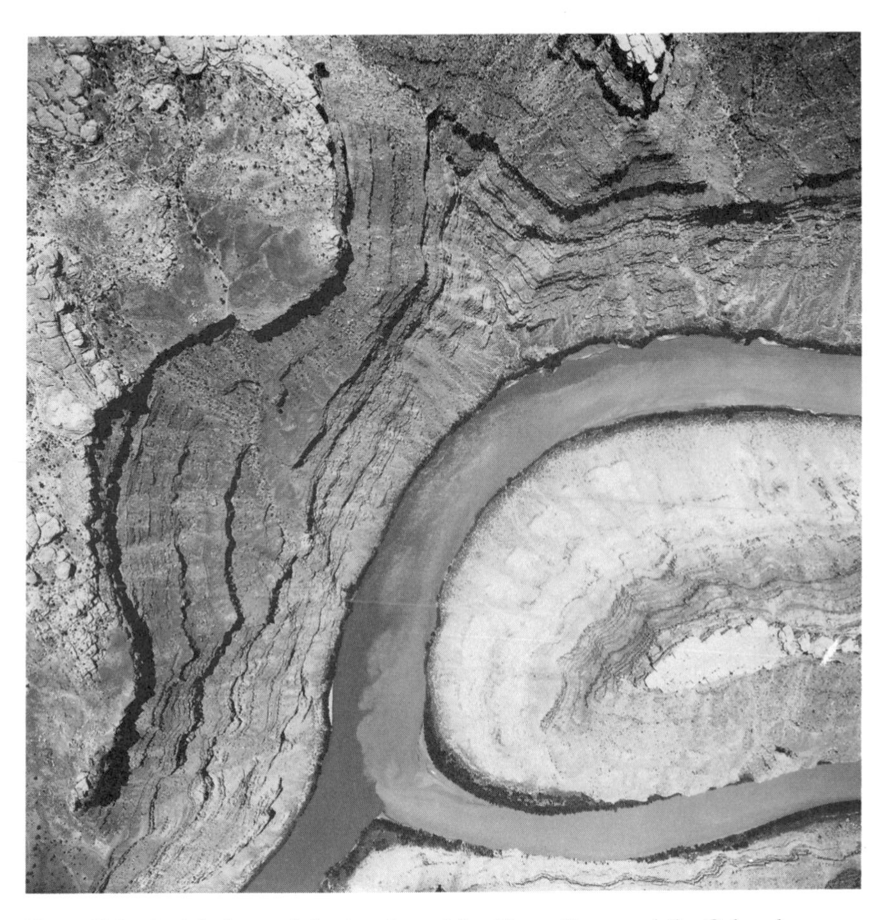

*Figure 10.8* Aerial photo of the junction of the Green River and the Colorado River, southeastern Utah. The muddy water of the Green mixes only slowly with the clear water of the Colorado.

A separate and quite independent theoretical derivation of the value of the exponents was presented by T. R. Smith, based on what he called conservation principles. His contribution shows that there is more than one way of computing the value of the exponents from theory. The theoretically derived exponents from our analysis and from Smith were plotted by Park on the triangular diagram of Figure 10.6. The theoretical values lie near the center of the cluster of observed data points.

CHAPTER ELEVEN

# Sediment Load

## Weathering and Its Products

The insoluble products of rock weathering, when moved by water, are generally called sediment. The source of sediment is, of course, the rocks that occur on the continental surface. Denudation, or lowering of the land surface by erosion, results from a number of processes, including solution, erosion, and transport by water; soil creep or downhill motion caused by gravity; erosion, and transport by ice, especially glaciers; freeze and thaw; landslides; and other types of flow.

Materials carried in solution by water (for example, calcium carbonate) are abundant in both number and volume. Solution of rocks and minerals and the consequent transport of solutes is surprisingly effective, and efficiency increases as climate is more humid and warm. In semiarid areas of the southwestern United States, dissolved load may be only 1 or 2 percent of total load carried by rivers; but in eastern states such as Pennsylvania, dissolved load may be as much as 64 percent of the total load.

Over much of the world the products of weathering carried toward the ocean by running water in creeks and rivers are composed principally of solid material or sediment. Sediment may be of many sizes, from very fine clay to boulders. The material fine enough to be called clay or silt, less than 0.06 millimeter in diameter, consists primarily of clay minerals, most of which are the result of the chemical alteration of feldspars. The material ranging in size from 0.06 millimeter in diameter, very fine sand, to 0.5 millimeter, medium sand, is predominantly quartz, that being the most common mineral not easily dissolved in water. Coarse sand tends to be a mixture of quartz and individual grains, often crystals, of various minerals. Larger particles, of size greater than 1.0 millimeter, are often bits of rock containing more than one mineral. Pebbles, gravel, cobbles, and boulders are rock fragments, usually somewhat rounded by abrasion during transport.

The lithologic or mineral character of the sediment carried seaward from its place of origin is determined by the types of rocks from which it is derived. The distribution of rock types in the continents of the world therefore determines the character of the sediment in transport.

## Rocks as a Source of Sediment

Rocks formed from sediment deposited in previous geologic epochs (sedimentary rocks) cover a much larger percentage of continental area than the igneous rocks that are the ultimate source. Five classes of rocks occupy about 90 percent of the continental area, distributed approximately as follows:

| | |
|---|---|
| Shale | 52 percent |
| Sandstone | 15 |
| Granite (and granodiorite) | 15 |
| Limestone and dolomite | 7 |
| Basalt | 3 |
| Others | 8 |

Combining mineralogic composition with the relative abundance of rock types gives the approximate proportions of common minerals exposed to weathering at the earth's surface:

| | |
|---|---|
| Feldspar | 30 percent |
| Quartz | 28 |
| Clay minerals and micas | 18 |
| Calcite and dolomite | 9 |
| Iron oxide minerals | 4 |
| Pyroxene and amphibole | 1 |
| Others | 10 |

The great preponderance of shale on the surface of continents is testimony to the importance of the fine-grained portion of total sediment and thus the clay minerals that are the origin of shale. The erosion of shale results in sediment of fine grain, usually easy to break down or comminute during transport. The fine-grained material, the principal object of measurement in the past, has been relatively easy to measure because of its large percentage bulk and because it is mixed in the water.

The coarse-grained material, however, makes up the bed and bars of many, if not most, rivers. Because it is not mixed through the whole column of flowing water, the coarse material has not been easy to measure. Indeed, its relative bulk and transport rate have not been measured except under special conditions. Scientists and engineers have turned their attention in recent years to this area of study.

## Suspended Load and Bedload

In a nonquantitative way, transport of sediment by flowing water is described by the terms "suspended load" and "bedload." Suspended load comprises the fine fraction of material in transport that is mixed intimately with the flowing water. It tends to make the water muddy. This fine material will settle through the water owing to its density, but it is sporadically and repeatedly caught in local turbulent eddies and lifted again and again into the body of the flow.

In contrast, larger particles are not swept up by eddies but are pushed along near the streambed, and for this reason are known as bedload. Whereas the concentration of the suspended particles decreases exponentially from bed to water surface, bedload never rises off the bed more than a few grain diameters. Bedload moves by a combination of sliding, rolling, and saltation. Saltation is defined as motion consisting of a series of short hops, often with temporary rests, before propulsion forward for another hop or short excursion.

Granular flow, including the transport of solids by wind or water, is a shearing motion in which successive layers of solids are sheared over one another. Such shearing cannot take place without some dilation or dispersion. It is well known that sand in close packing, that is, with no dispersion, behaves like a rigid or solid body. In contrast, quicksand experienced on margins of sandbed rivers or at springs owes its existence to a slight upward flow of groundwater that disperses the grains, with the result that man and beast sink into the quicksand.

In a river the dispersion must necessarily be upward against gravity. Keeping grains in motion requires that through any shear plane there must be an upward supporting force equal to the immersed weight of the solids above the plane.

Mechanical force per unit area, also called stress, can be transmitted by two possible mechanisms: (1) the transfer of momentum from solid to solid by continuous or intermittent contact; (2) the transfer of momentum from one mass of fluid to another and thence to the otherwise unsupported solid.

The support mechanism obviously must be maintained by the shearing motion, which in turn is maintained by the applied tractive force. The solid-transmitted stress owes its generation to the shearing of the solids over one another. The fluid-transmitted stress arises from the shearing of the fluid, for this maintains the internal turbulent motion.

Many individual solids are likely to be supported partially by solid-transmitted stress, but where a large number of solids is involved, one can treat them in two discrete parts. The distinction between the support mechanisms is then the logical division into bedload and suspended load.

Bedload is that part of the solid load which is supported by the unmoving bed, by intermittent contact among the moving solids, and ultimately by the layer of grains at the unmoving surface of the bed. When a tennis ball is dribbled up and down on the floor, though its contact with the solid floor is momentary and intermittent, the weight of the ball is supported by the floor. If a large number of balls are similarly bounced, their total weight must be supported by the floor. So sand grains moving near the bed in flowing water are bouncing, rolling, and colliding, and their excess or immersed weight is carried by the unmoving bed.

Suspended load, in contrast, is that part of the solid load the weight of which is transmitted by the fluid of the main flow to the fluid in the interstices of the grain bed. Airplanes give us an analogy. The weight of an airplane is ultimately carried by the air to the ground surface and thus the fluid pressure on the ground is increased within a cone below the flying plane.

The excess or immersed weight of the suspended load must then be equal to the mean upward flux of momentum by upward fluid currents in the turbulent eddies.

In an actual case—the river—the solid-transmitted stress and the fluid-transmitted stress cannot be separated by measurement, but physically the separation is a real one and provides a sound basis for defining bedload and suspended load.

Both mechanisms of support—solid to solid intermittent contact and upward flux of momentum due to fluid turbulence—are results of the tangential stress exerted on the bed by the flowing water.

### *Sediment Carried by Rivers*

In geologic terms, sediment deposited from water transport has covered such large parts of continents, and continental shelves surrounding the continents, that rocks formed from the consolidation of these deposits cover 67 percent of the continental areas of the planet. Shales, as shown previously, constitute the largest part of these deposits, and they are primarily the result of suspended load reaching oceanic areas where the load is deposited. But the geologic formations most important to man are those carried as bedload, coarser in grain size than the suspended clay, silt, and fine sand. The bedload, deposited and over time cemented, formed the materials that now are the aquifers holding groundwater, which is the water supply for at least half of the world's population. Sandstone, which generally is relatively porous and permeable, is the major rock type holding groundwater in its pores. Though shale is more extensive on the world's continents, it is quite impervious to the movement of water and therefore does not act as an aquifer.

Both the size and the composition of sediment are largely determined by the minerals of the original rocks. The mineral composition is responsible for the degree of susceptibility or resistance to the weathering processes of oxidation, reduction, and hydration. The weathering of feldspar, listed earlier as the most abundant mineral, is the origin of most of the sediment carried as suspended load.

Quartz, though less common in igneous rocks, is more resistant to weathering because of its low solubility. Quartz, then, is selectively concentrated during the process of rock destruction, transportation, and storage. While less resistant minerals are changed by weathering, broken down, and carried away in solution or traction, quartz is left behind and becomes an increasing component of the residue.

Interestingly, quartz can be dissolved in aqueous solution (though in low concentrations) and thus over time can be translocated and redeposited. Redeposition of quartz in the interstices of sediment mixtures produces a cement that can bind together grains or rock fragments to produce some of the most resistant and long-lived rocks known, quartzite and conglomerate. The filling of all the minute spaces between sand grains and the binding together of the sand is itself a wonderment. For as the deposition proceeds and the interstices fill, the movement of the water carrying the silica in solution becomes progressively more restricted. Such change, as in the formation of other sedimentary and metamorphic rocks, must take place under conditions of extreme pressure and temperature and requires millions of years.

Minerals of complicated composition are more easily weathered and changed than minerals of simple composition. The most common forms produce the clay minerals. The resistant materials and the rocks they comprise eventually become the main sediment moved as bedload in creeks and rivers. By far the most important are the quartz grains themselves, and the rocks containing or cemented by silica (quartz). Gravels and cobbles seen on streambeds do contain a wide variety of lithologic types, but when finally reduced to sand size by weathering and abrasion, the preponderance of grains are quartz. All these sizes down to that of medium sand are carried as bedload rather than suspended load.

Weathered and transported materials make up the sedimentary rocks that cover much of the continents of the planet; the volume of these deposits is immense. The carbonates carried by rivers to the ocean provide the materials used by marine organisms to produce the great deposits of limestone. Many sedimentary formations are thousands of feet in thickness.

Sediments carried by running water have immediate importance to humans. Soil erosion of agricultural and grazing lands has already had an impact on the life and welfare of millions of people. The clogging of irrigation works, the silting of reservoirs with consequent reduction in reservoir volume and efficiency, and the polluting of rivers and thus water supplies are among the ways sediment load is detrimental to man. Yet the overflow of river floodplains with consequent deposition of thin layers of sediment has, as in the case of the Nile, been a source of nutrients and organic material over thousands of years. And gravel in streambeds is essential for spawning of many fish species, as well as for the insect and microscopic forms of aquatic life.

Thus, sediment is an integral part of the total environment within which life forms have developed and flourished.

### *The Forces Propelling Bedload Grains*

The force of gravity that pulls an object downward can also push it forward. Imagine a car parked on a hill (Figure 11.1). The weight of the car is a force that acts vertically toward the center of the earth. But any force may be visualized as the sum of component forces that can be described by simple geometry. In the figure the arrow marked $W$ for weight represents the force exerted by gravity acting vertically. The force $W$ can also be expressed as the sum of two components: the arrow $N$ for normal is that part of the weight directed perpendicular to the ground surface, and the arrow $T$ for tangential is that part directed parallel to the ground surface or downhill. As in geometry, the weight $W = \sqrt{(N^2 + T^2)}$.

It is the arrow $T$ that is of particular interest here, for it is the force that pushes the car down the hill. The steeper the hill, the larger the $T$ force. Again, from geometry, $T = W \sin \alpha$, where $\alpha$ is the angle or slope of the hill.

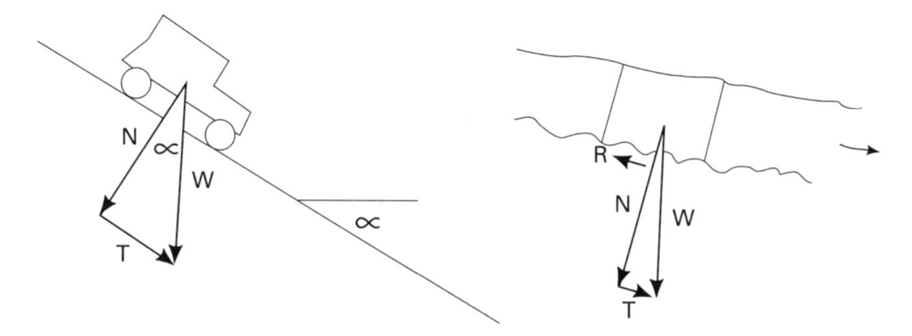


*Figure 11.1* The weight $W$ of an object, directed vertically, resolved into a normal component $N$ and a tangential component $T$. The resisting force $R$, because of friction, is equal to the force needed to drag the object.

Expressed another way, the car is pushed downslope by the downhill component of its body weight. This reason is also why water flows downhill. As in the right-hand diagram of Figure 11.1, a unit of water having weight $W$ is pushed downchannel by the tangential component $T$ of its weight.

It might seem logical that the transport of grains by a fluid is due to the velocity of the fluid. Indeed, the transport is related to, or roughly correlated with, the velocity of the fluid; but pushing a grain must result from a force, not a velocity. The force is the tangential component of the body weight mentioned above. The velocity of the flowing water is the result of the downhill component of the gravitational force, parallel to the bed and pushing the grains along.

This component of force propelling the water forward is resisted by friction at the boundary, which slows the parcels of fluid near the boundary. So the velocity of the fluid decreases toward the boundary in a predictable manner. Velocity increases upward as the logarithm of distance above the boundary. Thus, a plot of velocity of the water above the bed has the shape shown in Figure 11.2A. The shape of the curve is such that the mean value of velocity lies just six-tenths of the distance from the water surface to the bed.

This logarithmic decrease of velocity toward the boundary is seen in water flowing in a pipe, in natural rivers, and in air flowing over a surface (such as over a sand sheet in the desert). Of course, irregularities of the channel bed or banks cause the theoretical curve to be only approximated in most natural channels. In large and deep rivers the theoretical curve represents remarkably well the actual measured velocities.

One of the useful aspects of this relationship is the fact that Figure

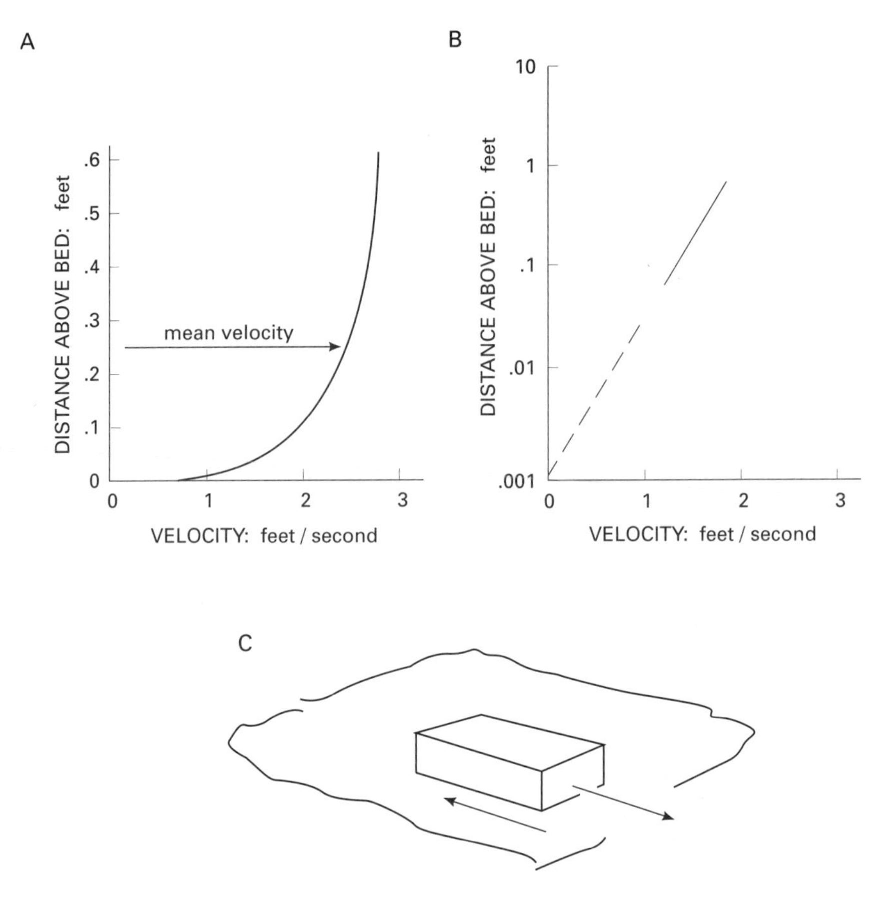


*Figure 11.2* Velocity in water as a function of depth, in Cartesian coordinates (A) and semilog plot (B). Diagram C shows the forces pulling and resisting a brick being dragged across a rough surface.

11.2A can be redrawn using a logarithmic scale of depth. In this case the profile is straight, as shown in Figure 11.2B. Although measurements very close to the bed are usually impractical, it has been found that the velocity profile, as in Figure 11.2B, continues straight to a distance above the bed where the velocity is zero. In the figure the velocity is zero at 0.001 unit above the bed.

Experiments in pipes and channels have shown that when the grains on the bed are immobile, the vertical position where the velocity is zero is one-thirtieth of the grain diameter. Some small velocity is still present around the grains in the top layer.

The velocity profile shows that, on average, each layer of water has somewhat larger velocity than that of the layer immediately below. The drag of one layer over another may be thought of as the upper layer tending to drag the lower layer along, and thus exerting a force in the downstream direction. Conversely, it may be visualized that the lower layer is a drag on the upper layer, creating a resisting force in the upstream direction.

This situation is in most respects the same as solid friction on a rough surface when one tries to drag a brick over the surface of a concrete street (Figure 11.2C). The force one exerts to pull the brick is in the direction of motion; it is resisted by a force acting in the opposite direction, exactly equal in magnitude if the brick moves at constant speed.

Thus the moving water exerts a force on the bed which is available to push sediment grains downstream. This force, the shear stress, can be expressed in pounds of force per square foot or kilograms per square meter. It is similar to the number of pounds of force one must exert pulling the brick across the concrete for each unit area of the brick.

In water a descending grain, even a heavy rock, does not hit the surface with sufficient momentum to crash into the bed and cause other grains to be catapulted upward. A descending grain can be bounced upward off the grain bed losing some of its momentum in the contact, but no crater or hole is formed in the grain bed.

A particle set in motion at the bed, and too large to be swept up in turbulent eddies as suspended load, makes its excursion by rolling, sliding, or saltation very close to the bed, usually hopping upward a distance only two or three times the mean diameter of the bed grains. Turbulence and eddy formation are not needed to help propel bedload grains. Flume studies by John Francis of Imperial College, London, have shown bedload motion in laminar flow, that is, flow when there are no turbulent eddies.

## Threshold of Movement

The conditions surrounding noncohesive grains on the streambed at the threshold of motion have been investigated theoretically and experimentally by several workers. The resulting theoretical statements enter into many of the equations used to predict the rate of debris transport. The average velocity profile in flowing water has been described, but when we deal with the threshold of grain motion, the temporal mean velocity at any height is the average of constantly fluctuating values caused by turbulent eddies.

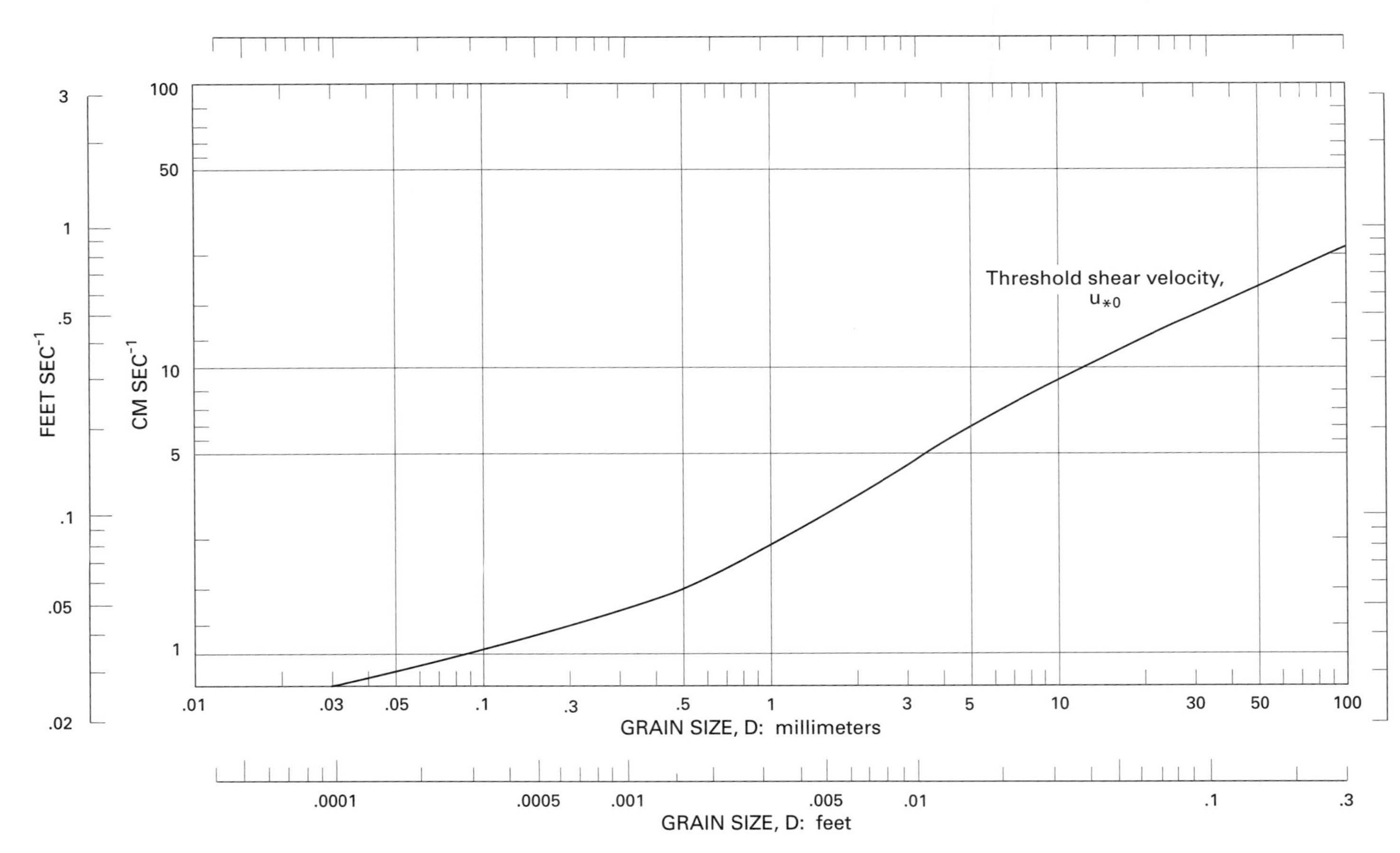


*Figure 11.3* Values of shear velocity at the threshold of grain motion, $u_{*0}$, as functions of grain size *D*. The Shields diagram has been transposed into the present form by R. A. Bagnold. It applies to quartz in water, shape factor 0.7 and temperature 10°C.

As the fluid passes over grains on the bed, the streamlines are deflected upward above those grains that are larger or higher than their neighbors. The fluid force on an exposed grain may be resolved into a tangential force, the drag, and a normal force, the lift. At low velocities the drag is primarily the result of viscous stresses, but at high velocities a pressure difference exists between the upstream and downstream faces of the grain and this constitutes the principal force. The eddies, shed downstream of such an obstacle, affect the time and spatial fluctuations of velocity acting on grains downstream.

The shear stress needed to set a grain in motion, starting from a bed of uniform size grains, was investigated by Shields. His diagram for initial motion is widely known. The relation is dimensionless shear stress $\theta$, as a function of grain Reynolds number $R_g$. The ordinate is $\theta = \tau_0/\gamma(S_s - 1)D$, where $\gamma = \rho g$ and $\gamma\ (S_s - 1) = \rho g\ (\sigma - \rho)/\rho$; $\gamma$ is specific weight of water, $S_s$ is ratio of grain to fluid density, $\sigma$ is grain density, $\rho$ is water density, and $D$ is particle diameter.

The entrainment function $\theta$ is plotted against Reynolds number of the grain, defined as $R_g = u_* D/\nu$, where $u_*$ is shear velocity, $D$ is grain diameter, and $\nu$ is kinematic viscosity of the fluid. Note that this differs from the usual Reynolds number, which incorporates depth of water rather than grain diameter.

Experiments on initial motion have been done with a variety of grain materials including amber, lignite, barite, and sand. Each experiment was conducted with a uniform size of grain. The data for all these materials of different density plot close to a single line on the Shields diagram.

For many purposes, however, where sand or gravel in water is the only concern, a modification of the same plot is more useful. What the investigator usually wants to know is the velocity at which motion will begin for quartz-density grains of various sizes. Ralph A. Bagnold converted the Shields diagram into these coordinates of shear velocity $u_*$ plotted against grain diameter (see Sagan and Bagnold 1975). The result is shown in Figure 11.3. Because the value of $u_*$ is the initial value or threshold value for grain motion, it is written $u_{*0}$.

The threshold curve of initial motion can be superimposed on another diagram where shear stress $\tau$ at the threshold of motion is plotted against grain size $D$, as compiled by Leopold, Wolman, and Miller (1964), and presented in Figure 11.4.

This diagram is a practical tool for approximating what size material on the streambed might move under specified conditions of flow. The plotted data are observed cases of particle motion from a variety of sources. The value of $\tau$ is computed as $\gamma ds$ or $\gamma Rs$ where $\gamma$ is unit weight

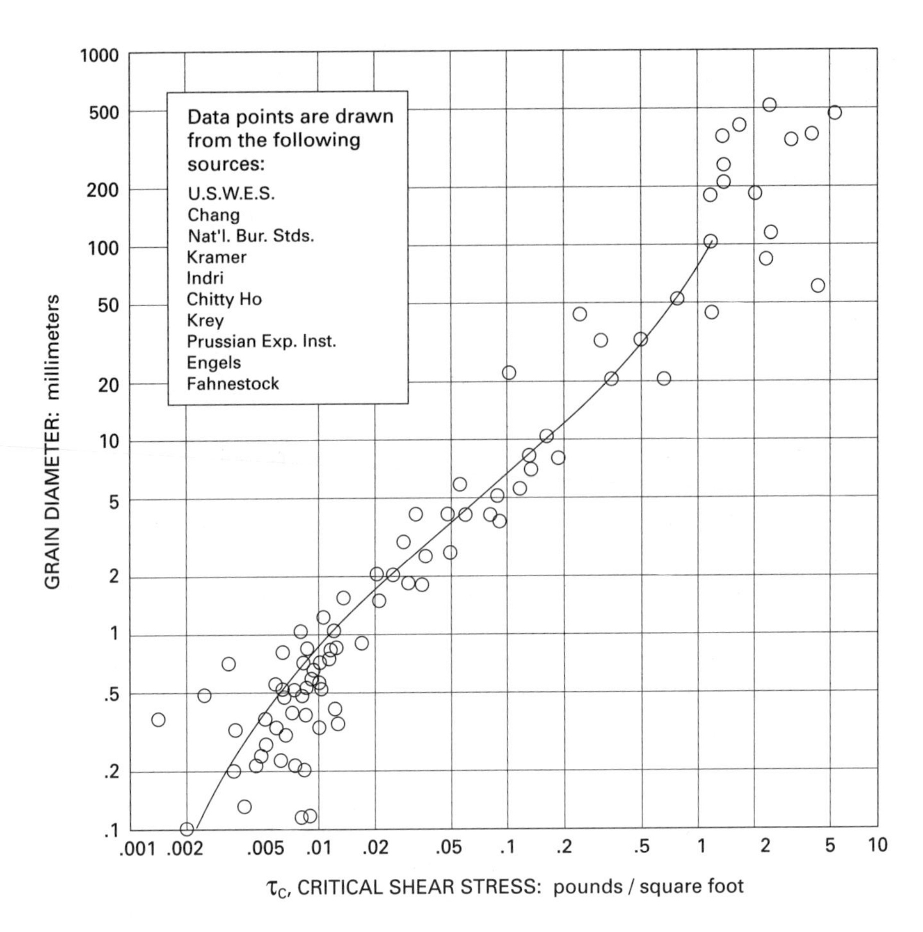


*Figure 11.4* Laboratory and field data on critical shear stress required to initiate movement of grains (Leopold, Wolman, and Miller 1964, p. 170). The solid line is the Shields curve of the threshold of motion transposed from the $\theta$ versus $R_g$ form into the present form, in which critical shear stress is plotted as a function of grain diameter.

of water (62.4 pounds per cubic foot), $d$ is depth (an approximation of hydraulic radius $R$) and $s$ is stream slope. The solid line drawn from the Shields diagram of initial motion goes through numerous observed values plotted on the figure. But in many river channels the mixture of grain sizes in the bed does not approximate the simple conditions expressed by the Shields experiments. Figure 11.4 should be used only as a first approximation; it may not apply to all field cases.

*The Sediment Particle in Motion*

The tractive force per unit area of the bed, or the tangential shear, tends to push grains of sediment in the surface layer of the bed in the downstream direction, proportional to the area of the grain surface exposed to this force and to the magnitude of the force per unit area. It is similar to the push you feel when standing in a strong wind. If you wear a bulky coat and carry an umbrella, the pressure you feel is greater than if you hunker down and are less exposed.

We have just discussed the force sufficient to set a grain in motion. What is needed to sustain motion of sediment particles near the bed? A sediment grain, initially at rest on the bed, is propelled up into a fluid moving at a velocity that increases with distance from the bed. A grain pushed into this field will be accelerated by the ambient fluid. At its maximum excursion from the bed it probably attains the same velocity as the fluid in which it is then imbedded. But as its upward motion is reduced to zero, it begins its downward fall commensurate with the usual speed of an object falling through a liquid. Throughout its fall it is carried forward by the fluid in which it is contained. Its vertical arc would be comparable to a baseball or a cannonball thrown upward, and its forward motion is determined by the effect of the various velocities of the fluid through which it is projected.

When shear stress associated with moving water is sufficient to dislodge grains on the bed surface, the surface of the grain bed must experience some degree of dilation in order for moving grains to pass over those still immobile. The stress exerted on the grains moves each one slightly, separating it from its neighbors by a small distance. This dilation decreases the linear concentration of grains, with the result that the grains, slightly separated, collide during their motion. These collisions impart some change of momentum to each grain, this momentum having a vertical as well as a horizontal component. The net sum of the upward-directed momentum component to any level must equal the weight of all the grains in the flow above that level if saltation is to be continuous. There is then, at virtually any level, an upward-directed force felt by any individual grain and proportional to the square of the diameter of the grain.

A similar downward-directed force is experienced by the unmoving grain bed. If the bed is composed of grains of equal size, when the topmost layer is put in motion the horizontal shear stress should be sufficient to set in motion the next layer, and the next, ad infinitum. But this does not occur; as more grains are set in motion, the downward-directed force increases and acts to hold in place the grains on the bed surface, with the result that only a finite thickness of the bed is peeled off and set in motion.

At the level of the unmoving bed, this dispersive stress directed upward and downward normal to the flow direction is equal to the weight of the immersed grains in motion. The upward-directed force is merely the normal component of a larger force, the tangential component of which is the grain shear stress at the bed. The ratio between the tangential and normal components is the coefficient of friction, tan $\alpha$, a quantity similar to and quantitatively approximating the solid friction coefficient for the same granular material.

The dispersive force exerted on a grain, being proportional to the square of the grain diameter, acts differentially on larger grains and pushes them toward the level of zero dispersive stress, or the top of the moving grain layer. In some instances this appears to be the explanation for larger than average rocks lying at the surface of the streambed.

### *Power Expenditure*

The force exerted on the bed grains is caused by the shearing of the fluid. Shear varies with depth (it is zero at the water surface) and increases linearly to a maximum at the bed. But the flow of water in a river is expending energy drawn from the potential energy of elevation. That potential energy is transformed to kinetic energy in the flowing water and dissipated into heat in turbulence and in friction with the channel margins. Some of this energy is used to perform work in transportation of sediment load. Seen in this light, the river is a transport machine and can be compared to any machine such as a locomotive. Power utilization results in work done, and power is the rate of doing work. Force is mass times acceleration, and work is force times distance. Power is the rate of work, or force times distance divided by time. The units of power are pounds per foot second. Weight is in units of force. Work done is weight moved through a distance, and power is work per unit of time.

In the fluid machine, power is the unit weight of water times discharge times slope. Water has a unit weight of 62.4 pounds per cubic foot, $\gamma = \rho g$. Slope is dimensionless. Power per foot of width is 62.4 $Qs/w$ in pounds per foot second.

Power is also the product of mean velocity times shear stress $\tau$, or $u \times \gamma ds$, where velocity is feet per second and stress is pounds per square foot. Then power is pounds per foot second.

Some conclusions can be drawn from consideration of the theoretical aspects of sediment transport.

The amount of mechanical work accomplished (in this case, sediment movement) is a function of power or time rate of doing work, the product of shear stress and velocity.

The size of the debris that can be moved depends on the bed shear stress or force exerted at the boundary. Thus it is dependent on the stress structure within the fluid and is related to boundary roughness.

Present theoretical analyses perforce deal with unigranular material, and the reality of heterogeneity of debris size must be approached through empirical relations. It is logical to relate sediment transport rate to the power available in the stream to transport it. Suspended sediment was sampled in many locations long before it was realized that power was the physically correct parameter to use in describing transport rate. Therefore, it was the practice to plot sediment transport rate against stream discharge at the time of sampling. The result is called the sediment rating curve. Many equations now use transport rate as a function of power rather than discharge.

## Sediment Measurement

In a joint effort of U.S. government agencies, a suspended-load sampling device was developed in 1948 that has become standard in many countries. Its design is based on the idea that the velocity of water in the entrance nozzle must be the same as the velocity of the ambient water flow. This equalization was accomplished by the design of the orifice for air escaping from the collecting bottle as the sediment-laden water entered. It is the practice to lower the sampler from the surface to the bed and back to the surface at just the rate that allows the collecting bottle to be filled. Thus, both the surface water of low sediment content and near bed water of higher concentration are sampled. Preferably, several vertical samples are taken across a river to represent various parts of the channel; but because suspended load is usually well mixed in the river, even a single sample may give a reasonable representation of the concentration.

The amount of load caught in the sampler must be determined by a not-so-simple laboratory procedure in which the water sample is weighed, then filtered; next, the filter paper containing the sediment is dried and weighed, and the weight of the filter paper is subtracted. Size distribution can be measured by one of several mechanical means.

*Figure 11.5* The East Fork bedload trap. Below the suspension bridge can be seen the concrete trough in the streambed that has a slot into which the bedload will fall. The endless belt in the trough carries the bedload to the far bank, where it is lifted by buckets and dumped into a hopper on a weighing scale. After being weighed, the sediment is returned to the river on an endless belt.

Size distribution of the collected sediment sample is easier to measure if the suspended sediment is fine sand or something coarser. A simple procedure is available whereby the sand is dropped into a column of still water. Because large particles fall faster than small, the measured rate of accumulation at the base of the column is a measure of the size distribution. The result of the laboratory analysis is usually expressed as concentration of sediment in the liquid in parts per million by weight. The published record ordinarily expresses the result as tons per day carried by the river. The computation is as follows: load (tons per day) = $3.46\,CQ$, where $C$ is concentration in parts per million and $Q$ is cubic feet per second.

Ability to make practical measurements of bedload is more recent. A sampling device was developed in Germany based on the concept that an entrance slightly flaring in cross section downstream would increase the pressure gradient just enough to overcome the effect of friction as water entered the mouth. The German device had some disadvantages that were overcome in a sampler of similar throat design made by Edward Helley and Winchell Smith. Any sampler needs field testing, but field facilities that actually measure all the bedload in a river are rare indeed.

*Figure 11.6* The East Fork bedload trap. The motor that drives the submerged endless belt is visible in the foreground at the end of the suspension bridge. The concrete trough in the streambed is partly out of the water at low flow. The gates to the slot are closed in this photo.

Just at the time this new sampler became available, I was designing a field installation that would measure all the bedload continuously for days at a time, during conditions of both low flow and flood. This apparatus, known as the East Fork bedload trap, successfully measured sediment during the runoff season for 7 years. Photographs of the installation are shown in Figures 11.5 and 11.6 and the nature of the river is seen in Figure 4.16.

During this period my colleague William W. Emmett was collecting several thousand samples using the Helley-Smith sampler lowered from the bridge at the bedload trap (Figure 11.7). The samples collected were compared with sediment measurements made in the bedload trap. In this manner, the field efficiency of the Helley-Smith sampler was established and its practical value proven. The sampler is now available in various sizes and weights, including a lightweight model that can be used by a person wading in a river, miniature models employed in research in very small channels, and giant models. Thousands of samplers are in use throughout the world.

Another trap of more limited use is a wire cage with a large open end facing upstream, placed on the channel bed to catch rocks of all sizes,

*Figure 11.7* W. W. Emmett lowering a Helley-Smith bedload sampler from the suspension bridge at the East Fork bedload trap.

especially those too large to enter the mouth of the Helley-Smith sampler. Other types of bedload measuring devices also are used for experimental purposes.

Much has been learned by means of the painted rocks technique. As explained in Chapter 1, a common process of rock movement is the plucking by flowing water of individual rocks off the bar surface. David Rosgen and his colleagues placed lines of painted rocks in mountain channels at various locations in the Front Range of Colorado. The rocks were observed each day over an entire snowmelt runoff season. Nine streams were involved. The rock sizes included a large size, the $D_{84}$ size for that stream, the $D_{50}$ size, and in some cases the $D_{35}$ size. A total of 769 rocks were placed. During the season 500 of them moved, despite the fact that none of the rivers reached bankfull that season. Our results confirmed the finding by Andrews that the $D_{84}$ size of a gravel-bed river is moved by a discharge equal to or in many instances less than bankfull.

Measurement of the size of rocks on the streambed is an important part of sediment evaluation. A bulk sample of sediment caught in either a suspended load or a bedload sampler is quite amenable to analysis. Yet a bulk sample from a riverbed of cobbles, rocks, and sand is formidable and requires a large backhoe or excavator. Therefore, the accepted manner of measurement is what we call pebble counting, a procedure proposed by M. G. Wolman. In an area of streambed, individual rocks are picked up at random. The B (intermediate) axis is measured with a scale, and the rock is discarded. Our early work showed that if more than 60 rocks are measured, the differences between sites exceed the differences among operators. It was decided then to measure 100 rocks and to use the sizes counted to construct a frequency curve of rock size. Such a curve expresses numbers of rocks of various size classes, not their weights. There has been discussion in the literature of which is better for computation purposes, but the general practice has been to report size distribution, not weight. The usual procedure is to record number of rocks in categories separated by $\sqrt{2}$, and thus the sizes in millimeters of: less than 4, 4, 5.6, 8, 11, 16, 22, 32, 45, 64, 90, 128, 180, 256.

The main problem in recording size distribution is not the technique of measuring but the choice of the area to be sampled. Many alternatives are possible. Our experience indicates that to sample the size distribution of a gravel-bed river, the following is a reasonable approach, approximately as first outlined by Wolman.

Select a typical gravel-covered point bar. The surface often is somewhat more coarse on its upstream portion than its downstream portion. Choose an area in mid-bar region, with about one-third area under water and two-thirds the exposed mid-bar region. Visualize the area to be sampled as a rectangle. Walk over it in parallel lines, picking up a rock at every stride (about 5 feet). The total number of rocks measured should be 100, so that in each category the number measured is also its percentage of the total.

Plotting the data is an important part of size distribution analysis. Whether the data were obtained by pebble counting or by weighing sieved fractions, it is highly desirable to plot them in two ways, not merely one. The usual plot of size distribution is a cumulative curve of percentage by weight or number versus size, the latter on a log scale. In addition, a plot of percentage by weight or number not cumulated is prepared and plotted on a logarithmic ordinate scale.

A sample of two such plots is shown in Figure 11.8. The upper diagram permits a much more sensitive picture of the mode, or most frequent size, than the cumulative graph, because the changes in slope of the latter are often too subtle for visual recognition.

### *Sediment Rating Curves*

The most common sediment rating curve is one in which transport rate of suspended sediment or bedload in tons per day is plotted against

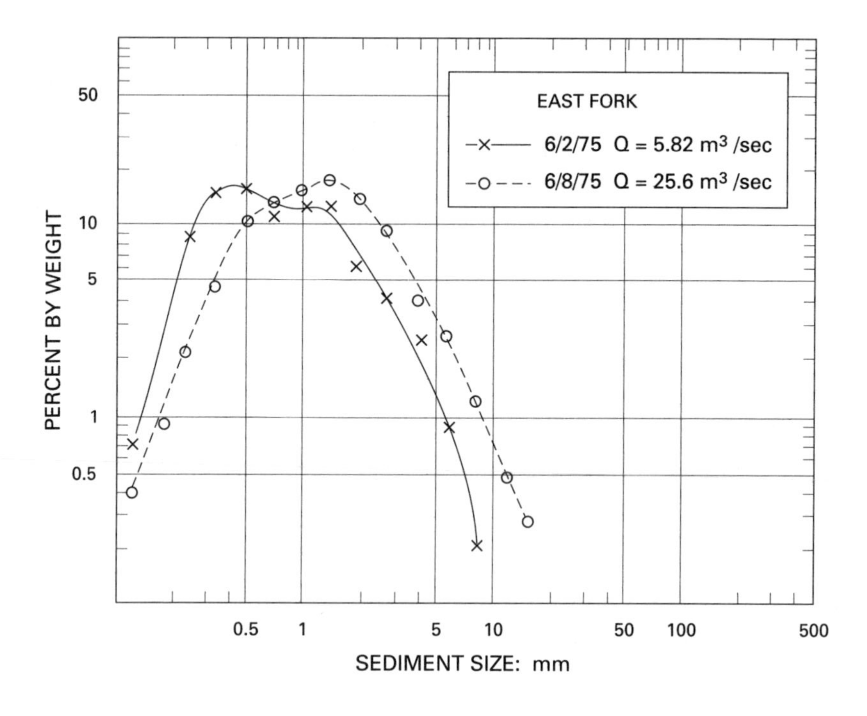


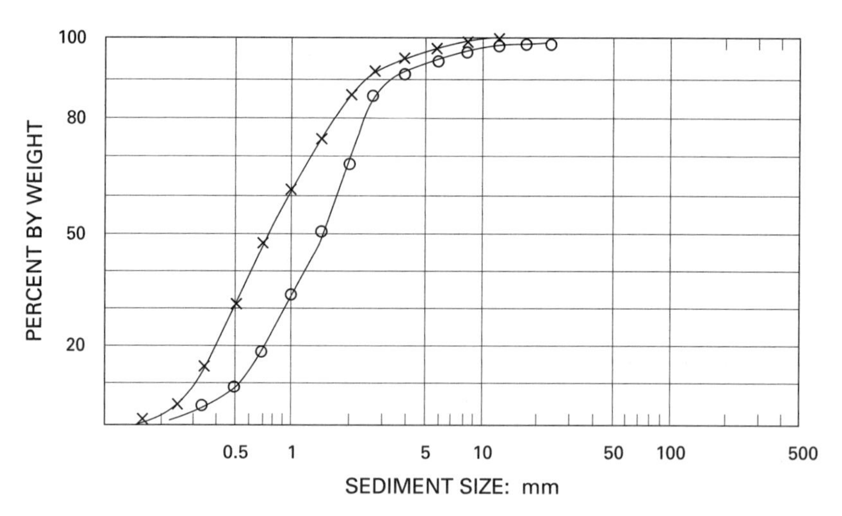


*Figure 11.8* The size distribution of bedload caught in the East Fork bedload trap during the periods of highest and lowest discharge during the 1975 runoff season.

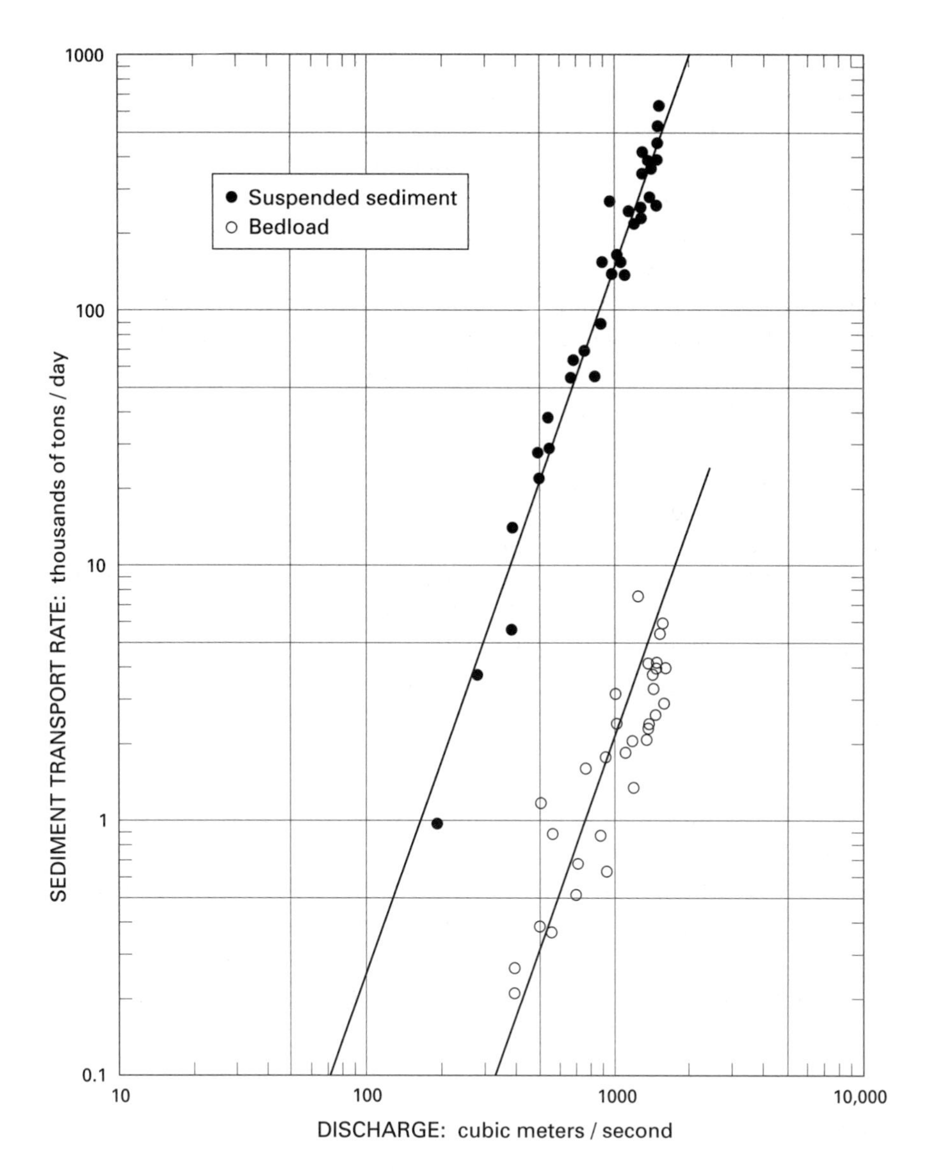


*Figure 11.9* Emmett's suspended sediment rating curve and bedload rating curve for the Tanana River near Fairbanks, Alaska. The data include measurements at two sites known as "at Fairbanks" and "near North Pole," for the years 1977–1979. The sites are 15 miles apart, and the drainage area is 20,500 square miles.

discharge in cubic feet per second on double-log paper. In nearly all such plots the slope of the mean line is steep, between 2.0 and 3.0.

Although a very large number of sediment rating curves have been prepared, most have been for suspended load. Very few indeed are the data for both suspended and bedload at the same station for the same time period. Some of the best data sets in existence are those of Emmett. He supervised collection of data on the Tanana River near Fairbanks, Alaska (Figure 11.9), during 1977–1979, and similar data are now available for later years. A long, unbroken record of both suspended sediment and bedload is available from Emmett's station on Little Granite Creek near Bondurant, Wyoming. Rating curves for the individual years 1982–1988 are shown in Figure 11.10. The line for each year extends over the range of discharges experienced in that year.

In the year-to-year comparison in Figure 11.10, the Little Granite Creek data show only slight changes in the slope of the lines. The slope of the suspended-load curves has an average value of 1.62 and that of the bedload curves is 2.24, so at high flows the bedload transport rate is approximately that of suspended load. At a discharge of 20 cfs, the transport rates of suspended and bedload are respectively 2.0 and 0.02 tons per day, but at 200 cfs they are 90 and 4 tons per day. At the lower flow suspended load was 100 times the bedload, but at high flow 22 times the bedload.

The annual amounts of suspended sediment and bedload are shown in Figure 11.11, expressed in tons per year. A more meaningful plot is in units of bedload per unit width in kilograms per meter second or pounds per foot second as ordinate, and power per unit width as abscissa, in the same units as above. This curve approaches a uniform slope at high discharges equal to $n = 1.5$ in $L = a\omega^n$, where $L$ is transport rate and $\omega$ is power per unit width.

## *Amounts of Sediment Transported*

To describe how much sediment is transported by rivers is a difficult task because of variability in space and time, region to region, and year to year. Data are usually inadequate; they may deal with bedload, but not suspended load, or the opposite. Records are not long. Nevertheless, I shall try here to provide some quantitative information and to indicate various ways of presenting sediment volumes.

Table 11.1 is a regional comparison compiled by the U.S. Water Resources Council. It gives orders of magnitude in tons per square mile per year. Table 11.2 is similar, but presents data from individual gaging stations rather than regional averages. This brief table makes it clear that

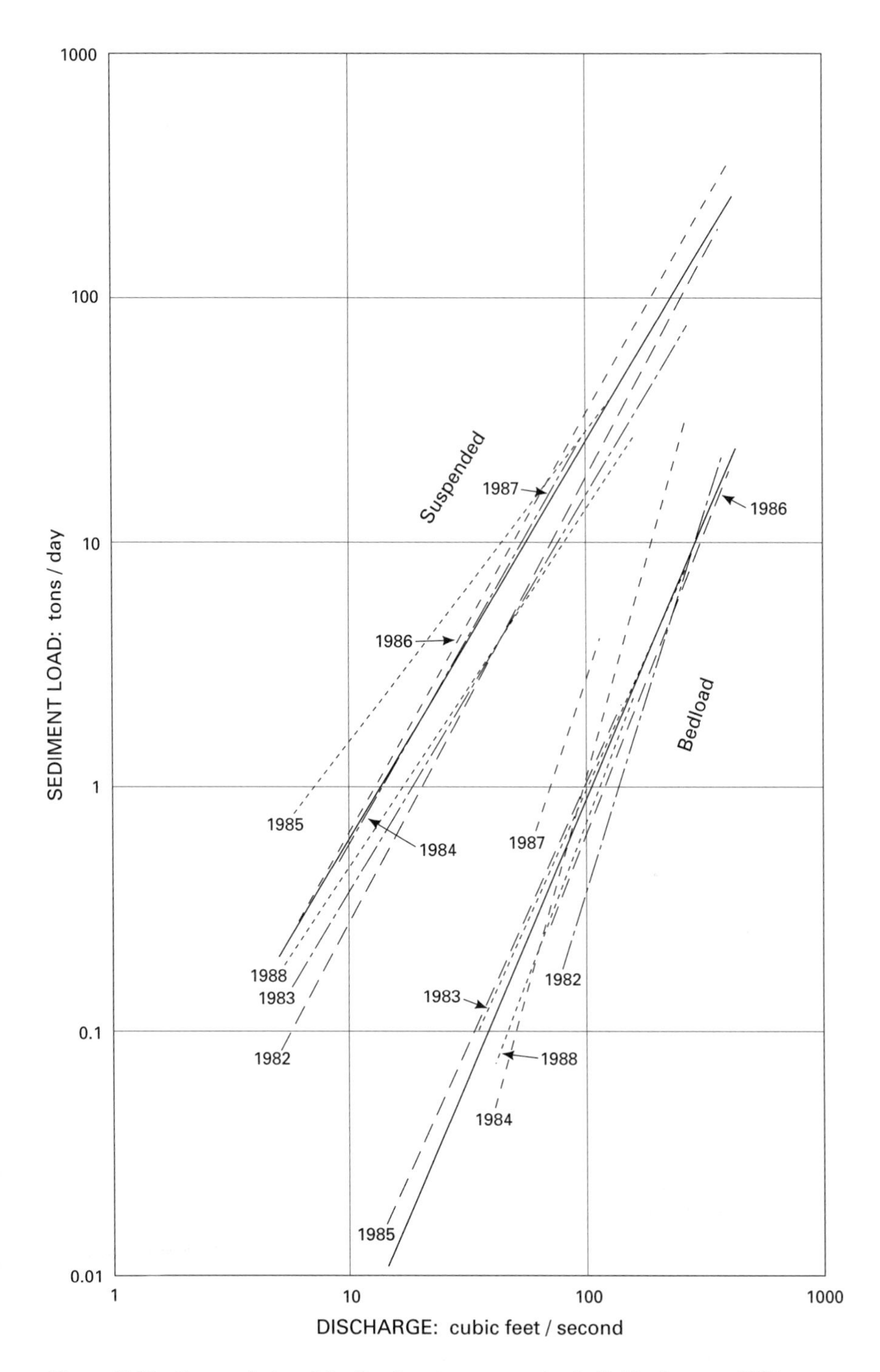


*Figure 11.10* Suspended and bedload rating curves for individual years, 1982 to 1988, at Little Granite Creek near Bondurant, Wyoming. The drainage area is 21.1 square miles. (Data from Emmett 1990.)

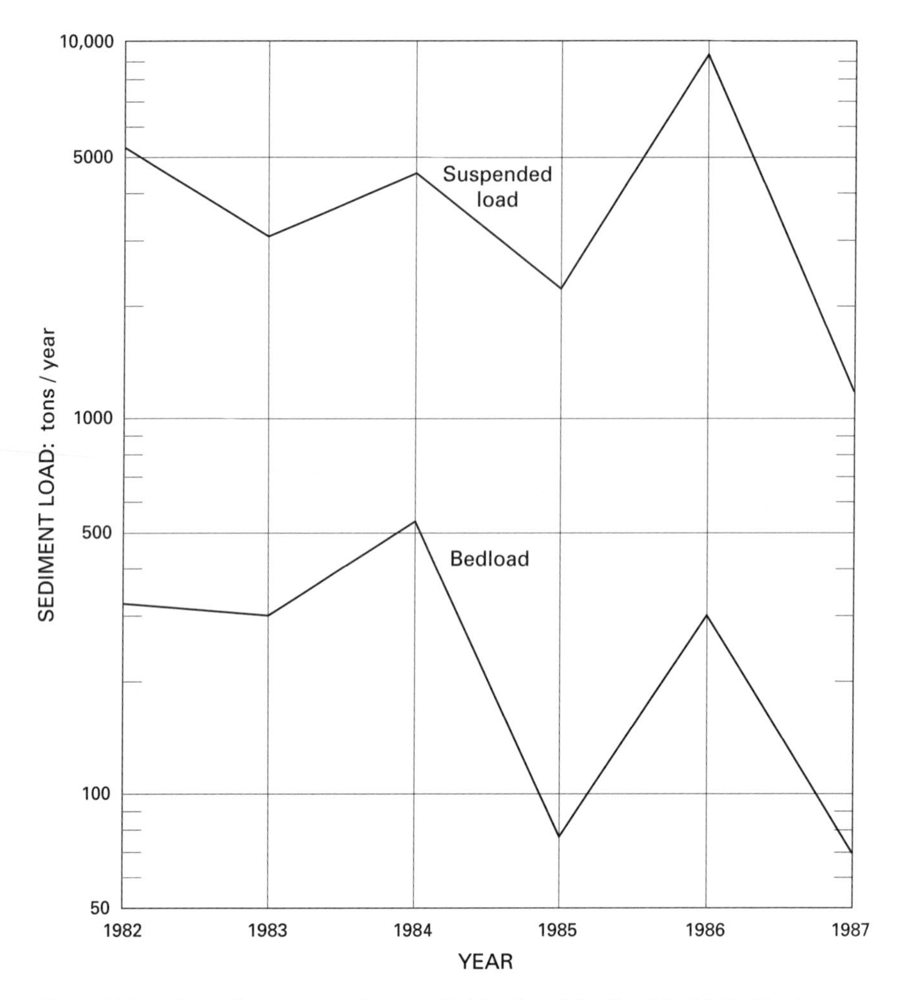


*Figure 11.11* Annual amounts of suspended load and bedload in Little Granite Creek, Wyoming, for each of six years. The weight of suspended load is 15 to 25 times that of bedload.

*Table 11.1* Sediment yield from drainage basins of 100 square miles or less in the United States (from U.S. Water Resources Council 1968, pt. 5, chap. 5, p. 4)

| | Estimated sediment yield (tons/sq mi/yr) | | |
|---|---|---|---|
| Region | High | Low | Average |
| North Atlantic | 1,210 | 30 | 250 |
| South Atlantic—Gulf | 1,850 | 100 | 800 |
| Great Lakes | 800 | 10 | 100 |
| Ohio | 2,110 | 160 | 850 |
| Tennessee | 1,560 | 460 | 700 |
| Upper Mississippi | 3,900 | 10 | 800 |
| Lower Mississippi | 8,210 | 1,560 | 5,200 |
| Souris-Red-Rainy | 470 | 10 | 50 |
| Missouri | 6,700 | 10 | 1,500 |
| Arkansas-White-Red | 8,210 | 260 | 2,200 |
| Texas—Gulf | 3,180 | 90 | 1,800 |
| Rio Grande | 3,340 | 150 | 1,300 |
| Upper Colorado | 3,340 | 150 | 1,800 |
| Lower Colorado | 1,620 | 150 | 600 |
| Great Basin | 1,780 | 100 | 400 |
| Columbia—North Pacific | 1,100 | 30 | 400 |
| California | 5,570 | 80 | 1,300 |

*Table 11.2* Samples of sediment yield at specific gaging stations in different parts of the United States

| Basin | Drainage area (sq mi) | Sediment load (tons/sq mi/year) | | |
|---|---|---|---|---|
| | | Bedload | Suspended | Total |
| Coon Creek, Medicine Bow, Wyoming | 6.5 | | | 17 |
| East Fork Encampment Creek, Medicine Bow, Wyoming | 3.5 | | | 2.6 |
| Front Range, Colorado | | | | |
| Goose Creek No. 1 | 81 | 6 | 3 | |
| Goose Creek No. 2 | 81 | 3 | 5 | |
| Goose Creek No. 4 | 81 | 3 | 6 | |
| Left Hand Creek | 52 | 23 | 17 | |
| Little Beaver | 12 | 4 | 6 | |
| Lower Trap | 5 | 4 | | |
| Middle Boulder | 29 | 5 | 9 | |
| South Fork, Cache la Poudre | 88 | 2 | 11 | |
| Eastern United States | | | | |
| Watts Branch, Maryland | 4 | | 227 | |
| Juniata, Pennsylvania | 3,354 | | 265 | |
| Delaware, New Jersey | 6,780 | | 270 | |
| Western Plateau | | | | |
| Green River, Utah | 40,600 | | 530 | |
| Colorado River, Cisco, Utah | 24,100 | | 808 | |
| Bighorn River, Kane, Wyoming | 15,900 | | 114 | |
| Little Granite Creek, Wyoming | 21.1 | 264 | 4,170 | |

mountain areas such as the high country of Colorado produce only very small amounts of sediment compared with the Piedmont in Maryland and Pennsylvania, a more humid region. Records of soil erosion from small agricultural areas are another way of expressing sediment production. Soil loss is different from sediment yield, because much of the material eroded from farmland is deposited nearby and remains in storage for long periods. A sample of data on soil loss from agricultural areas is found in Dunne and Leopold (1978, table 15).

Still another insight into sediment transport is gained by citing the transport rate in a river at bankfull stage, when most of the transport actually takes place. Analyses were made of the suspended-load rating curves and the flood-frequency data for 28 gages in the Coast Range of California. I tabulated the suspended load in tons per square mile per day at bankfull discharge. This quantity for individual days ranged from 20 to 2,000, with an average of 534. Of course, bankfull stage may or may not last a full day, but these figures give an impression of the magnitude.

Those stations with the highest average discharge, located in the northern counties of California (tributaries to the Eel River are examples), had an average of 652 tons per square mile per day, only slightly higher than the average of the stations. Interestingly, the two highest values at bankfull included Bull Creek, a tributary of the Eel River with a basin area of 28 square miles; its value was 1,997 tons of discharge per square mile per day. The other was Wildcat Creek in the hills behind Berkeley, 7.8 square miles of drainage area, and 1,793 tons of discharge per square mile per day. Both basins are prone to massive landslides, a probable cause of the large values of sediment flux.

## Sources of Sediment

The gaping wounds in the surface topography where gullies eat into the soil suggest that arroyo trenching is perhaps the most important source of sediment in some streams. Geomorphologists have compiled very few sediment budgets using actual measurement over time. The few that exist indicate that gully erosion is producing far less sediment than surface sheet erosion by sheetwash, rills, and rain splash.

Except where terrace remnants confine a channel and are eroding, and where landslides are principal sediment sources, sheet erosion is the most ubiquitous process putting sediment in rivers. A sediment budget we compiled by field measurement of various processes in a semiarid basin in New Mexico showed that surface erosion was producing 13,600 tons per square mile per year, gully erosion 200 tons, and mass movement 90 tons. The overriding importance of surface sheet erosion was evident.

Rivers tend to gradually erode the concave banks of channel bends. Vertical unvegetated banks seen on such bends appear to be a major source of sediment. As erosion occurs on the concave bank, the point bar on the convex bank grows correspondingly, so that channel width remains constant over time. Moreover, material eroded from a stream bank tends to be deposited on the next point bar downstream on the same side of the channel. This trading process contradicts the apparent importance of channel bank erosion as a primary source of sediment.

Bank erosion is, however, a major source of sediment in the many river

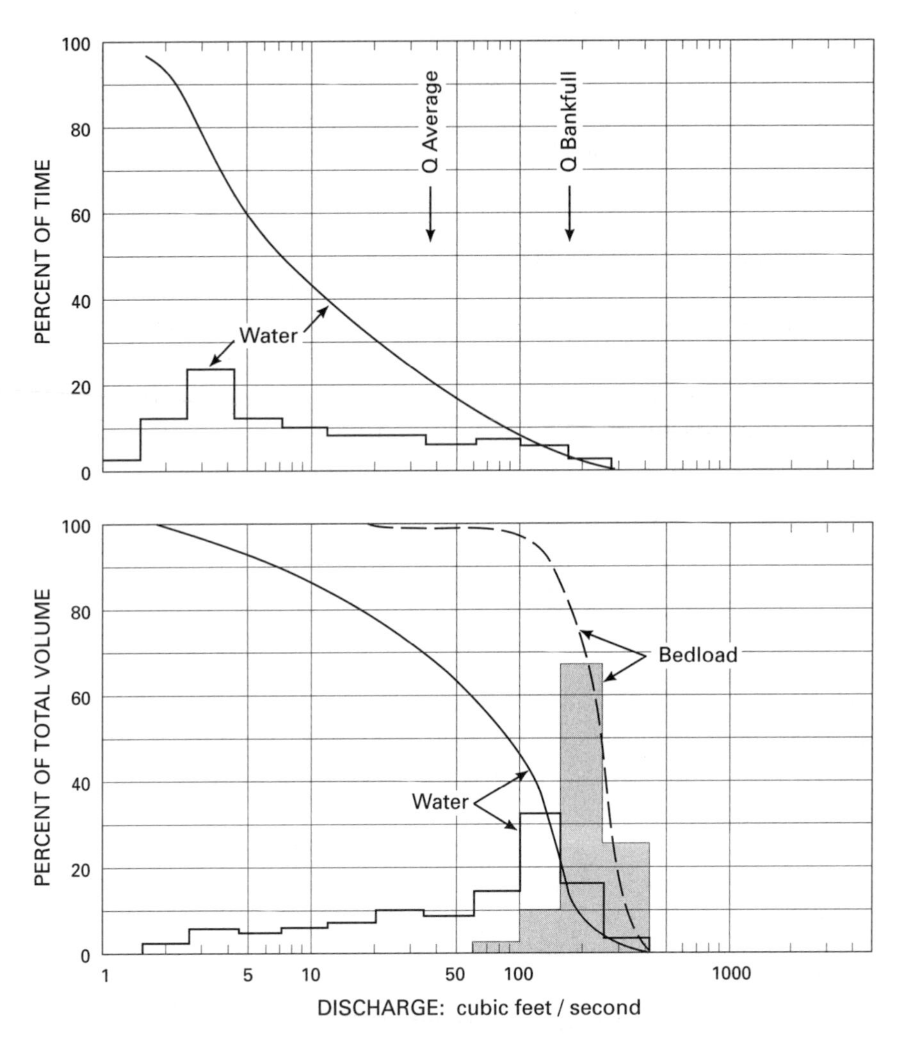


*Figure 11.12* The percentage of time various discharges occur *(above)* and the percentage of annual water volume and bedload volume contributed by various discharges *(below)*, Left Hand Creek, Colorado. In both diagrams the quantities are shown as cumulated and as noncumulated amounts.

valleys that have one or more terraces. These terraces present relatively high banks to a channel moving laterally against them. When a river erodes a terrace that stands above the floodplain, the volume of sediment eroded per unit of lateral erosion is proportional to the height of the terrace. This terrace sediment is larger in volume than that involved in the usual trading of material between cut bank and point bar, and thus represents a net increase in the sediment load of the stream.

In some areas, including the northern coastal ranges in California and some of the steep land in western Oregon and Washington, landslides are an important sediment source—particularly where they were active in the wet period of the late Pleistocene and have remained dormant for long periods. When such slides are reactivated by undercutting at the toe, immense amounts of sediment are delivered to channels whose dimensions have been developed under present conditions in the absence of massive slides. Road construction and the growth of channel bars due to deposition downstream of forest clear-cutting have been responsible for channel impingement on formerly stable hillslopes, with resulting reactivation of ancient slides.

In his years of geologic mapping in the Front Range of Colorado, Richard Madole came to a similar conclusion for mountain valleys, most of which were glaciated only at their extreme upstream end. Glacial melting in the early Holocene was influential in moving sediment down into the valley fill. Madole found, however, that most streams he studied in their mountain valleys were flowing on and moving late Holocene and recent alluvium.

## Relation of Bedload to Water Quantities

In many streams bedload does not begin to move until the discharge is greater than the mean annual flow. Average discharge in California rivers fills the channel half full; in Wyoming rivers, one-quarter full. In other words, discharge must be appreciable before bedload is in motion. Furthermore, bedload moves only a small number of days per year.

To give quantitative meaning to this statement, two types of analysis are presented in Figures 11.12 and 11.13. The graph for Left Hand Creek in the Colorado Mountains shows that on 24 percent of the days in a year, the discharge is between 3 and 5 cfs, but these flows contribute only 5 percent of the total annual discharge. Discharges less than the mean annual value occur on 80 percent of the days and contribute 30 percent of the annual volume. Average annual discharge, 40 cfs, is exceeded on

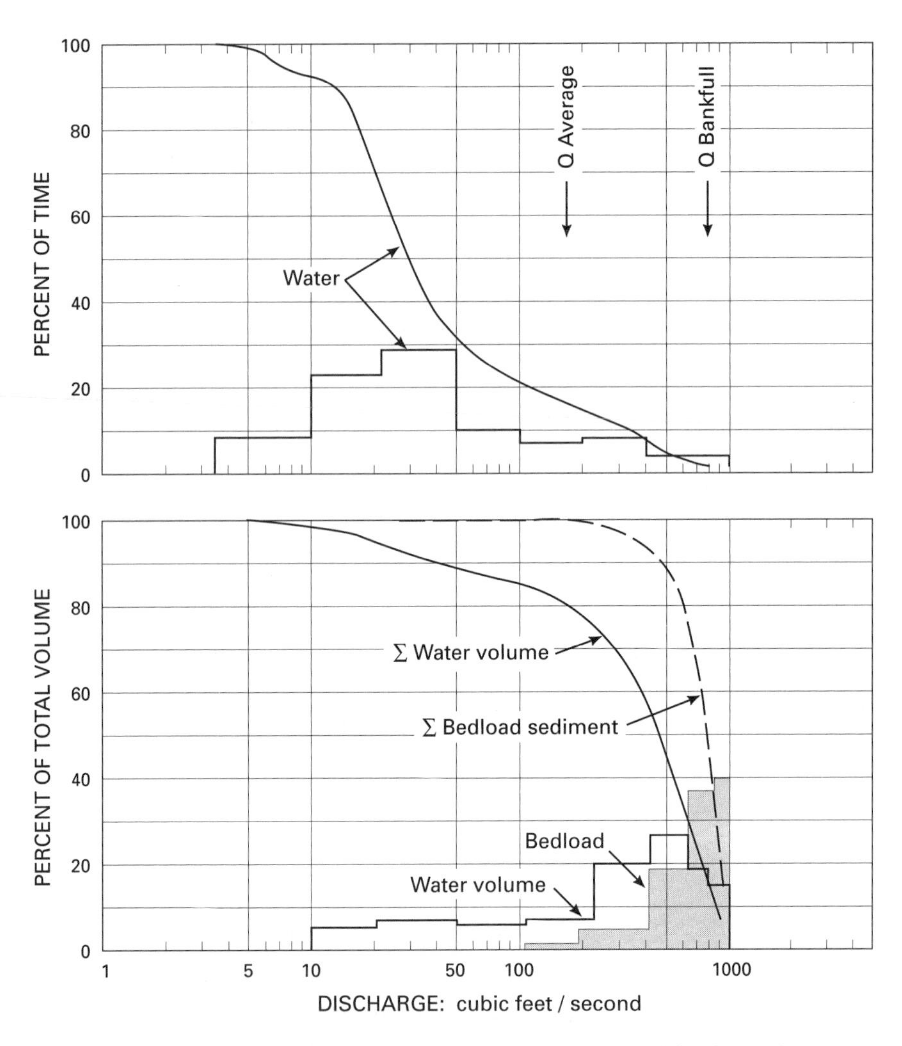


*Figure 11.13* The percentage of time various discharges occur *(above)* and the percentage of annual water volume and bedload volume contributed by various discharges *(below)*, East Fork River, Wyoming. In both diagrams the quantities are shown as cumulated and as noncumulated amounts.

only 20 percent of the days in a year. Yet the greatest volume of water, 32 percent of the annual total, occurs in the discharge range 100 to 150 cfs, a value of just less than bankfull. Indeed, 60 percent of the annual water volume occurs in the discharge above mean annual discharge, or in 20 percent of the days in a year. Bedload is not moved by flows less than 60 cfs in this stream, and by far the largest volume is moved in discharges near bankfull. These data agree with the statements made earlier that effective discharge, that which carries the greatest sediment load, closely coincides with bankfull.

Nearly identical conditions exist on the East Fork River, Wyoming (Figure 11.13), where we constructed and then operated our bedload trap for nearly ten years. Average discharge is equaled or exceeded only 19 percent of the time but in that time 80 percent of the annual volume of water is discharged. Again, bedload hardly moves at discharges less than mean annual, and most bedload is transported by flows just below bankfull.

### *Relation of Bedload, Suspended Load, and Dissolved Load*

Rocks appear hard and solid, so it is not easy to visualize rock material dissolved in water like sugar in a coffee cup. But dissolved material is an important part of the load carried by rivers. As percentage of the total load (dissolved plus sediment), the dissolved portion varies from 2 to 60 percent, increasing with increased precipitation and runoff.

The transport of material dissolved in water is more effective than one might intuitively surmise. Six scientists independently compiled river records for the world and came to surprising agreement. They found that the dissolved load per unit of drainage area, including data from rivers on all continents, is between 32 and 37 metric tons per square kilometer per year (90 to 104 tons per square mile per year). All these authors agreed that the total dissolved load from all continents is about 3.50 billion metric tons per year (3.86 billion tons per year). Maybeck (1976) in addition computed the probable amount of both dissolved and solid (bedload and suspended) load for the major rivers of the world. The ratio of solid to dissolved load is 46 for the Colorado River and 10.5 for the Brahmaputra. Low values of the ratio include 0.3 for the Ob River, and 1.7 for the Amazon. The Ganges has a ratio of 6.9; the Congo, 1.1.

The total of dissolved plus suspended load varies from about 100 to 800 tons per square mile per year in rivers in much of the United States. Webb and Walling (1982) studied a 105-square-mile basin of the river Creedy in Devon, Southwest England. In this basin, with a mean annual

*Figure 11.14* John P. Miller, the distinguished geologist who mapped many quadrangles in the Sangre de Cristo Mountains of New Mexico and was an early researcher on mountain stream channels.

precipitation of 35 inches comparable to the midwestern United States, dissolved load exceeded suspended load. The average annual value for dissolved load was 230 tons per square mile per year. The total, 354 tons, was comparable to the total for the Mississippi and somewhat larger than that of the Delaware River in New Jersey.

The chemical species, such as silica, dissolved in river water were measured in accordance with the usual chemical analysis of dissolved solids. John P. Miller studied the areal and lithologic source of various chemical species in the Sangre de Cristo Mountains of New Mexico, where he had mapped the geology in more than six quadrangles (Figure 11.14). He was able, therefore, to sample streams emanating from only a single lithology and to compare the solution products of granite, quartzite, and calcite-cemented sandstone.

## Materials in the Streambed

The size of rock material in a riverbed is larger in the headwaters and decreases downstream in roughly exponential manner. Several factors are involved. The size of rocks initially contributed at the headwater is dependent primarily on the jointing in the rock outcrops, which determines the size of blocks made available. The jointing may be quite different in various lithologies and differs among sedimentary, igneous, and metamorphic rocks. But the material available in the headwaters is strongly affected by past geologic conditions, especially glaciation, weathering processes, or previous episodes leading to transportation and deposition.

The downstream decrease in size is due to a combination of abrasion (rounding and breakage during transport), solution in place, and sorting (the movement of finer particles, leaving coarser fragments as a lag deposit). The relative importance of abrasion and sorting cannot be determined in the field, but by the study of separate lithologies useful inferences can be drawn. In the headwaters of the Shenandoah River, J. T. Hack found that the number of boulders of sandstone decreased downstream of an outcrop from 50 to less than 10 in a few hundred feet. Limestone cobbles persisted as a major component in the streambed for no more than 5 miles, and in one formation for only 1 mile below the source outcrop. Hack concluded that limestone blocks are destroyed or worn down within a short distance of the source.

In the river Noe in Derbyshire, England, A. D. Knighton found that sandstone, one of only two lithologies present, decreased in size exponentially from a median size of 80 millimeters to 35 millimeters in a distance of 20 kilometers (12 miles). He also found that the rate of increasing roundness, which provides some indication of abrasion effectiveness, was greatest in the steep headwater zone. In the high mountains of New Mexico, John P. Miller found that the mean particle size of granite decreased from 150 millimeters to 75 millimeters in 60 miles; that of quartzite dropped from 130 millimeters to 70 millimeters in 79 miles.

Of course, the decrease in size of a particular lithology does not clearly differentiate between abrasion and sorting, but some other information implies that the above-mentioned changes in size are due to abrasion, breakage, or wear. The lines of painted rocks placed in mountain streams in Colorado reported by Rosgen and me involved rocks of sizes $D_{84}$, $D_{50}$, and $D_{35}$ averaging respectively 150, 70, and 50 millimeters. All 769 rocks placed were observed each day for a full snowmelt season. Of the 500 rocks that moved, the percentage moved was nearly the same for the three size classes. Furthermore, the average distance moved in one excursion was less than 2 feet and did not vary with rock size. In one stream 64 rocks of $D_{84}$ size were placed; 22 moved once in the runoff season, 17 moved twice, 7 moved three times, and 2 moved four times. Thus, the time spent exposed to weathering in the streambed is great and presumably prepares a rock for eventual abrasion.

Another line of evidence is that of Leopold, Emmett, and Myrick in the painted rock data for an ephemeral channel in New Mexico. The hundreds of rock movements recorded showed that the distance traveled did not depend on rock size. In fact, differential movement by size was shown to be absent. Because both bed material size and channel gradient decrease exponentially downstream, there is an intuitive tendency to assign the flattening of slope to the decrease in bed material size. Several studies have shown this to be simplistic and misleading. John P. Miller found no consistent relationship between slope and bed material size. Both are related to other factors such as areal distribution of lithologic types and processes operating in the geologic past. Hack showed that only when drainage area is specified does there exist a relation of slope to bed material size.

Numerous studies lead to the conclusion that channel slope decreases downstream primarily as a result of the increase of discharge downstream.

## Computation of Sediment Transport Rate

It is desirable for many purposes to be able to compute transport rate of sediment from flow parameters. For example, it is often necessary to compute the sediment transport rate at a location where no sediment measurements have been made. Available theory seems adequate, but the choice of flow parameters and sediment size is not simple because of the many combinations of variables that are possible in rivers. Generally, computation of transport rate utilizes the average of past measurements in the form of a sediment rating curve. The calculation of individual conditions without using such an average is a formidable task.

Sediment rating curves are drawn from measurements made by a sampler under various flow conditions. For suspended sediment, the curve is a plot of either concentration or transport rate as a function of discharge. For bedload, the curve is usually expressed as tons per day as a function of either discharge or stream power. The tons per day for an individual measurement is already an extrapolation of the weight caught in a sampler, expanded to a chosen width and to an assumed constant rate for some period of time. For example, the mouth of a Helley-Smith sampler is about 3 inches wide. The sampler is held on the bed for 30 seconds, or perhaps up to 3 minutes, at about 20 sites across a channel. To express the average of this sample as tons per day is, as indicated, an extrapolation to a large width and to a long time period and can result in serious error.

Nevertheless, if data collection by sampler, whether on suspended sediment or bedload, is carried out according to tested practices, the sediment transport rate for the river as a whole can be determined with acceptable accuracy. Certain facts must be considered. In contrast to suspended load that tends to be well mixed across the channel, bedload moves in only a portion of the channel width and not usually in the deepest portion as might be expected. The main bedload movement tends to be on the sloping surfaces of point bars or central bars, where depth is intermediate between the deepest and shallowest zones. Data for mountain streams in Colorado show that 80 percent of the bedload moves in 40 percent of the channel width. Therefore, if bedload is computed per unit of width, great circumspection is required in translating the result into total bedload transport for the entire river.

This problem does not arise if the bedload rating curve presents the load in tons per day for the whole river. But to arrive at that unit, the samples must be taken at uniformly spaced distances across the channel. Those locations where the load is zero must be given the same weight as other locations in computing the total.

Various equations for computing transport rate from hydraulic measurements consume a large part of the literature on sediment transport. Several of the equations proposed are based on the idea that transport rate per unit of width is a function of bed shear stress computed from depth and slope, less a critical shear stress needed for initial motion. The latter involves the grain size of the sediment, often approximated by the mean grain size $D_{50}$. Some equations compute transport rate as a function of mean velocity.

In a more general sense, formulas all involve some combination of weight of fluid and solids, kinematic viscosity of the water, particle diameter, water depth, and shear velocity (which includes slope as well as depth). According to W. R. White of the Wallingford Hydraulic Research Station in England, the above parameters can be grouped in four basic quantities: a dimensionless grain size, a mobility number, a dimensionless flow depth, and relative density. Each river or flume experiment has a nearly constant value of the grain size parameter, but a range of values of the flow depth and mobility parameters.

In an extensive analysis, White, Miller, and Crabbe of Wallingford compared eight equations proposed by different authors by computing the transport rate that had been measured in 1,020 flume experiments and 260 field experiments. The best formulas gave results that differed from the measured values by less than 50 percent. That is, the computed values lay within the range of one-half to twice the observed transport rate. The formula that gave the best results is that of Ackers and White, published in 1973. This formula computed 68 percent of all the experiments within the limits just defined, and 49 percent within the narrower limits of two-thirds to three-halves the observed transport rate. The Ackers-White formula specifically computes the dimensionless transport parameter as a function of the dimensionless grain size and a computed mobility number.

To give some idea of how such calculations proceed, let us use one of Bagnold's equations. He expressed sediment transport rate as a function of stream power, which is $\Omega = \gamma Qs$ for the whole stream, and $\Omega/w = \gamma Qs/w = \omega$ for unit width. He found, in a well-designed set of experiments testing the effect of flow depth, that at constant stream power $\omega$ the bedload transport rate per unit width $i_b$ decreases with increased flow depth $d$ as $i_b \propto d^{-2/3}$. Analyzing the famous flume experiments of G. K. Gilbert, Bagnold (1980, 1986) found that when depth was held constant, transport rate $i_b$ decreased as grain size $D$ increased, according to the relation $i_b \propto D^{-1/2}$.

It has long been known that at high discharge the transport rate reaches a constant relation to stream power expressed as $i_b \propto \omega^{3/2}$, where $\omega$ is power per unit width $\gamma vds$, in kilograms per meter second or pounds per foot second.

If the power required for initial motion is $\omega_0$, then the excess power to move sediment is $\omega - \omega_0$. This led to Bagnold's empirical equation $i_b = (\omega - \omega_0)^{3/2} d^{-2/3} D^{-1/2}$. This simple equation includes those parameters appearing in most transport relations: width, depth, velocity, slope, grain size, and experimental values of flow needed for initial motion. The equation predicts transport rate under some conditions, perhaps with less dependability than the Ackers-White equation but with the advantage of clarity.

CHAPTER TWELVE

# The Drainage Network

## Joining of Channels

The drainage basin is the area that contributes water to a particular channel or set of channels. Precipitation falls on the basin and finds its way over or through the ground to surface streams. The amount of water derived from precipitation and potentially feeding to a given point along the channel system is generally proportional to the size of the source area, at least in regions where precipitation is more or less uniform over the area.

Because the magnitude of a river is a function of the contributing area, the size of the basin feeding any point in a channel system is strongly correlated with nearly every hydrologic and morphologic parameter—average flow, flood magnitude, width and depth of the channel, to name a few. Within any basin the relevant processes can be categorized as (1) those occurring on the unchanneled area or slopes and (2) those occurring in the channel and floodplain. The present discussion concerns the integration of the incipient water courses on the slopes, rills, draws, and channels.

As water accumulates on the ground surface during a storm event, the direction of overland flow acquires a component toward the deepest rill or channel, enhancing the channel discharge during storm periods and promoting its ability to collect water from adjacent hillslope areas. Because the process is self-reinforcing, channels once developed tend to persist.

The importance of random chance has been fully justified by many analyses of channel networks. We introduced the idea of using random walks as an analogue to channel network analysis. The procedure has been widely used in different forms. One consists of developing random walks that begin from a series of evenly spaced points. The direction of

*Figure 12.1* A hypothetical channel network produced by a random-walk procedure. The direction of each downward step is determined by chance.

every step in the random walk is dictated by a random number table or by drawing cards from a deck. The direction of a step is determined by chance, five possibilities being available: left, right, down, or one of two diagonals. Backward and uphill steps are excluded. A sample is shown in Figure 12.1. It is evident that the quantitative characteristics of such a network closely resemble those of river basins.

When two channels join, the drainage area below the junction is the sum of the areas of the two tributaries. Thus drainage area increases downstream in a series of discrete jumps. Between the successive tributary junctions, drainage area increases only by the addition of the unchanneled zone draining directly into the channel. An example of the increase in drainage area along the length of a river is plotted in Figure 12.2. The data represent Watts Branch near Rockville, Maryland. The drainage area of each individual tributary was measured with a planimeter, and the accumulated area was plotted as a function of linear distance along the main channel from the headwater divide downstream.

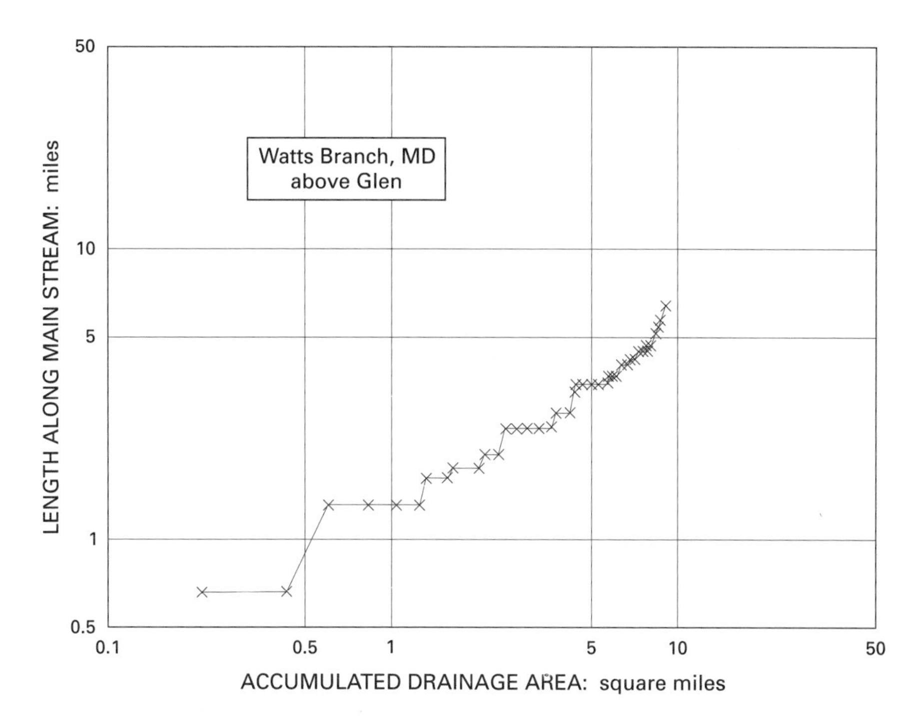


*Figure 12.2* The increase in length of the main channel as the drainage area increases. Watts Branch, Maryland.

The stepped nature of accumulating drainage area is expected. Yet, to generalize, a smooth curve can be drawn through the points with the result that channel length increases as the 0.59 power of drainage areas for this basin. Such a relation was first stated by John T. Hack, who found that for streams in the Shenandoah Valley channel, the channel length along the main stream could be expressed as

$$L = 1.4D_A^{0.6}$$

where $L$ is length in miles and drainage area $D_A$ is in square miles. For Watts Branch the coefficient is 1.3, very similar to Hack's coefficient in the Shenandoah. The same relation holds for large rivers as well. Important increments of drainage area contribute directly to the main channel. But the largest increments of drainage area are those that are added at tributary junctions.

If, as basins of larger size are considered, average basin width and length were to increase in the same proportion, the relation of length of

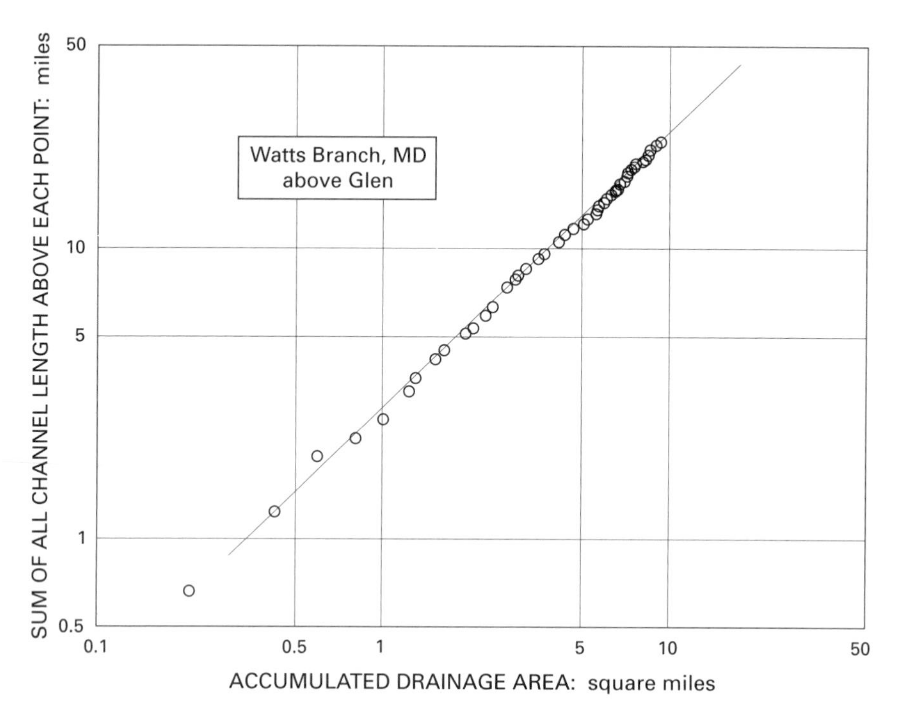


*Figure 12.3* Total channel length, including all tributaries, as a function of drainage area. Watts Branch, Maryland.

the longest stream to basin width would increase as the square root of basin area. That the exponent is not 0.5 but 0.6 to 0.7 indicates that, with increasing size, the length of drainage basins increases faster than width. This relationship holds through a wide spectrum of basin size.

Drainage density is the total length of all channels in a unit area. The average distance between two adjacent channels is the reciprocal of the drainage density. One-half of this distance represents the length of flow over or through hillslopes from a divide to a channel.

The total channel length in each separate drainage area, measured and accumulated downstream, gives total channel length as a function of accumulated drainage area. Such a relation for Watts Branch above Glen is plotted in Figure 12.3. The relation is $\Sigma L = 2.6D_A^{0.94}$, where $L$ is accumulated channel length in miles and $D_A$ is drainage area in square miles. If the exponent were exactly unity, the basin would have the same drainage density for any portion of the basin, drainage density being defined as $\Sigma L/D_A$ at any point along the main channel. For this basin a unit drainage area of 1 square mile is drained by 2.6 miles of channel.

## The Drainage Network

The joining of channels as more and more tributaries enter a main or master stream means that the drainage area increases downstream, so channels get progressively larger as discharge is enhanced by continued additions. The result is a network of channels much like the veination of a leaf. Indeed, the same words used to describe leaves have been applied to drainage networks.

The network of drainage channels has been variously described as trellis or palmate, and by other terms descriptive of veination of different sorts. The use of such descriptive terms implies an organization of the net, and indeed this organization has been shown to be unexpectedly simple. Nature's frequent simplicity of pattern reflects the operation of a few dominant physical processes.

The drainage net is the pattern of tributaries and master streams in a drainage basin as delineated on a planimetric map. In theory, the net includes all the minor rills that are definite watercourses, even all the ephemeral channels in the most distant headwaters. In practice, the amount of detail that can be defined is dependent on the scale of the map used to trace the channels.

If the contributing drainage areas of individual stream segments are outlined on a topographic map, further characterization is possible. Such contributing areas are seen to have a variety of shapes. Although most are oval or perhaps pear shaped, a number are subrectangular and others almost circular. This feature adds another measure of description of the drainage characteristics.

Although such qualitative expressions of dissection and shape tell something about the differences between sample areas, they are of limited use when it comes to investigating the causes of such differences, or to distinguishing basins where the differences are less marked. Recent studies have made possible more comprehensive quantitative descriptions that provide useful tools in the study of the origin and processes of land sculpture.

## Numbers, Lengths, and Orders of Stream Channels

Among the many ideas contributed by the late Robert E. Horton, the one that more than any other has caught the imagination of students of fluvial processes is the relation of lengths and numbers of streams to stream order. To define the last term first, stream order is a measure of

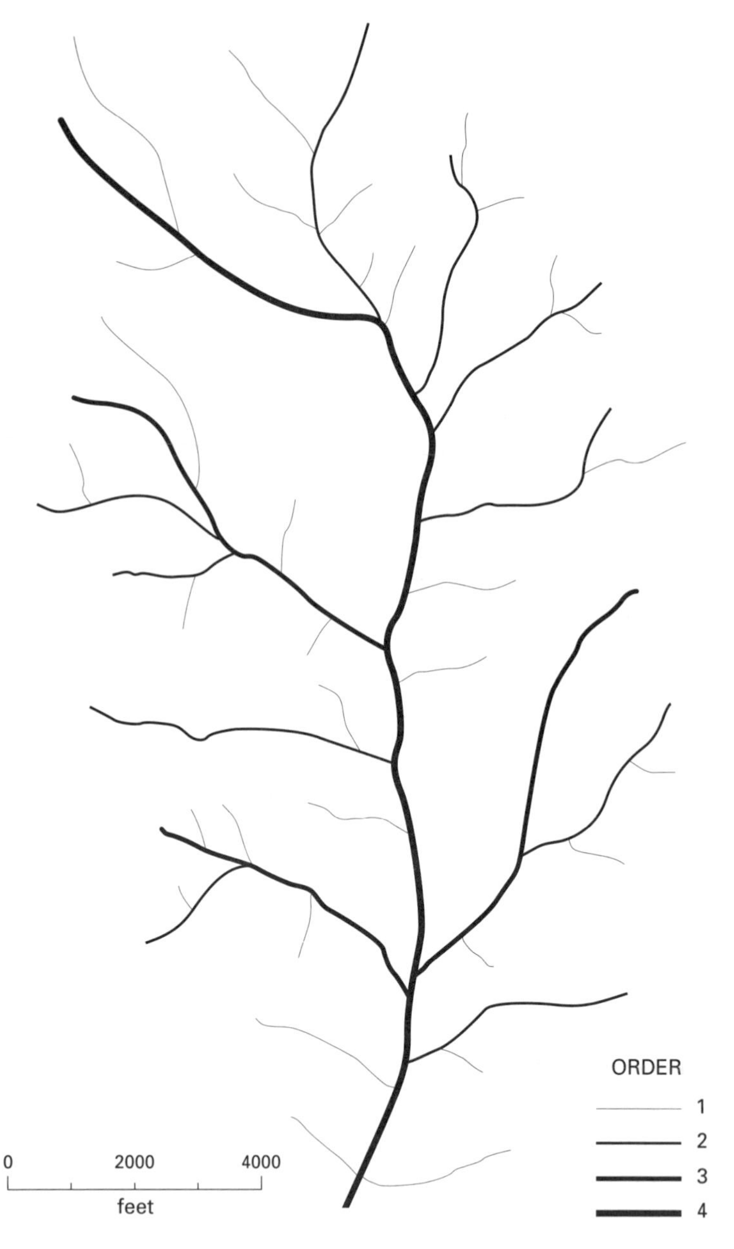


*Figure 12.4* Network classification using the Horton definitions of stream order. Watts Branch above Glen, Maryland.

the position of a stream in the hierarchy of tributaries. Given a map of a certain scale, the first-order streams (Order 1) are those which have no tributaries. The second-order streams (Order 2) are those which have as tributaries only first-order channels. Each second-order stream was considered by Horton to extend headward to the tip of the longest tributary it drains. Which tributary to call the headward extension of a given second-order stream, where differences in length are insignificant, is a matter of choice. The one that seems to be the linear extension of the second-order stream is usually chosen—although when we correlate discharge with drainage basin parameters, the tributary with the largest drainage area is probably most significant.

The third-order stream (Order 3) is formed by coalescence of two second-order streams, and it receives as tributaries only first-order and second-order channels. It is also considered to extend headward to the end of the longest tributary. It can be seen, then, that in practice, after the second-order channels are labeled and the third-order channels are identified, one of the previously marked second-order streams is renumbered to make it the headwater extension of the third-order stream.

An example of the Horton classification system is shown in Figure 12.4. Note that a channel of any order extends headward to the place the most distant tip ends, near the basin divide. For each stream order Horton plotted the mean length, total number, and mean slope of channels of that order. These plots invariably approximate a straight line on semilog paper, as shown in Figure 12.5. Data on average slope are the least consistent and the most affected by the particular way a channel is terminated at the headwater tip. Horton defined two ratios that describe essential elements of a network. The bifurcation ratio, or branching ratio, is the average number of channels of Order $x - 1$ divided by the number of channels of Order $x$, that is, the number of channels of a given order tributary to the next-higher order.

A comparable ratio is the average length of channels of Order $x + 1$ divided by the average length of channels of Order $x$.

Arthur Strahler suggested a different way of designating stream order to overcome the difficulty of renumbering one headwater tributary as necessary in the original Horton scheme. He restricted the designation of order to stream segments.

In the revised scheme the Order 2 stream segment is defined as beginning where two unbranched, or Order 1, channels meet, and extending downstream to the place where that segment joins another channel of Order 2. Thus in the same basin, streams of higher order are shorter in the Strahler system than in the Horton system. The two schemes are

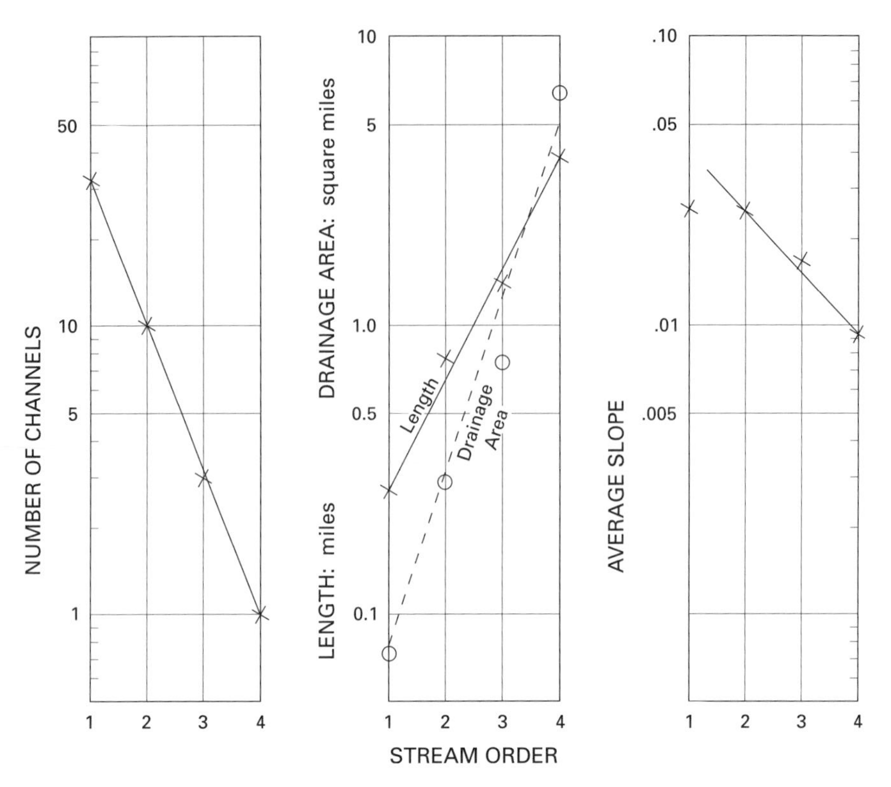


*Figure 12.5* Plot of average number, length, drainage area, and slope as functions of stream order, using the Horton classification. Watts Branch above Glen, Maryland.

similar in that a channel of any given order has as tributaries only streams of lower order.

Classification by segments is easier in that one can start at the upstream end and assign order numbers without any required change. Because of the ease of application, most workers use the Strahler system rather than the original Horton scheme. The disadvantages of the Strahler system, seldom acknowledged, arise from two sources.

First, the delineation of a channel as a blue line on a topographic map is an incomplete picture of the occurrence of channels on the ground. If this delineation is improved by adding definite channels to those shown as printed blue lines, then the Strahler ordering invariably leads to an unrealistically short length of the highest stream order.

Second, the data on the number and length derived from the Strahler ordering system are sensitive to the inclusion or omission of ephemeral tributaries. The addition or omission of a minor channel can materially change the order designation of segments downstream. The Horton scheme is rather insensitive to such a minor change in the mapped network.

Whichever system of nomenclature is used, the actual lengths and numbers of channels in each order depend significantly on the scale of the map used, and on the detail to which topographic draws or swales are scrutinized as possible channels of Order 1. On maps of small scale, regardless of the contour interval, the detail in which the channels of smallest order can be shown is less than on maps of more enlarged scale. Thus the drainage area contributing to channels designated Order 1 is larger on maps of small scale than on maps of large scale.

But the detail in which the network will be shown depends also on the care with which the investigator studies the contour line configuration. On nearly all maps the network shown by the blue lines representing streams will be too sparse, and additions should be made by the investigator to include definite draws or swales not carrying the printed blue symbol.

### *Channels Described by Blue Lines on a Map*

Horton recommended that streams designated as ephemeral (dashed blue lines) as well as those called perennial (solid blue lines) be included in the network. Such differentiation, even on the excellent 1:24,000 maps of the U.S. Geological Survey, is based on appearance only, not on any hydrologic criterion.

The actual practice in mapmaking regarding the use of a solid blue line, an interrupted blue line, or no blue line at all may be summarized as follows from the topographic instructions of the U.S. Geological Survey: "Perennial water features are symbolized by solid blue lines. . . . Intermittent water features are mapped with broken blue symbolization." The definitions of "perennial" and "intermittent" are given in the instructions as:

> Perennial: Those hydrographic features which contain water throughout the year except for infrequent and extended periods of severe drought.
>
> Intermittent: Those hydrographic features which contain water only part of the year.
>
> . . . The field man doing field completion is responsible for completing and correcting the drainage compiled, classifying drainage, and indicating streams to be published.
>
> . . . In addition to his direct observations, the field man must obtain supporting evidence to judge whether observed conditions are characteristic

> of the stream systems throughout the year. The most useful supporting evidence is the information obtained from local residents.
>
> . . . All perennial streams are published regardless of length.
>
> All intermittent streams are published that are longer than 2,000 feet. . . . In applying these rules they should be modified where necessary to produce a consistent portrayal, especially in the extension of streams in headwater drainage.
>
> . . . Streams at the source and upper part of a drainage system are an integral and important part of a complete drainage system. In general, headwater drainage shown on the published map should terminate no higher than about 1,000 feet from the divide, or at the upper confluence of streams, whichever appears most appropriate.

These instructions to the staff preparing a topographic map show that the headward limits of the blue lines do not reflect any statistical characteristic of streamflow occurrence. The specification that the blue line terminate no higher than about 1,000 feet from the watershed divide does not reflect differences in hydrologic performance among various combinations of climate, topography, and geology. Rather, the choice of what is to be shown as an interrupted blue line is based on "consistent portrayal," as the instructions state. The geomorphologist must provide a personal rationale and evidence for designation of first-order tributaries in any given area.

It would be desirable to have some criteria resulting from field studies that would give specific statistical or physical significance to the type of line used on a topographic map. The criteria might stem from a study on the frequency or duration of flow in channels of different sizes or drainage areas. For example, it would be useful if one knew that the solid blue line became dash-dot where a streamflow changed from 90 percent of the time to 80 percent. This change might be a function of drainage area within a given physiographic or lithologic unit. The determination would have to be specific, but also easy in mapmaking practice.

I tried to devise a way of defining hydrologic criteria for the channels shown on topographic maps and developed some promising procedures. None were acceptable to the topographers, however. I learned that the blue lines on a map are drawn by nonprofessional, low-salaried personnel. In actual fact, they are drawn to fit a rather personalized aesthetic. It is surprising that geographers, long interested in mapmaking, have not considered this problem and devised some useful and simple rules based on generalizations from field facts.

In Figure 12.6 appear three versions of the drainage network of Watts Branch above Viers Pasture near Rockville, Maryland. The first map

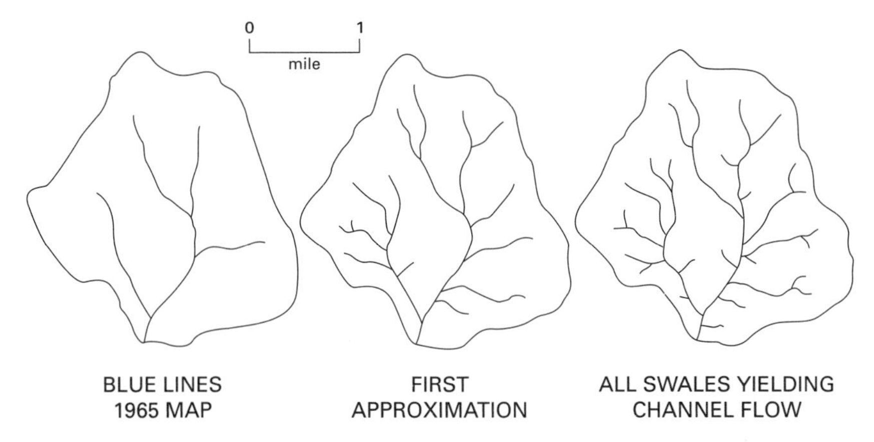


*Figure 12.6* Three versions of the channel network. At left, blue lines printed on a 1965 topographic map; at center, a more detailed net; at right, careful analysis of the contour pattern to include all significant swales or draws. Watts Branch, Maryland; drainage area 3.7 square miles.

shows the drainage net depicted by blue lines on the 1965 edition of the topographic map. This net includes both solid blue and dot-dash blue lines, indicating streams as depicted on the published quadrangle map.

The middle version is the drainage net derived from study of the contour lines of the topographic map, in an effort to draw a stream channel in the principal contour reentrants. The channels are drawn in a consistent manner, so that the same contour configuration applies to each draw or swale considered important enough to be represented as a channel.

Finally, the third version is an attempt to add to the drainage net all swales considered prominent enough to yield channel flow during storms, an admittedly subjective choice. For consistency, headwater tips of tributaries shown in the second version are extended headward to be comparable in drainage area and distance from divide to the smallest newly added channels.

The three maps show the differences in drainage network that might logically be derived from the same topographic map. From both field experience and network analysis, preference would be for the third or most detailed version. To construct such a complete map, however, requires considerable time in excess of what is needed for the second, intermediate version. Such extra effort may be precluded if a network map is required for a large area. Either the second or the third version is preferable for nearly all purposes to the first. Comparable maps for some European countries, especially Belgium, are close to the second version, having a more consistent relation of the channel symbol to contour configuration than do U.S. maps of nearly the same scale.

### *Horton and Strahler Classifications*

Each of the three maps shown in Figure 12.6 has been subjected to a stream-order analysis by both the Horton and Strahler schemes to elucidate the differences in both the network drawn and the method of analysis. The more detailed drainage network results in more channel segments of each order and in the assignment of a higher order for the trunk or master stream. As a consequence, it makes the average length of each segment shorter. These results apply whether the analysis is by the Horton or the Strahler method.

Consider the comparison of numbers and average lengths for the three networks used in this simple sample, analyzed by the Strahler method. Version A uses only blue lines on the map. The length of the Order 4 channel in Version C is only part of the Order 4 segment and is therefore shorter than the other segments. Such an incomplete length of the highest-order channel is typical when an arbitrarily chosen drainage basin is subjected to a Strahler-type analysis.

For the same basin, the difference in the way the network is drawn changes the number of Order 1 streams from 4 to 23 and decreases their average length from 0.9 to 0.3 mile. The relation among orders remains quite constant, even when the net is made much more detailed, though the numbers of orders increases. The inference is that if the net is drawn consistently with respect to the contours, the relation among orders remains, even if the net is not detailed. The more detailed the net, however, the larger the sample becomes within the same basin. The greater number of examples of segments of lower orders gives a more stable average value of the measured parameters.

An appreciable number of channels join rivers of considerably higher order. For example, the basin of Watts Branch above Meeting House Road has a drainage area of 16.3 square miles. There are 651 Strahler Order 1 tributaries; of these, 20 enter channels of Order 3 or Order 4.

A considerable portion of the area of any basin discharges directly into high-order channels without sufficient drainage area to maintain a first-order channel. In Watts Branch above Glen Hills, with a drainage area of 6.97 square miles, 9 percent flow directly into channels of Order 3 or larger. This percentage usually is between 10 and 20.

In the example shown in Figure 12.6 the branching (or bifurcation) ratios are as follows:

| Between orders— | Version A | Version B | Version C |
|---|---|---|---|
| 1 and 2 | 4.0 | 4.0 | 2.9 |
| 2 and 3 | | 3.0 | 4.0 |
| 3 and 4 | | | 2.0 |

It must be stressed that a basin larger than the one used in the example would give a more adequate sample for the determination of branching ratio. But even with this small sample, the average value for version B is 3.5 and for version C is 3.0. It is typical that the value of branching ratio is between 3.0 and 4.0, and 3.5 is common. Thus there are on the average three or four branches of smaller order, tributary to a channel of given order. Experience has shown that the bifurcation ratio is partly dependent on the relief ratio, the ratio of topographic elevation at the headwater divide to basin length.

It is usual that the number and lengths of Order 1 tributaries do not plot on the mean lines of order versus number and length. The bifurcation ratio is generally more stable and constant if determined from the higher orders rather than from Order 1 or Order 2.

First Broscoe, then Strahler, noted that there is an equal drop in elevation for each stream order in a well-developed network of channels.

Geomorphologists have inquired diligently into the processes and constraints that make the bifurcation ratio so relatively ubiquitous. The many studies of Horton diagrams have, with few exceptions, demonstrated that there are either three or four channels of a given order tributary to a channel of the next-higher order. These inquiries have centered on drainage density and length of overland flow.

Kirchner (1986) makes an alternative argument. He says that the observed regularity of channel networks does not reflect an underlying uniformity of channel structure, but instead reflects a choice of how to observe and classify that makes this regularity of observations all but inevitable. He is of the opinion that the near-constant values of bifurcation ratios and of length ratios are merely an artifact of the way the Strahler scheme defines orders. This provocative hypothesis might be considered in light of growth models such as trees.

### *Effect of Map Scale*

The need to complete and unify the drainage net beyond the blue-line network of standard maps applies to all map scales. The three principal scales of published maps of the United States are 1:250,000, 1:62,400, and 1:24,000. Network analysis can be applied to any scale and any contour interval. The main difference is that the size or drainage area of the smallest tributary shown will, of course, be larger for small-scale than for large-scale maps.

The minimum size of channel shown on a map, whether by printed blue line or by insertion on the basis of contour lines, depends on the map scale. The 1:24,000 scale, or 2,000 feet to the inch, is available for most areas of the nation. On such a scale, the smallest tributary easily discernible on the basis of 10-foot contours has a drainage area of about 0.7 square mile and a length of about 1,500 feet.

Our everyday experience suggests that definite channels smaller than that are common. Therefore, if actual channels are mapped on the ground, a far larger number will be found than those discerned on a published map. The scale of map, and thus the number and size of channels, should be chosen in light of the particular need to be served.

To show the effect of map scale, I analyzed a drainage net on 1:24,000 maps for Arroyo de los Frijoles in New Mexico. The channel of Order 1 had a drainage area of 0.3 square mile and a length of 0.3 mile (1,600 feet). A small portion of this basin was mapped, and every discernible channel was walked and measured. This miniature watershed was found to have a master stream of Order 5 on the Horton system. The smallest basin, Order 1 on a map at 1:24,000 scale, was Order 5 on the larger-scale map. One can estimate that the order of the largest basin would be 11 on an enlarged scale map. If there is one basin of Order 11, projecting upward within it are about 200,000 small basins of Order 1 having the length of 0.01 mile.

## Synthetic Networks by Random Walks

The channel network is guided by tendencies toward a most probable state. This stochastic element can be demonstrated in various ways, such as the close relation of random-walk analogues to field relations. Earlier it was shown that random walks proceeding in any direction within an arc of 180 degrees eventually meet. The joining resembles the joining of river channels. We introduced another random-walk model that has, with modifications, been widely used.

On a sheet of rectangular cross-section paper, each square is presumed to represent a unit area to be drained. The drainage from each square has an equal chance of leading off in any of four cardinal directions, subject only to the stipulation that flow in the reverse direction is not possible. Under these conditions one or more streams may flow into a unit area, but only one may flow out. Random selection is used to place an arrow

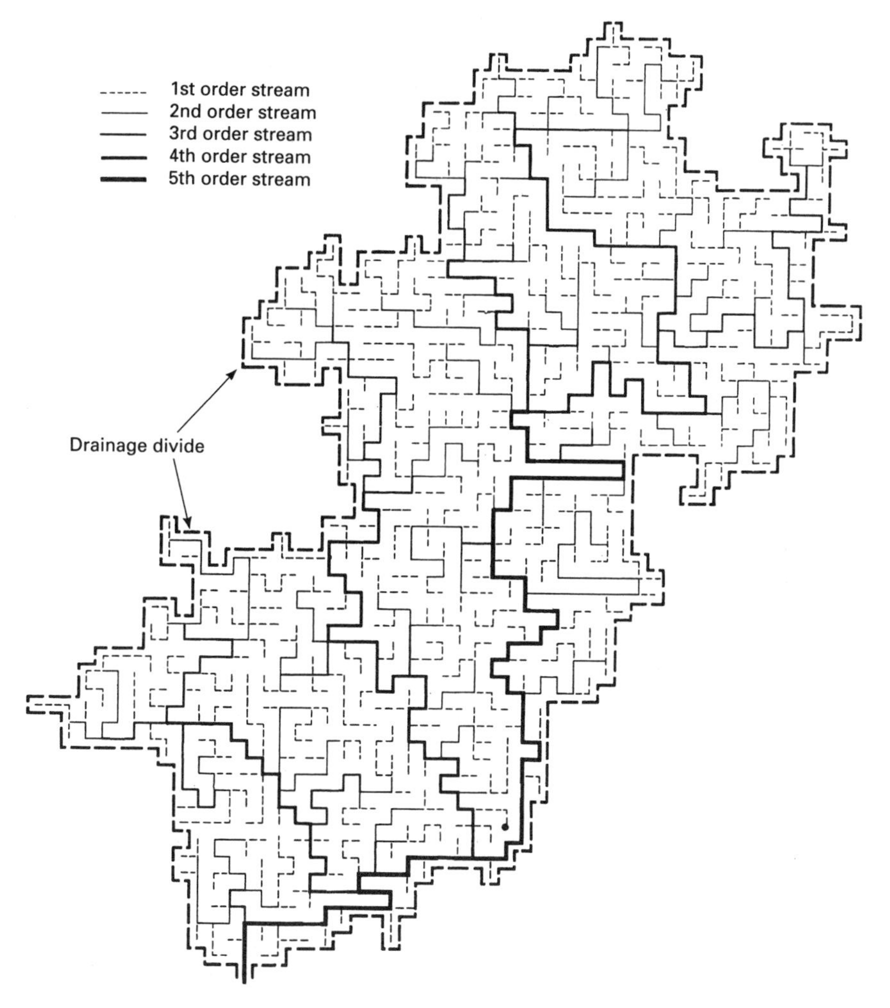


*Figure 12.7* Development of a random-walk drainage network.

indicating flow direction from each square. An example of a network so developed is shown in Figure 12.7. In the figure all networks leading to other master streams have been eliminated.

In Figure 12.8 are plotted the graphs of average stream lengths for two random-walk models, the networks of Figures 12.1 and 12.7, and a sample of a river basin. The master streams are orders 3 to 5. In all three, the number of streams approximates an inverse geometric series, as Horton's law of stream numbers postulates. The near-parallelism of the graphs is worthy of attention.

The fact that networks closely resembling actual river nets can be

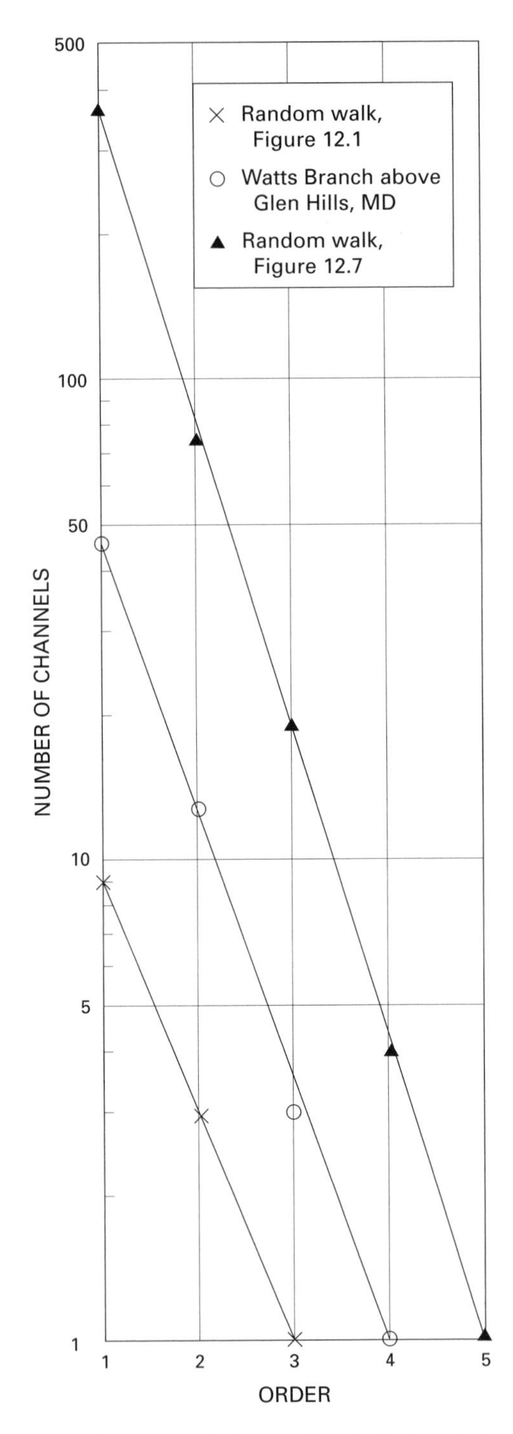


*Figure 12.8* The number of streams of given order plotted against order for two random-walk models and for the basin of Watts Branch, Maryland, using the Strahler ordering system. The random-walk data are drawn from Figures 12.1 and 12.7.

constructed by random procedures led us to conclude that river networks are themselves random. Ronald L. Shreve independently concluded that channel networks are topologically random. Networks developed by random walks suggest that random factors affect the size of drainage basins. One might ask, for example, why on the North American continent there is only one basin the size of the Mississippi River Basin. Why should there not be two or more?

I drew an outline map of the North American continent on cross-section paper and used the procedure described for Figure 12.7 to assign from random numbers the direction of outflow from each unit area. By the nature of the procedure, each basin drained to a point on the coast of the continent. The basins so developed were outlined and the drainage area expressed in square miles. A sample is shown in Figure 12.9. The purpose of the exercise was to examine how random processes result in a hierarchy of basin size and to determine whether such processes result in a size distribution of basins comparable to what is observed on the ground.

The sizes of the basins developed by random walk were arranged in rank order. Similarly, the real rivers on the North American continent that drained to the coastline were outlined and their drainage areas measured. These were arranged in rank order of size. In Figure 12.10 two random trials are compared with the reality of basin size existing on the continent. The graph plots (1) the drainage areas in rank order for two random trials and (2) the reality. The two random trials produced a maximum basin size of 1.8 million and 1.9 million square miles, whereas the largest real basin, the Mississippi, has an area of about 1.2 million square miles. The distribution curves of basin size in rank order are of similar slope and somewhat parallel through a large number of ranks.

The random-walk development suggests that chance alone would tend to develop a variety of basin sizes, the distribution of which is similar to actual basins and therefore rivers. This reinforces the inference drawn from Figure 6.1 that the sizes of drainage basins in all the continents tend toward a log-normal distribution indicating the role of random chance in geomorphic forms.

## The Branching of Trees

The preceding discussion demonstrates that branching networks tend to be random, but the constraints operating to circumscribe the random operation of processes are not obvious. We may gain some additional

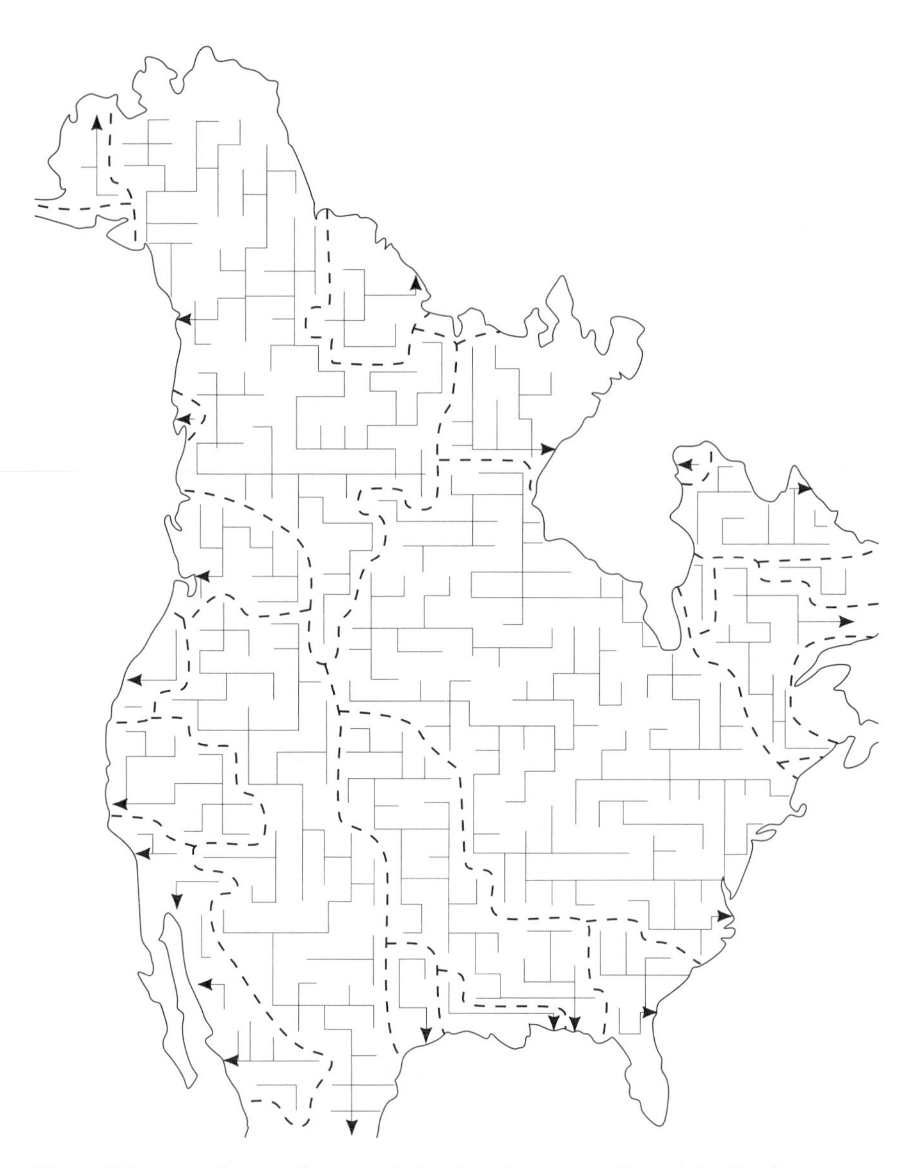

*Figure 12.9* A random-walk network developed on an outline of the North American continent to examine the basin area sizes that might result from a random process.

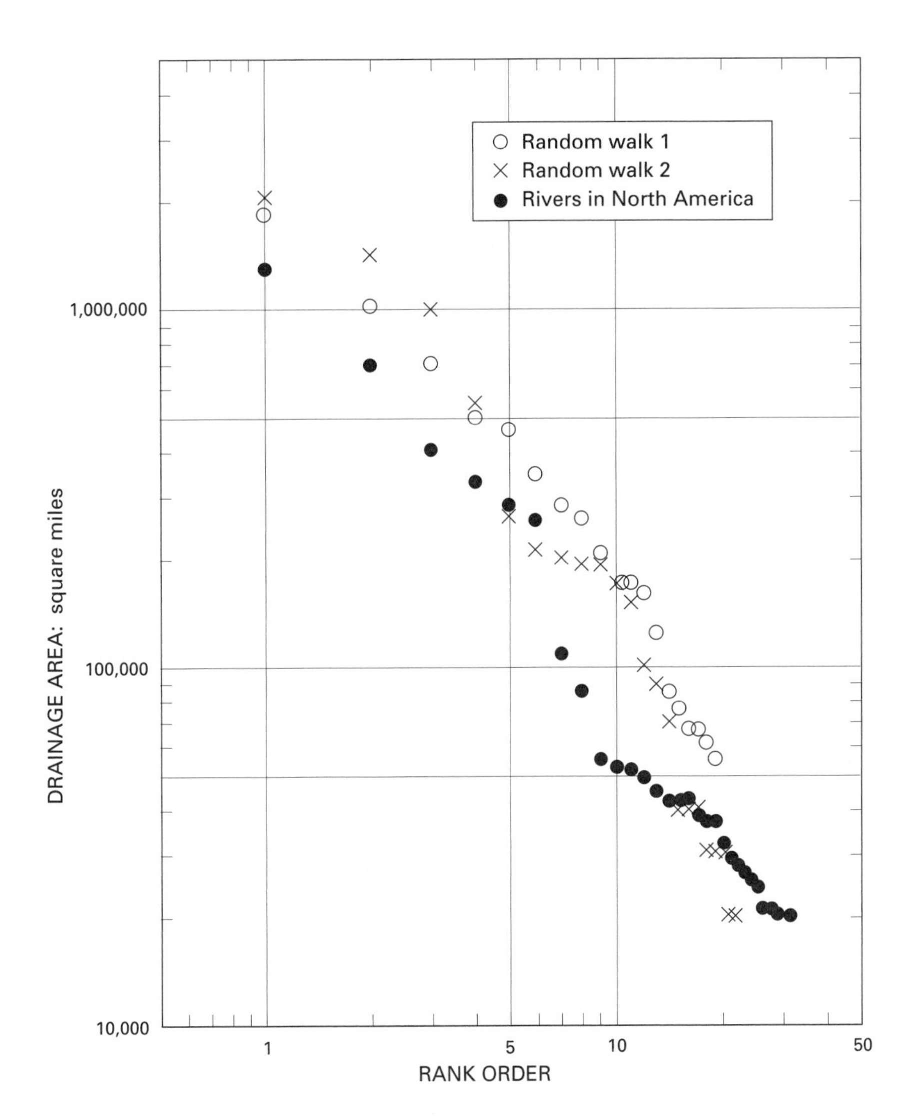


*Figure 12.10* Drainage areas of two random-walk trials plotted against actual basin areas in North America. Basin areas are arranged in order of size of drainage area and are plotted against rank. The Mississippi Basin, largest in the continent, has an area of about 1.2 million square miles. The two random-walk trials have maximum basin sizes of 1.8 million and 1.9 million square miles.

insight by considering other hierarchies in nature. The branching of plants is useful for this purpose.

The stem system of a plant forms a structural support that exposes photosynthetic organs to sunlight and at the same time provides the route for removing photosynthetic products from them. In the most common types of plants, the stems support the leaves and hold them up and out to the sunlight. The leaves are unit areas of photosynthetic surface and must be supplied with water and nutrients. The starches and sugars derived from photosynthesis in the leaves are carried to various parts of the plant to be stored or to support growth.

The branching form that carries the starches and sugars to the rest of the plant is analogous to the drainage system of a river, in that the network of streams and tributary rills serves various parts of the drainage area as routes for carrying surface runoff and erosion products from the basin, eventually to the ocean. The network of stream channels is composed of trunk streams and tributary branches similar to those of a tree.

Service to a large number of individual leaves or, in the case of a river drainage basin, a large number of unit areas can be accomplished by a variety of stem patterns. In both the plant and the drainage basin there might be, for example, an individual branch going to each unit but all branches converging to a single central locus. Such a pattern would eliminate stems of intermediate size. The plant world contains such examples, one being Ocatillo, a thorny desert shrub that has long stems emanating from a main root; leaves are distributed all along each stem and the stems are unbranched.

Another possible configuration would be a large number of petioles emanating from a single tip of stem, but no leaves distributed along the small branches. This configuration would be like a spoked wagon wheel placed atop a flagpole. Still another possibility would be a tree like a flagpole with many spoked wheels located alongside its length.

Interestingly, one can imagine a variety of possible configurations that would fit the basic purpose of serving each leaf or unit area of a drainage basin but, with rare exceptions, trees and rivers exhibit only one of these. The nearly universal pattern is a large central trunk that successively divides into more numerous small branches or tributaries. The tree is a three-dimensional structure. The river system as analyzed in plan view on a map is a two-dimensional structure. This difference does not contravene their basic similarity and the virtual absence of other basic forms.

To show this similarity we can perform a Horton analysis of branching on a tree. The definitions and plotting procedure discussed earlier can be applied. A typical set of data are plotted in Figure 12.11, including both

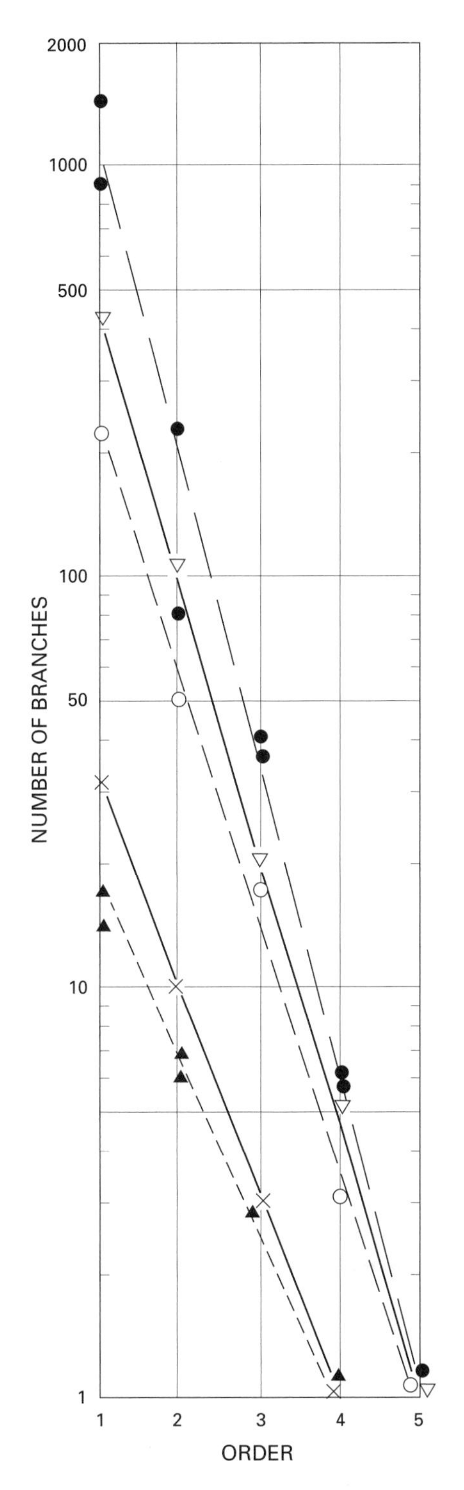


*Figure 12.11* Number of branches plotted against their order for five networks analyzed by the Horton method: two ash trees (*Fraxinus* sp.); a 10-foot fir tree (*Abies* concolor); a random-walk network; the channel network of Watts Branch, Maryland; and three-dimensional trees constructed of Tinker Toys.

a deciduous species (ash, *Fraxinus* sp.) and a conifer (fir, *Abies* concolor). The graph includes data on a river and a couple of random-walk models in order to show the similarity. The numbers of branches as functions of order are remarkably similar for the fir tree, the ash trees, and the random-walk networks. Similar graphs for the basin of Watts Branch and a three-dimensional random-walk tree made with Tinker Toys are slightly different from the other three.

It is perhaps easier to visualize the constraints imposed on the branching pattern of a plant than of a river. The plant stem must provide structural support as well as a route for the translocation of fluids. The structural system must be flexible to withstand external forces, must be expandable as the plant grows, and is self-manufactured. With these requirements it appears logical that there is a genetic advantage to minimize the volume and thus the total length of branches. Some economy in length is achieved by the branching network and by the angle of joining of branch to main stem.

Consider the problem posed in Figure 12.12, in which three points, A, B, and C, are to be joined in such a way that points A and B are to be fed by, supported by, or otherwise served by a single point C. Three diagrams represent some of the alternative lengths and angles at which branches could be constructed, depending on the choice of the junction point D. The graph below in Figure 12.12 shows the relation of position of junction and total length of branches. A minimum point exists, not very sensitive but nevertheless real, in which point D is about a third of the maximum distance down the length of branch D–C.

Thus, to minimize total branch lengths the branches in the example would not meet in a T as in the left diagram, nor in a V as in the right diagram, but in a Y in which the stem of the Y is about two-thirds of its total weight.

Consider now the case of a square, 8 units on each side, or 64 unit areas. The goal is to serve each unit area from a point centered in the square. One possible pattern would be to have a separate branch proceed from the central point to each of the 64 units, as suggested in Figure 12.13A. Another would be to have four principal branches connected to the center; from the ends of each branch would radiate 16 shorter stems, as in Figure 12.13B. A further subdivision involving subbranches is pictured in Figure 12.13C. The total lengths of stems in the three patterns are approximately 200, 124, and 88 length units, respectively. The order of the largest branches in the three patterns are respectively first, second, and third.

Clearly, an increase in the maximum order of the branching pattern serves the same number of unit areas with a progressively smaller total

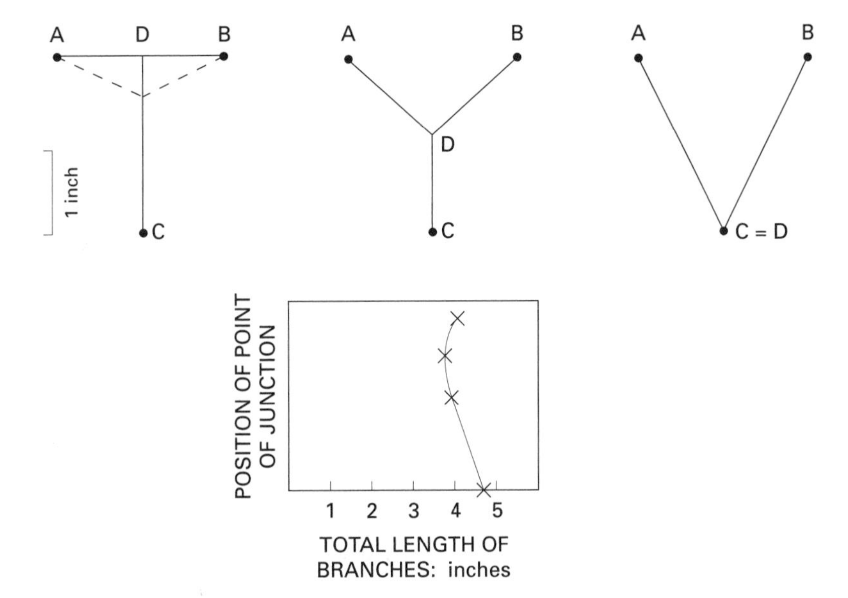


*Figure 12.12* Effect of the position of the junction of a Y on total stem length; three possible arrangements. Below is a graph of the position of the junction point in relation to total stem length. The ordinate scale is identical to the length of the ordinate in the diagrams above.

length of stem. Branching, therefore, represents an increase in efficiency in considering total stem length. (It is understood that constraints limit the bifurcation.)

Random-walk models of river networks closely approximate natural field systems. Whether a similar analogy could be used in the three-dimensional case of tree branching is a logical question.

Three-dimensional models are too complicated to visualize in detail on paper, so actual models were built from a child's set of Tinker Toys. These are basically wooden wheels each drilled with holes around the circumference and in the plane of the wheel. Wooden sticks of various lengths can be fitted into the holes so that they project radially from the wheel. There are eight radial holes in each wheel and one in the axial direction.

Many rules for the construction of a branching structure might be devised, but the following are typical of those chosen for the models here described. The face cards are removed from a deck, leaving aces to tens inclusive. After each draw of a card, it is replaced and the deck is shuffled.

The game begins with a single vertical stalk set on a base for convenience. A wheel is placed on its upper end so that the stalk sticks into one

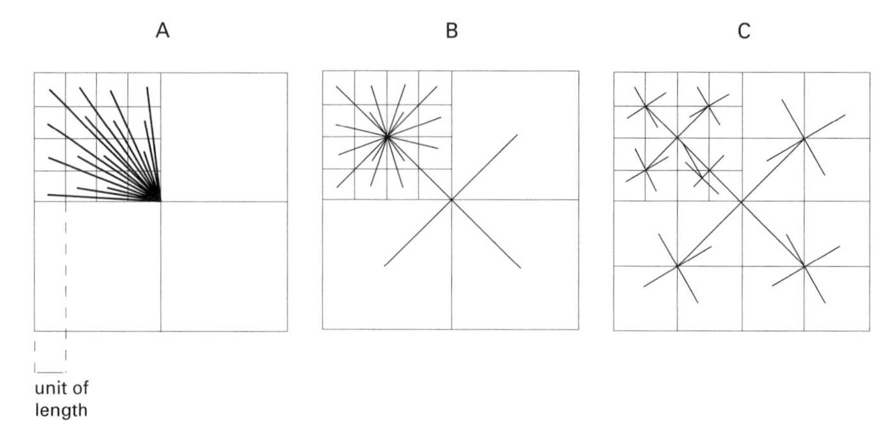


*Figure 12.13* Possible branching patterns to serve 64 unit areas from a central point. A. Separate individual stems from the center to each unit; the total length of stems is 200 units. B. Four principal branches, from each of the 16 radial stems; total length of stems, 124 units. C. Four principal and 16 subbranches and radiating stems; total length of stems, 88 units.

of the radial holes. A card is drawn. If it is black, two sticks are placed in the wheel radiating out at 45 and 135 degrees from the horizontal. The black draw, in other words, gives two stems or a bifurcation. The length of the added stems is determined by the odd-even rule, and the orientation of the whole wheel relative to a predetermined direction is governed by the value of the card. The rule just described involves equal chance for adding two 45-degree sprouts or no sprouts at all.

The end of each sprout is taken consecutively over the whole tree, a card being drawn for each existing sprout. Then the process is repeated, and in each repetition there is a larger number of sprouts. The model is considered ended when the total number of available wheels or sticks is used, usually 40. Analysis of numbers and average lengths of each order was carried out as in the drainage basin analysis described.

Other rules were made up, primarily by changing the number of added stems on the draw of a black card. Equal probabilities were tested for zero and two sprouts, zero or three sprouts, zero or one or two or three sprouts, zero or a long sprout consisting of two alternative sprouts and one terminal short sprout. The data plotted included one or several trials of each rule. It was found that successive trials under the same rules give very comparable results, for the data converge rapidly with replication.

Table 12.1 gives the average values of the bifurcation and length ratios for a variety of field examples and random models. The values of each ratio are quite comparable among the various types of examples. The

*Table 12.1* Average values of the branching characteristics of several field examples and random models

| Model | Bifurcation ratio[a] | Length ratio[b] |
|---|---|---|
| Two-dimensional | | |
| River networks | 3.5 | 2.3 |
| Random models of river networks | 4.1 | 2.5 |
| Three-dimensional | | |
| Trees | 5.1 | 3.4 |
| Roots | 3.2 | 2.4 |
| Tinker-Toy random models | 3.3 | 2.3 |

a. Defined as the average number of branches of a given order per branch of the next-higher order.
b. Defined as the ratio of the average length of branches of a given order to the average length of the next-higher order.

bifurcation ratio averages about 3.8, only slightly larger than the value 3.5 for river networks (for which a large number of data are available, compared with a small number of examples of all other categories). The length ratio averages 2.6.

## Branching Patterns—A Summary

Systems in which paths connect points distributed in space to some particular point or locus of points may take many forms. I postulate here that the form that is most probable tends to minimize the total length of all paths within the applicable constraints. The most probable pattern will consist of a progressively branching system in which there is at all potential points of branching some specific probability that branches will develop.

Earlier, in the case of river drainage networks, and here, in simple three-dimensional networks, it has been demonstrated that random models develop branching systems in which the branch lengths and the number of branches increase logarithmically with order number. This logarithmic relationship is one of optimum probability.

Geometric relationships of at least some branching forms show that minimum total length of all branches is associated with high values of bifurcation ratio and low values of length ratio. The few models tested suggest that when the former exceeds 3.8 and the latter falls below 2.6

further change has little effect on total branch length. I infer, then, that values taken from nature for these ratios, being in close agreement with the test numbers cited, typify the most probable branching configuration.

That the organization of the drainage net described by Horton is guided by tendencies toward a most probable state is implied by the fact that networks having characteristics similar to river networks can be derived by stochastic models. Several simple random models have been tested. One involved random walks proceeding from equally spaced points along a line, the random walks tending sooner or later to meet or join, thus developing a branching tree. The other consisted of directional arrows whose orientation was randomly chosen, entering into squares of a rectangular grid. Connecting the arrows, within given rules or constraints, resulted in a network. For such models stream order was linearly related to logarithm of length and number. It has been concluded that the logarithmic relationship is one of optimum probability.

In addition to streams and trees, other networks existing in nature have branching patterns with similar properties tending toward minimum work and/or maximum probability. Michael Woldenberg has studied the branching structure of bile ducts and blood vessels in bovine livers and other mammalian body structures, and has compared them with tree branching.

He has also studied the areal distribution of market centers in many countries. The location of market centers relative to one another is analogous to the way drainage basins are served by streams. The arrangement of market centers resembles nested hexagons, dictated by the fact that demand for a product sold at a center declines with distance, that is, with transportation costs. The hexagonal organization is a least-work solution. Woldenberg found that the hexagonal nesting has properties also exhibited in drainage basin arrangement, although superficially basin areas on a map do not look like hexagons. His analysis showed that Horton plots of stream order against properties such as lengths, numbers and slopes, can be analyzed in the same way as nested hexagons, with comparable results.

Woldenberg concluded that "the equilibrium state in fluvial systems reflects a combination of least work and maximum possible entropy. . . . The least work derives from a compromise between least overland work loss and greatest gains due to economies of scale in the channel. . . . The maximum possible entropy [maximum probability] derives from the fact that basin areas of Order *u* are reasonably equal in size and have reasonably equal channel slopes and relief" (1969, pp. 97–98).

These comments give significant support to the hypothesis presented in this book.

## CHAPTER THIRTEEN

# Energy Utilization

### The River as Machine

The operation of any machine might be explained as the transformation of potential energy into the kinetic form that accomplishes work in the process of changing that energy into heat. Locomotives, automobiles, electric motors, hydraulic pumps all fall within this categorization. So does a river. The river derives its potential energy from precipitation falling at high elevations that permits the water to run downhill. In that descent the potential energy of elevation is converted into the kinetic energy of flow motion, and the water erodes its banks or bed, transporting sediment and debris, while its kinetic energy dissipates into heat. This dissipation involves an increase in entropy.

The process involves all the dynamic aspects of water flow—the shearing process that gives rise to the tangential shear stress at the bed that moves sediment, and the turbulence that supports the suspended load. Considering the fact that the sedimentary rocks that constitute the main surficial material on the world's continents all were carried and deposited by these forces, some elementary consideration of such forces and such deposition is essential to an understanding of the river. Frictional resistance is a key element in generation of the stress structure that produces and maintains channel form. This consideration of such forces and deposition leads to a further elaboration of the tendencies in mechanical systems toward minimal work and uniform distribution of work rate.

### Velocity and Resistance to Flow

The shear stress at the bed propels the bedload. The tangential component of the weight of the water pushes it downhill. This force is resisted by the friction at the boundary. Because the velocity is in general not increasing, the propelling force and the resisting force are equal. The force can be thought of as the number of pounds per square foot exerted on the bed parallel to the bed surface, just as with the solid friction exerted by the rough floor over which a brick is being dragged (Figure 11.2). The force is indeed the tangential component of the body weight.

In mathematical terms, $\tau_0 = \gamma Rs$, where $\tau_0$ is the tangential force at the bed in pounds per square foot; $\gamma$ is the weight of a unit volume of water (62.4 pounds per cubic foot); $R$ is the hydraulic radius (area divided by wetted perimeter), approximately equal to depth of water in feet; and $s$ is the energy slope, but it is well approximated by the channel gradient.

The shear stress $\tau_0$ in the nonaccelerating fluid is equal to the resisting force and is proportional to the square of the velocity:

$$\tau_0 = ku^2 = \gamma Rs$$

where $k$ is the flow resistance.

The derivation of the well-known Chezy equation for velocity follows from the above statement. Let $\sqrt{(\gamma/k)}$ be called the Chezy coefficient of resistance, $C$. Then velocity $u = C\sqrt{(RS)}$, which is the Chezy formula.

Because

$$\tau_0 = \gamma Rs = \rho g Rs$$

then

$$gRs = \frac{\tau_0}{\rho} \quad \text{and} \quad \sqrt{(gRs)} = \sqrt{\left(\frac{\tau_0}{\rho}\right)} = u_*$$

where $u_*$ is the shear velocity, a measure of shear stress. It is a useful quantity because it can be related to flow resistance as follows.

In pipes, a friction factor **f** called the Darcy Weisbach friction factor was found to be $\mathbf{f} = 2g\mathbf{d}s/u^2$, where **d** is the pipe diameter and **f** is a measure of the uniformly distributed wall friction per unit length of pipe. Because $R$, the ratio of area to wetted perimeter, is

$$R = \frac{\pi r^2}{2\pi r} = \frac{r}{2} = \frac{\mathbf{d}}{4}$$

where $r$ is pipe radius and $\mathbf{d} = 4R$, then in channels

$$\mathbf{f} = \frac{2g\mathbf{d}s}{u^2} = \frac{8gRs}{u^2}$$

where $R$ is the hydraulic radius of the channel.

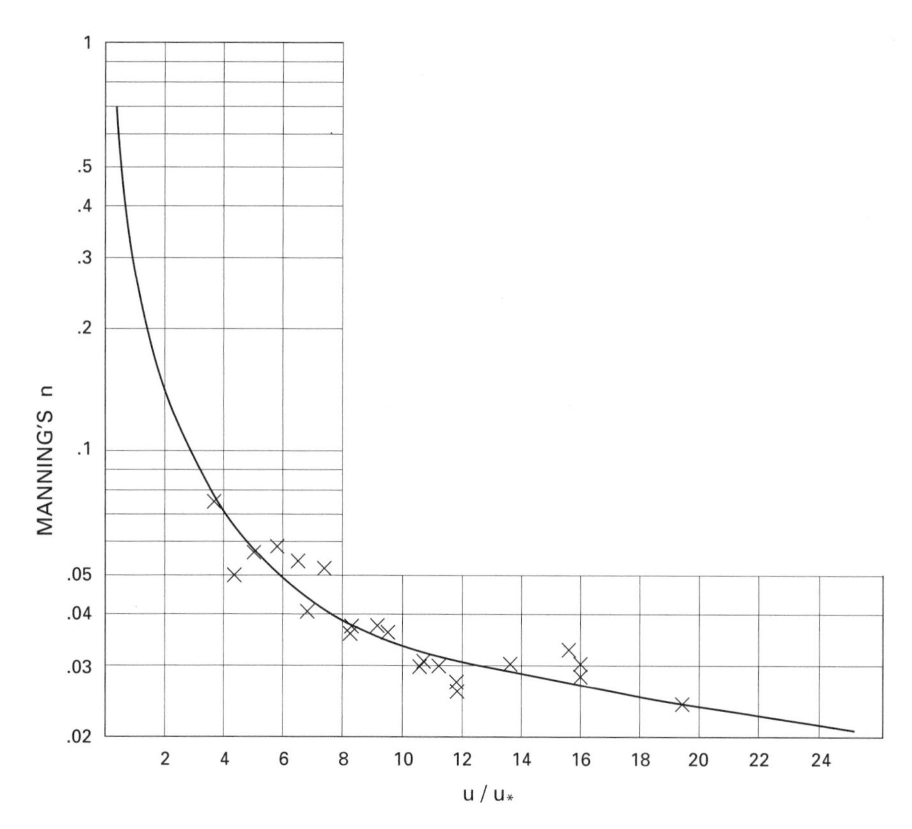


*Figure 13.1* The relation of the friction factor $u/u_*$ to Manning's $n$. The plotted measurements were taken by the U.S. Geological Survey and published in Water Supply Paper 1849.

The friction factor can be written as

$$\frac{1}{\sqrt{\mathbf{f}}} = \frac{1}{\sqrt{8}}\frac{u}{u_*} \quad \text{or} \quad \frac{u}{u_*} = \sqrt{\left(\frac{8}{\mathbf{f}}\right)}$$

Thus there are various forms of the dimensionless friction factor:

$$\frac{u}{u_*} = \sqrt{\left(\frac{8}{\mathbf{f}}\right)} = \frac{C}{\sqrt{g}} = \frac{u}{\sqrt{(\tau_0/\rho)}}$$

These variants appear in many places in the literature of hydraulics. For present purposes, friction will be expressed as $u/u_*$. This form has the advantage of being dimensionless, and in rivers varies from about 2 to 16, the larger number being for smooth riverbeds and the smaller values for rough.

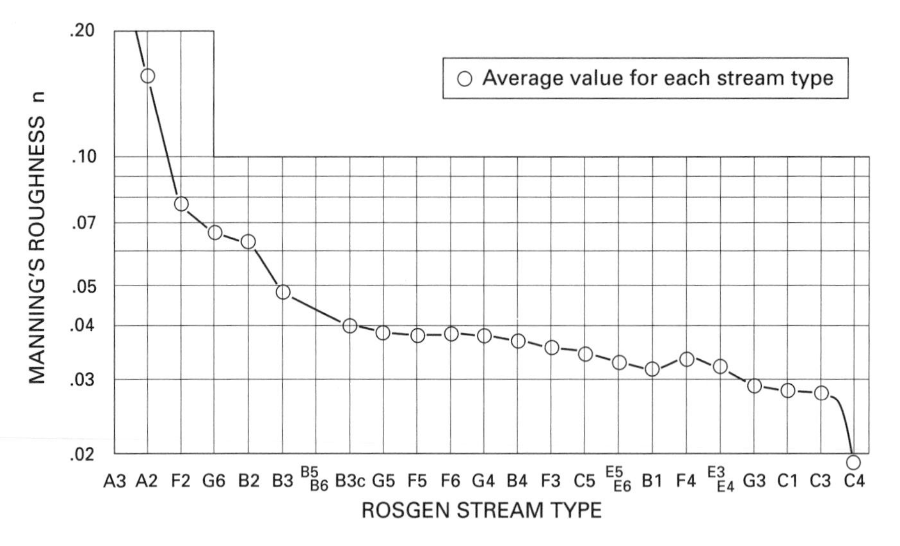


*Figure 13.2* The relation of Manning's $n$ to the Rosgen stream classes. Vegetated channels having greater friction than those unaffected by vegetation are not shown. (Data for $n$ from individual measurements by David Rosgen and from data published by the U.S. Geological Survey (Barnes); data for New Zealand streams from Hicks and Mason 1991.)

The Manning formula for velocity, widely used in American practice, is

$$u = \frac{1.5d^{2/3}s^{1/2}}{n}$$

The resistance factor $n$ varies slightly with depth and varies over a large range from about 0.01 to 3.0. The relation of Manning's $n$ to $u/u_*$ is shown in Figure 13.1. The Rosgen stream classification system provides an estimate of the friction coefficient, as can be seen in Figure 13.2.

## Effect of Wall Roughness

Stress and velocity distribution are related. If the roughness of the walls of a pipe is changed and the discharge is then adjusted to give the same hydraulic gradient (slope), the respective velocity profiles in the central zone are parallel, as shown in Figure 13.3. The abscissa is, in dimensionless terms, the position between the pipe center line ($r/r_0 = 0$) and the

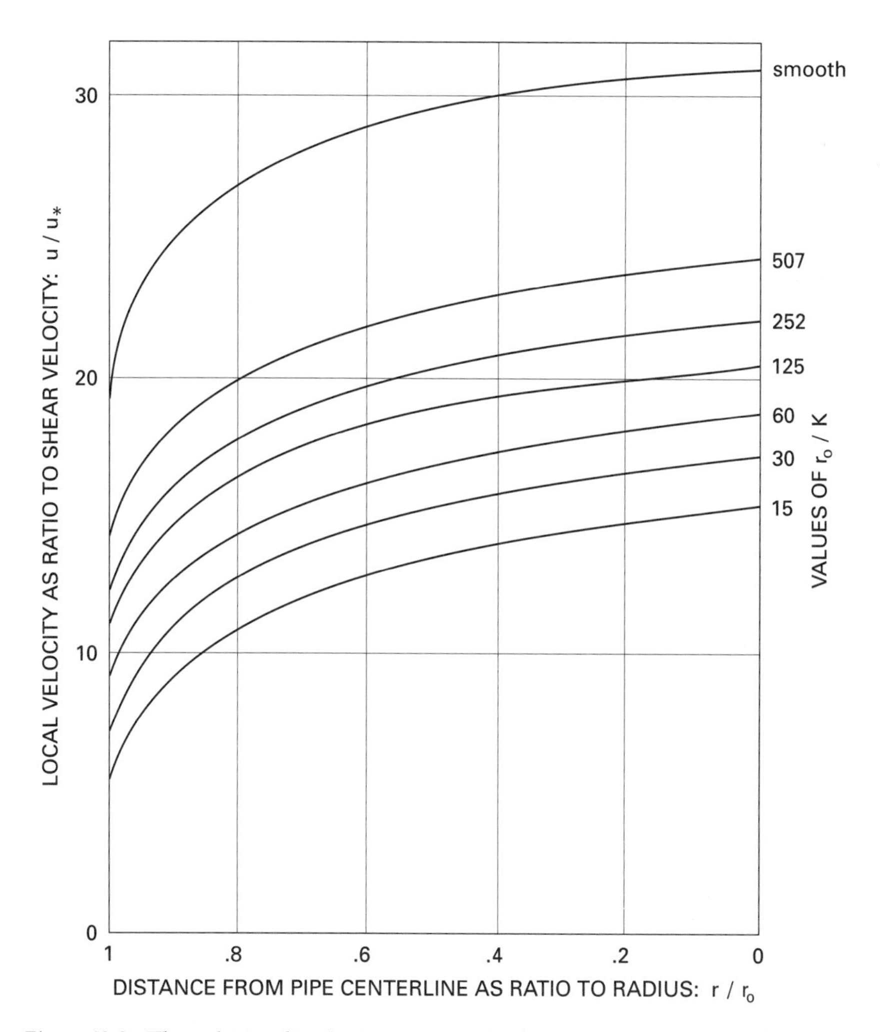


*Figure 13.3* The velocity distribution in pipes having the same hydraulic gradient but differing in the relative height of the boundary roughness elements. The abscissa is the position relative to the pipe wall, $r_0$ being the pipe radius. The ordinate is the value of local velocity, $u$, as ratio to shear velocity, $u_*$. The family of lines designates different ratios of pipe radius to roughness element height $K$. (From Bakhmeteff and Allen 1945; by permission of ASCE.)

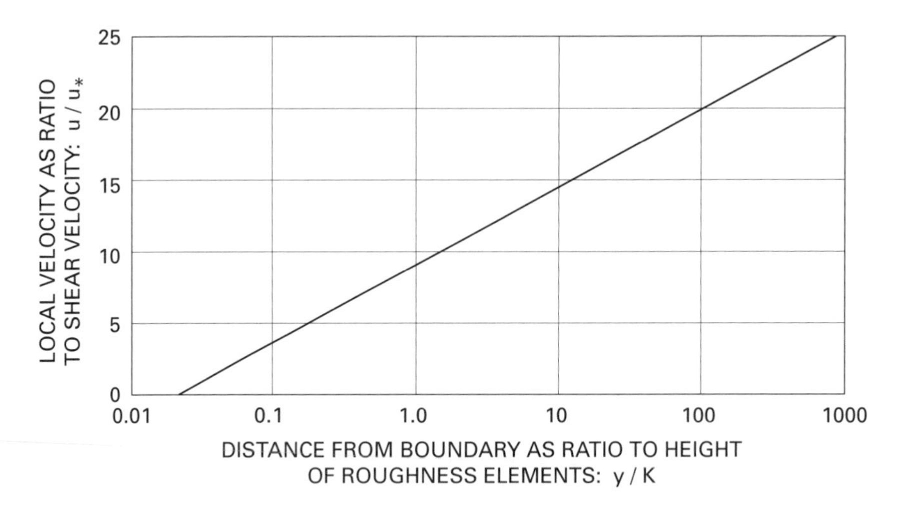


*Figure 13.4* Velocity profile of pipes of various roughness. The local velocity $u$ is expressed as a ratio to the shear velocity $u_*$. The distance is expressed as a ratio of the distance from boundary $y$ to the height of the roughness element, $K$. The applicable equation is $u/u_* = 8.5 + 5.75 \log y/K$.

pipe wall ($r/r_0 = 1$), where $r_0$ is the radius of the pipe. The ordinate is velocity, expressed as local velocity $u$ divided by shear velocity $u_*$, or

$$\frac{u}{u_*} = \frac{u}{\sqrt{(gRs)}}$$

The various curves represent velocity distributions associated with different sizes of the roughness elements or protuberances $K$. These are expressed as the ratio of height of roughness element to pipe radius, $r_0/K$, called the relative smoothness. Note that the curves are parallel, except for a zone very close to the wall.

The data are derived from the famous experiments of Nikuradse, who glued sand on the inside of pipes to give a standard roughness or known height of roughness element, then tested the effect of change in the ratio of pipe diameter to roughness size.

The distribution of velocity at different distances from the wall for pipes having various values of relative roughness and the same value of hydraulic gradient (Figure 13.3) shows the similarity of shape among the curves. This implies that they are identical if put in dimensionless form by plotting the same data in a graph having the same ordinate, $u/u_*$, but the abscissa measuring distance from the boundary, $y$, as a ratio to roughness height, $K$. This plot of the Nikuradse pipe data is shown in Figure

13.4, in which the data establish a single straight line on semilog paper. It then provides a general equation for velocity distribution in rough pipes in the form

$$u/u_* = 8.5 + 5.75 \log y/K$$

On this semilog plot, where $y/K$ is 1 and $\log y/K$ is 0, the value of $u/u_*$ is 8.5. When $\log y/K$ goes from 0 to 1, $y/K$ goes from 1 to 10. The slope of the line, then, is the amount of change in the value of $u/u_*$ for a unit change in the abscissa. The ordinate changes by 5.75 when $y/K$ is increased tenfold.

The equation for the velocity profile can be written in another useful form, as

$$u_2 - u_1 = 5.75\, u_* \log \frac{y_2}{y_1}$$

where $u_1$ and $u_2$ are the velocities at distances $y_1$ and $y_2$.

When the line in Figure 13.4 is extended downward to the point at which velocity is zero ($u/u_* = 0$), the corresponding value of $y/K$ is 0.033. This distance from the boundary is equal to one-thirtieth of the roughness height $K$. Thus, the velocity becomes zero at $1/30\ K$, or at a level deep within the interstices between the roughness elements—when the flow is turbulent, the boundary rough, and the roughness elements not moving. This situation can be of practical importance if, when the flow conditions satisfy the requirements of steady uniform flow over a rough boundary, the vertical velocity profile is measured and the mean line is projected to the place where velocity became zero. That depth or distance value should be one-thirtieth of the height of the effective roughness elements. Thus, the effective roughness height can be ascertained from the measured velocity distribution away from the boundary.

## Velocity Distribution in the River Channel

The natural river channel is formed and maintained by the river flow and its associated sediment through differential action along the boundary of the full three-dimensional velocity distribution. The friction between the flowing water and the channel boundary is appreciable, whereas that between the water surface and the air is small. The friction at the bed is felt by all levels of the flow, for each level in the flow is dragged by the level below and in turn drags the level above. Under ideal conditions the

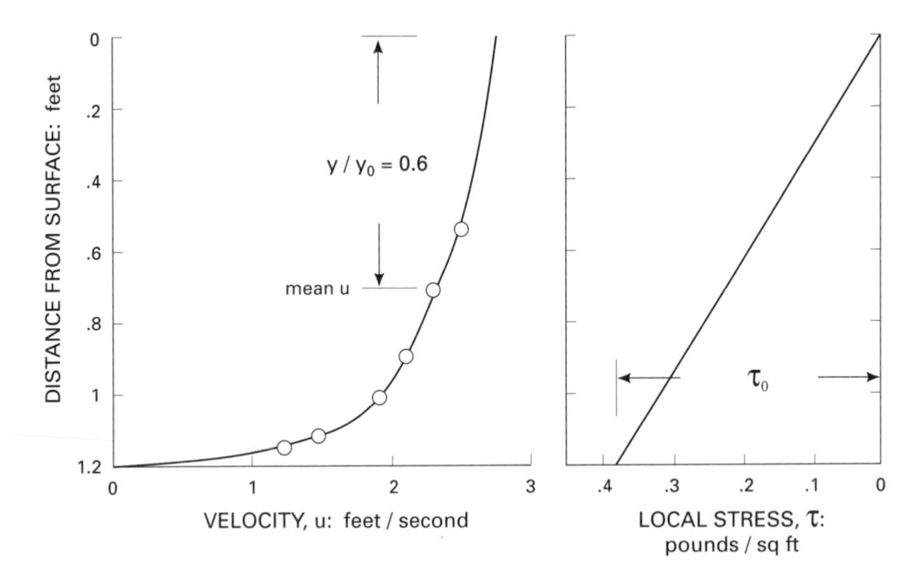


*Figure 13.5* The vertical velocity curve measured on the Hoback River, near Bondurant, Wyoming; slope 0.0051, depth 1.2 feet. At the right is the computed variation of local shear stress.

velocity is maximum at the water surface and decreases toward the bed in a logarithmic manner, as expressed in the preceding equation. For the same reason, the velocity decreases from the center of the stream to the banks.

The logarithmic decrease of velocity toward the boundary observed in pipes and uniform conduits is approximated in natural channels. An example is a profile measured in the Hoback River near Bondurant, Wyoming (Figure 13.5). Also shown is the computed relation of shear stress to depth. In this example the level of the average velocity occurs very close to six-tenths of the distance from the surface to the bed, as expected from theory.

Rivers have (1) an organized variability of bed, as in the alternation of pools and riffles, (2) an organized variability in flow direction, as in meandering reaches, and (3) a random variability resulting from a host of heterogeneities of bank and bed conditions. Protuberances and reentrants lead to many local deviations in velocity distribution, expressed as variance from the ideal logarithmic relation of velocity to distance from the boundary. These irregularities cause eddies, reverse currents, and circulation in the cross-sectional plane.

A quantitative picture of such variability can be obtained from the velocity isopleths drawn in Figure 13.6. The figure shows velocity distribution in cross section taken in a straight reach of a small river. The example is Baldwin Creek near Lander, Wyoming, a river that rises in the high mountains of the Wind River Range and emerges into a broad, flat valley. The measurements were made in a portion of the stream in the alluvial valley only a few miles from the mountain front. The river here is about 15 feet wide, the banks stand about 4 feet high, and except for the snowmelt season in early June the flow is small and the water clear. High stages occur only during the snowmelt season, the period during which the data reported here were taken. The velocity pattern is shown for three stages of flow, the highest of which is near bankfull.

Several details are worth noting. The stage increased 0.47 foot from the condition shown in the bottom diagram to that in the top one. This increase reflects a change in discharge from 33 cfs to 86 cfs (2.6 times as much), but an increase in depth of 16 percent. The same change in discharge caused the velocity nearest the surface and slightly to the right of the center line to increase from 4.40 to 6.65 feet per second, or 51 percent.

Despite the generally straight alignment of the river in the vicinity of the section, neither the cross-sectional shape nor the velocity distribution is symmetrical. The left bank is sloping, whereas the right bank is nearly vertical. The gradient of the velocity away from the bank is small on the left bank but large at the right boundary. Though there is a perturbation in alignment 50 feet, or 4 widths, upstream of the section, it might be expected that the channel would have become somewhat more symmetrical than it actually has.

Experience in dealing with natural rivers suggests that the asymmetry of cross-sectional form and velocity distribution caused by even a minor change of channel alignment will persist downstream for a considerable distance, seldom less than 5 widths and often twice that distance. It is not known, in fact, over what distance upstream from a particular point the bed and bank roughness are integrated to provide the effective flow resistance that determines the velocity at the point in question.

The depression of the thread of maximum velocity to a position below the water surface is commonly noted in isopleth patterns and is exemplified in the measurements of Baldwin Creek. This depression is often assumed to be due to the friction of the water and the air, but it must be primarily caused by circulatory motion in cross section. Because vertical and cross-channel components of velocity are difficult to measure in natural channels, very little of a quantitative nature is known about the circulatory motion in rivers. The depression below the surface of the

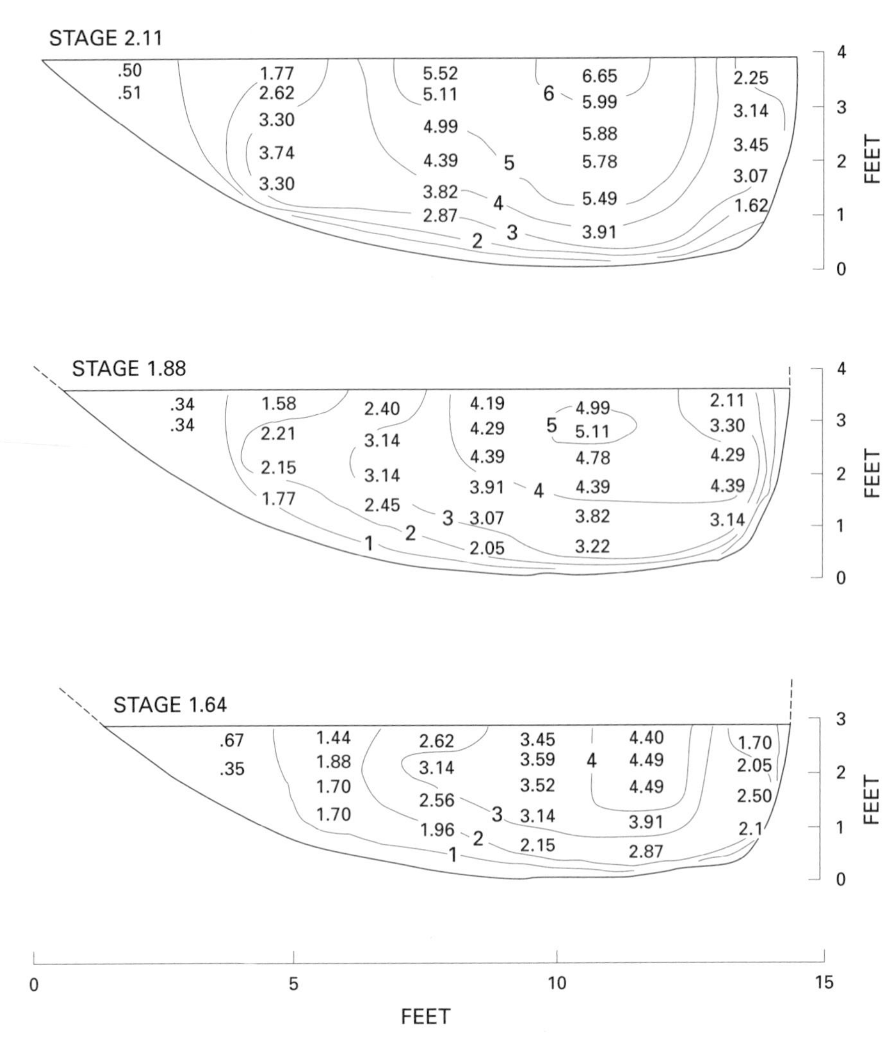


*Figure 13.6* The distribution of flow velocity in a cross section on a straight reach of Baldwin Creek, near Lander, Wyoming. The point velocities measured with a current meter, are written numerically in feet per second; the velocity isopleths are drawn at intervals of 1 foot per second. The ordinate is depth in feet. The three cross sections represent three stages of flow during the period June 10 through June 13, 1959.

maximum velocity is very marked in meander curves where the helical motion is expected. A similar but less dramatic depression of the maximum velocity observed in straight channels has a similar origin.

Some insight into the net result of these circulation patterns is provided by an old practice on big rivers. My uncle Frederic Leopold, of Burlington, Iowa, spent his life near the Mississippi River. He recalls that when he was a young boy in the first decade of the century, river hoboes eked out an existence in shacks on the riverbank. To avoid the rigorous winter, these hoboes each fall would build a simple, small houseboat or raft and float down the river to Louisiana, abandoning or disposing of the craft in the spring. Such a boat had no motive power except a scull oar. The guidance system for the thousand-mile cruise consisted of a large, nearly submerged log, to which the boat was tied. The log had essentially no wind resistance and its course was governed by the streamlines of flow at the depth of the log. In any long, straight reach the cross circulation apparently pushed the log to the center of the river. As the craft entered a curve, it would be carried toward the concave bank, but long before reaching the shore would have passed the point of inflection of the meander curve and thus be carried toward the opposite bank. The log described a sinuous course alternately on one side and then the other of the river centerline, tending to keep quite clear of the river banks by responding to the helical motion of the flow.

Even if the nearly submerged log had closely approached one of the banks, it would seldom strike it. Richard J. Russell, a geographer of great experience in river matters, once remarked to me that you will never see a log impinging on the concave bank of a meander bend. Rather, as it approaches the shore, owing to the cross circulation of surface water toward the concave shore its path will become more and more parallel to the bank as the cross-channel component becomes damped to zero at the boundary. Bank erosion on the concave bank is due only in part to impingement of high-velocity water by a component orthogonal to the boundary. An additional influence is exerted on the falling stage of the hydrograph when pore pressure in bank materials is directed toward the free face, with resultant bank sloughing and caving of wetted material.

Several people have observed that at times of high flow in a straight reach of channel, the water at the channel centerline is higher than along the banks. We have made measurements in several rivers and confirmed this phenomenon. It apparently is the result of two circulation cells with water moving toward the centerline from each bank, thence descending toward the bed and, at the bed, flowing toward each bank. It is a subtle phenomenon and has been measured in only a few instances. These

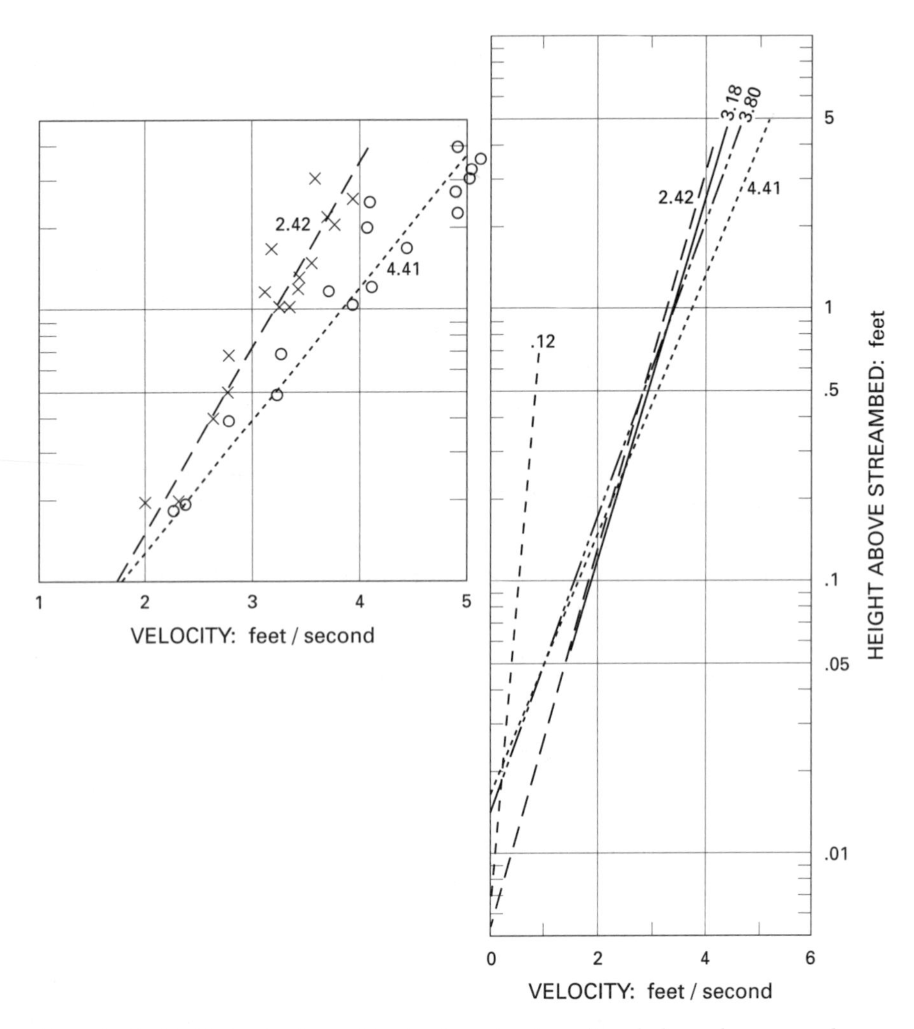


*Figure 13.7* Vertical velocity profiles in the natural unaltered channels measured from a cable at Station 54 of the East Fork River bedload project near Boulder, Wyoming, June 1971. The left-hand graph shows individual velocity measurements with a mean straight line drawn through them for two stages of flow; the graph at right gives mean lines for five stages, extrapolated to a position where the velocity is zero. The numbers are values of stage (gage height) for each set of measurements.

*Table 13.1* Vertical velocity curves from a cable at East Fork River near Boulder, Wyoming, Station 54, 1971, $s = 0.0007$

| Date | No. of meter readings | Stage (ft) | Depth, $d$ (ft) | Slope of water surface | Graphed slope,[a] $m$ (ft/sec/cycle) | $u_*$ (ft/sec) | $5.75u_*$ | Velocity at 0.1 ft above bed (ft/sec) |
|---|---|---|---|---|---|---|---|---|
| August 10 | 6 | 0.12 | 0.53 | 0.0003 | 0.45 | 0.07 | 0.41 | 0.55 |
| June 8 | 16 | 2.42 | 3.2 | .00064 | 1.41 | .26 | 1.48 | 1.8 |
| June 9 | 21 | 3.18 | 3.8 | .0008 | 1.45 | .31 | 1.80 | 1.8 |
| June 10 | 22 | 3.80 | 4.3 | .00095 | 1.93 | .36 | 2.08 | 1.6 |
| June 15 | 22 | 4.10 | 4.7 | .00103 | 1.56 | .39 | 2.27 | 1.9 |
| June 17 | 20 | 4.41 | 4.6 | .0011 | 2.10 | .40 | 2.32 | 1.7 |

a. Measured on semilog plot of data.

measurements support Frederic Leopold's observation that the houseboat tied to the log tended to be moved toward the center of the channel.

When field data are collected by placing a current meter at various distances above the streambed, the vertical velocity curve may be drawn through the points plotted on semilog paper. The slope $m$ of the vertical velocity curve at a particular place in the channel increases at high flow because it approximates $5.75u_*$. If water-surface slope remains constant with stage, then the value of $m$ should increase as $\sqrt{d}$, the square root of depth, because $u_* = \sqrt{(gds)}$.

These considerations indicate what might be expected in a series of vertical velocity curves representing a specific position or location in the cross section of a channel, at different values of discharge. We have made a concerted effort to obtain such a set of measurements, including reliable values of water-surface slope. Our data are plotted in Figure 13.7, with pertinent values listed in Table 13.1. Slope of the water surface was surveyed over a reach 700 feet long on both banks at each of 12 different discharges, and a relation of slope to stage developed. Current-meter readings were taken at the same location from a cable, with no disturbing influences nearby. The measurements were taken with a pygmy current meter close to the bed and with a Type A (standard) meter at greater distances from the bed. Note in Table 13.1 and Figure 13.7 that most of the curves are drawn through more than 20 points in the vertical.

Theory suggests that the curves pass through a single point if the bed is immobile. In a gross way, the curves do nearly pass through a constant velocity value of 1.8 feet per second at a distance of 0.1 foot above the bed. The one exception is the very low value of depth, at which stage it is nearly certain that little or no bed debris was in motion. For the five curves representing depths of 3.2 to 4.6 feet, the expected relations are approximated. The slopes $m$ vary only slightly, the highest being 1.5 times the lowest because the range of depths was small.

The observed slope of the graph, $m$, is compared to the expected value, $5.75u_*$, in Table 13.1; the two are of the same order though not equal. Few other published data show the change of velocity profile with stage at a specific location in the channel.

The extrapolated velocity curves reach zero velocity at about 0.01 foot above the bed. The height of the roughness elements controlling velocity should be 30 times this height or 0.30 foot, which is approximately the height of sand dunes observed on the streambed. It is presumed that these dunes are the effective roughness elements.

To indicate the range in shape of vertical velocity curves, Figure 13.8 presents curves for two large rivers. The Columbia River at The Dalles has a bed covered with large angular blocks, so the roughness must be very great. The Colorado River at Grand Canyon must have a bed composed primarily of gravel, but covered with patches of sand that move in the form of dunes. In the Colorado, there is only a small difference in velocity 1 foot above the bed and at the surface. In the Columbia, all velocities are low and there is a large difference between the surface and near-bed velocities.

Values of slope are not known for these two examples. The shear velocity does not vary much from small to large rivers, because slope tends to become smaller as depths become larger. For example, the lower Mississippi has a slope of about 0.00009, a depth of 80 feet bankfull, and thus a value for $u_*$ of 0.49 foot per second. In Table 13.1 for the East Fork, where the depth was 4.7 feet and the discharge 1,500 cfs, $u_*$ is 0.39 foot per second.

These examples indicate some of the ways in which the irregularities of the natural channel and the associated convergence, divergence, and nonparallelism of stream lines make the hydraulic action in rivers different from that in straight uniform pipes. Added to these problems of uniformity are the complexities introduced by a movable boundary.

Bank irregularities are larger than bed roughness elements. Bank friction therefore offers greater flow resistance per unit of perimeter area than bed friction. Because the wetted perimeter of the banks is usually smaller than that of the streambed, however, the contribution of bank

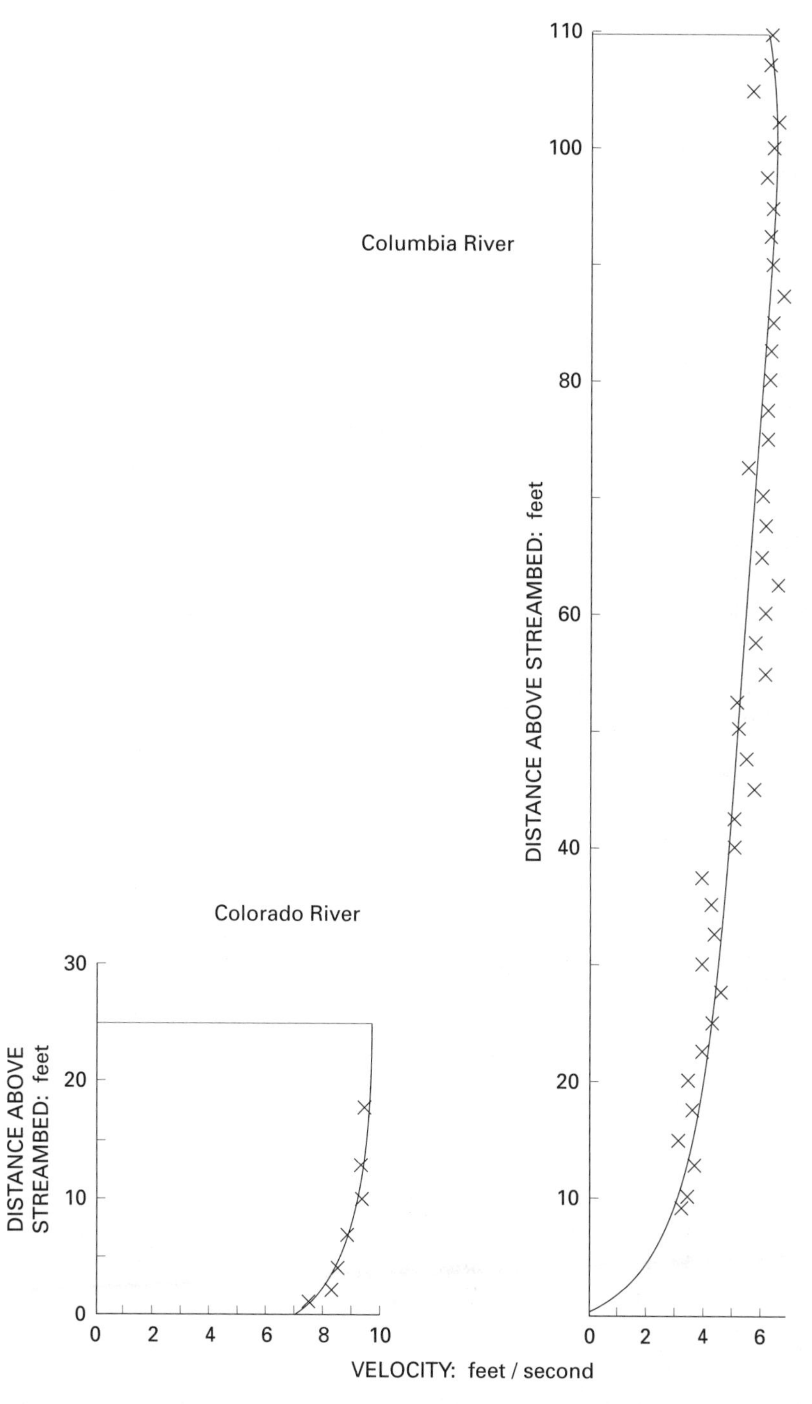


*Figure 13.8* Vertical velocity curves for two large rivers: the Colorado River at Grand Canyon at a depth of 25 feet, May 26, 1924, Station 300, discharge 62,200 cfs; and the Columbia River at The Dalles at a depth of 109.7 feet, June 3, 1971, at a station 500 feet from the left bank, stage 84.18 feet.

friction to total channel friction is less than half. When Thomas Lisle measured bank resistance as stress per unit area times perimeter area applicable, it contributed an average of 37 percent of the total channel resistance. When curves in the channel exist, as in meanders, they are usually the dominant contributor to energy dissipation.

The pattern of velocity isopleths in a given river cross section implies that the vertical profiles differ from one place to another across the channel. At each point the shape is determined by the local shear distribution, which in turn is affected by the local boundary roughness. This pattern differs from pipes, where the flow is symmetrical and the stress structure is the same at all points of the uniform boundary.

In most practical hydraulic computations for rivers, the total roughness and the total flow are the quantities available. The discharge for the entire channel is the product of the cross-sectional area and the mean velocity. Therefore, it is the mean velocity that is computed by the use of flow equations such as those of Manning or Chezy. In these equations, then, the resistance factor, $n$, $C$, $u/u_*$, or $\mathbf{f}$, applies to the whole channel on the average. The resistance factor does not, however, describe the effect of the particular size of gravel, cobble, or boulder constituting the bed of many natural channels. There is a continuing interest in finding empirical relationships between bed material size and a total resistance factor.

The elements involved in computing mean velocity are the same in all the formulas—depth or hydraulic radius, energy gradient or slope, and a resistance coefficient. Mean depth is the easiest parameter to determine in the field, by surveying a cross section from which depth can be measured, even for high stages not being observed. Except for the effect of bed scour during high flow, the measurement of depth is simple.

The appropriate slope is more difficult, because at any stage of flow it is not obvious over what distance the mean elevation drop should be measured. At high stage, it is assumed that the mean slope of the water surface is parallel to that of the channel bed, if the value is an average over 10 to 20 channel widths.

The resistance factor is made up of the roughness offered by: the bed material, presumably a function of its size; the irregularities of the channel due to bars; pools and riffles; bed forms; and channel alignment. For simplification it is assumed that if a channel reach is fairly straight and apparently uniform, size of bed material is the principal determinant of total resistance. Therefore, the computed resistance from measured values of mean velocity, mean depth or hydraulic radius, and slope should correlate empirically with size of bed material. The difficulty is that the bed material is not of uniform size, and therefore a choice must be made from the size-distribution graph. Further, it is unclear what kind of sample of bed material should be used to obtain a size-distribution graph. It is common practice to assume that the resistance is governed by the surface material rather than the material found at some depth below the surface of the bed.

The Wolman pebble-counting procedure is widely used to obtain a size-distribution graph of bed material. One hundred rocks are picked at random from the bed surface near the head of a riffle bar (downstream end of a pool). The measurement on each rock is the B-axis, or intermediate axis. This technique is applicable only to a channel whose bed is predominantly gravel exceeding 4 millimeters in size; it cannot be used on sandy or fine-grained channel beds, where the bed roughness is dominated by the bed forms.

Even when a rock-size distribution is obtained, it is necessary to choose a particular size for correlation, commonly the median size $D_{50}$. Other reasoning suggests that the effective size is larger than the median. The value $D_{84}$ is most appropriate.

We have developed an empirical relation of bed particle size and resistance for gravel rivers:

$$\frac{1}{\mathbf{f}} = 1 + 2 \log\left(\frac{d}{D_{84}}\right)$$

The relation was reviewed by J. T. Limerinos, who compiled hydraulic and bed-material size data for 11 study reaches and involved 55 current-meter measurements of discharge. His study confirmed the equation with only slight variation. The above formula is perhaps easier to use if the resistance factor is $u/u_* = \sqrt{(8/\mathbf{f})}$, or

$$\frac{u}{u_*} = \sqrt{8} + \sqrt{8}\left[2 \log\left(\frac{d}{D_{84}}\right)\right]$$

$$= 2.83 + 5.7 \log\left(\frac{d}{D_{84}}\right)$$

The relation is plotted in Figure 13.9.

A nearly identical empirical formula was developed by D. I. Bray for Canadian rivers, in which $u/u_* = 6.25 + 5.75 \log d/3.5\mathbf{k}$. In this instance $\mathbf{k}$ is the diameter of the sand grains.

A similar equation was found by the Hydraulics Research Station at

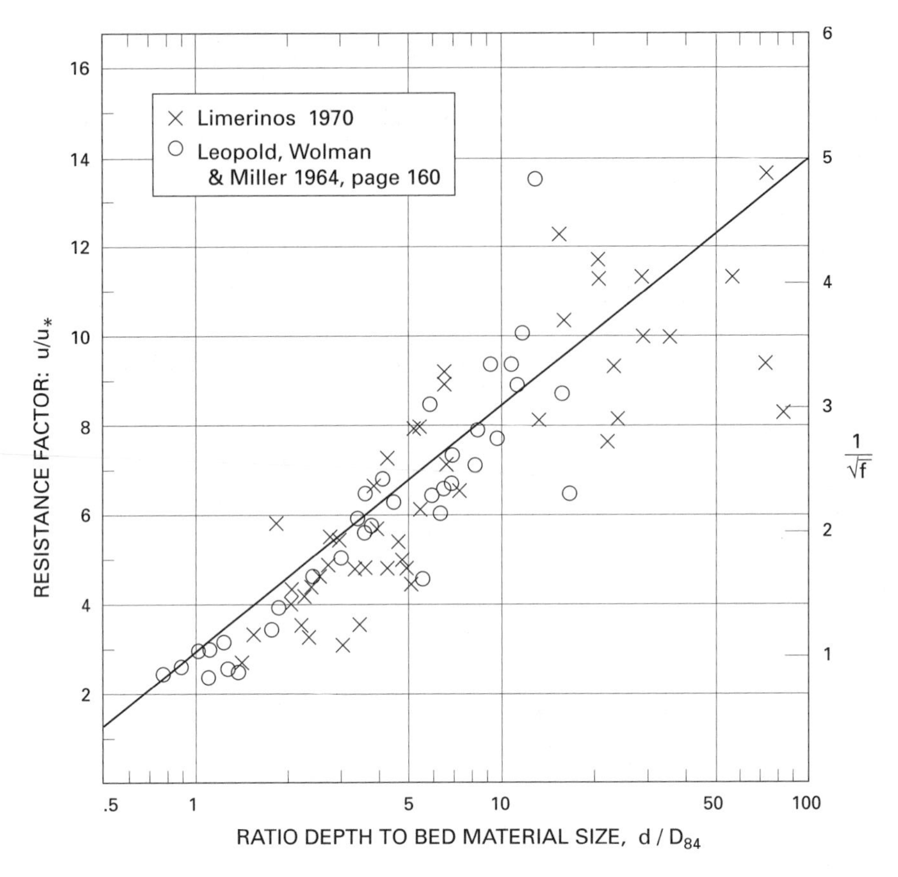


*Figure 13.9* The relation of bed-particle size to hydraulic resistance, with river data from a variety of eastern and western streams. Resistance factor $u/u_*$ and $1/\sqrt{f}$, as a function of relative roughness; ratio of water depth to bed material size $d/D_{84}$, from field measurements.

Wallingford (1977, p. 41) to apply to gravel-bed streams in the United Kingdom:

$$\frac{u}{u_*} = 6.29 \log 1.82 \left( \frac{d}{D_{90}} \right)$$

As in many relations derived from field data, there is considerable scatter in the graph, and the plot should not be used for values of $d/D_{84}$ close to unity.

Another practical approach to the estimation of flow resistance is the use of comparative photographs exhibiting the visual character of various channel reaches for which the resistance coefficient has been computed from measured values of velocity, depth, and slope. Over a period of years the U.S. Geological Survey has compiled such a series of photographs and measurements. Selected examples have been published (Barnes, 1967) in a handbook useful for the estimation of river channel resistance values. A similar handbook for New Zealand rivers has been published (Hicks and Mason, 1991). In attempting to estimate the resistance factor in a field situation, one needs to consider both the Geological Survey photographs and the equation above. The Rosgen classification and its relation to roughness (Figure 13.2) are also useful.

## Bed Forms and Resistance

With the increase in sediment transport rate accompanying increased discharge in rivers, there is the tendency toward lower flow resistance as entrained sediment alters the turbulence structure. But a far greater effect on resistance may be generated by the formation and progressive alteration of bed forms with changing discharge as sand in traction assumes various forms. More is known about bed forms in medium and fine sand than in gravel and cobble beds, because most laboratory flume experiments have dealt with sand rather than gravel.

With increasing discharge and thus load on a sandbed stream, the bed usually goes through a progressive alteration. First ripples are formed, differentiated from dunes by their small size. Wave lengths are usually about 0.5 foot and seldom as much as 2 feet. Height is usually less than 0.2 foot. Ripples do not form if the sand-particle size is greater than 0.6 millimeter.

At higher discharges, dunes will form on a sandbed, with lengths from 2 to several feet and heights of 0.2 to 1 foot in large laboratory flumes. Forms on the bed of the Mississippi River are several hundred feet long and as much as 40 feet high. Dunes can build up to a height that constitutes an appreciable proportion of the mean depth, and often they have ripples or smaller dunes riding over their backs. Both dunes and ripples are characterized by a relatively long back, sloping upward in the downstream direction. The downstream end has a steep face, over which the sand avalanches. Thus, the ripples and the dunes move progressively downstream. The dunes offer a large resistance.

In sandbed channels at some relatively high stage, often unpredictable in a natural river, the shear stress is sufficient to wash out the dunes and the bed becomes essentially plane. The change from dune to plane bed is associated with a marked decrease in flow resistance and a consequent increase in flow velocity.

At still higher rates of flow, near or coincident with critical flow (Froude number equal to unity, $\mathbf{F} = u/\sqrt{(gd)} = 1$), the bed may develop antidunes that cyclically build up in height, causing waves at the water surface in phase with the bed forms. In this situation the water surface tends to be parallel to the underlying bed form, with the water surface trough over the trough between antidunes, and a wave crest over the dune crest. The surface waves then break with the wave curl falling upstream. The dunes quickly wash out and the water surface becomes smooth again. The cycle of antidune buildup and breaking in sandbed rivers may be as short as one minute between the alternation of smooth water and breaking waves. The breaking antidune washes away the downstream face and deposits on the upstream face, so the dune tends to move upstream as it grows, even though the sand is moving downstream through the bed form. Nevertheless, I have seen standing waves of the antidune type that do not go through a buildup-breaking cycle and persist for days. An example is a reach of the Colorado River at the lower end of the Grand Canyon within the backwater influence of Lake Mead reservoir. For a distance of half a mile, the river is often an uninterrupted line of surface waves 6 to 10 feet high, which do not break but remain quite stationary for long periods of time.

Considerably less is known about bed forms for either fine or coarse gravel. Cobble to coarse gravel beds of rivers do not exhibit forms comparable to the dunes in sandbed flume experiments. Rather, they form alternating or en-echelon lobate bars, without the distinct sloping back and steep face characteristic of sandy dunes. In cobble rivers the central gravel bar is a common feature—that is, a lobate whaleback bar extending down the channel near its center line, long in comparison to its width. In natural rivers these alterations of bed configuration result from changes in the amount of sediment being brought from upstream into the particular reach.

A distinction must be made between the recirculating flume and the natural river. Most hydraulic flumes are of the recirculating type, in which both the water and the sediment in motion are propelled by a water pump through a closed-pipe system. Sediment is available on the bed of the open-channel part of the flume and will be picked up by the moving water. In a recirculating flume sediment size and width are fixed. Discharge can be determined by the operator, and either slope or depth can be chosen. But there is no unique relation among sediment transport rate, depth, and slope. The velocity is not uniquely determined by shear stress, as it would be if the flow were in a rigid-bed channel with a fixed friction factor. Depth and velocity can be used by the investigator as independent or predetermined variables, with slope, shear stress, and friction factor as dependent variables. Their values will be determined by the mutual interaction of water and sediment, after adjustment of the depth and velocity to the values selected.

In the river channel, the dependent and independent variables can be and usually are in another combination. Channel width is not adjustable over any short period such as the passage of one flood. Neither is slope, although in a pool and riffle sequence the slope may have a small but orderly change as discharge varies. Discharge is independent in the sense that the basin upstream provides an amount of water and its time distribution based on the storm characteristics or the snowmelt. The difference between river and flume is that the sediment transported to the river reach in question is governed not only by what sediment is washed off the basin, but also by what is available or what is picked up from the beds of channels upstream. Thus, sediment is an independent parameter in a natural river reach, whereas it is a dependent factor in the recirculating flume.

In the river, then, the dependent factors in a given reach are depth, velocity, and flow resistance. The flow resistance is altered by the bed configuration that results from the interaction of the dependent variables and the independent ones. If in a given reach the incoming sediment is less than the flow might carry and movable sediment on the bed is available, the river reach may act in a manner intermediate between a fully loaded river and a recirculating flume. It may scour its bed in response to imposed conditions of discharge and incoming sediment load at the slope and width inherited from the past.

Thus, a natural river reach is considerably more complicated than a recirculating flume. The point to emphasize is that bed configuration—bed form and thus bed roughness—is a dependent factor, adjustable mutually with depth and velocity and influenced by the magnitude of the incoming sediment load from upstream.

## Effects of Bed Form Changes on Hydraulic Parameters

The change in flow resistance resulting from bed alterations attributable to the inflow of sediment is exemplified by measurements made at certain gaging stations, especially in the semiarid West, where sediment loads are high. Consider, for example, the passage of spring snowmelt flood past the Grand Canyon station near Bright Angel on the Colorado

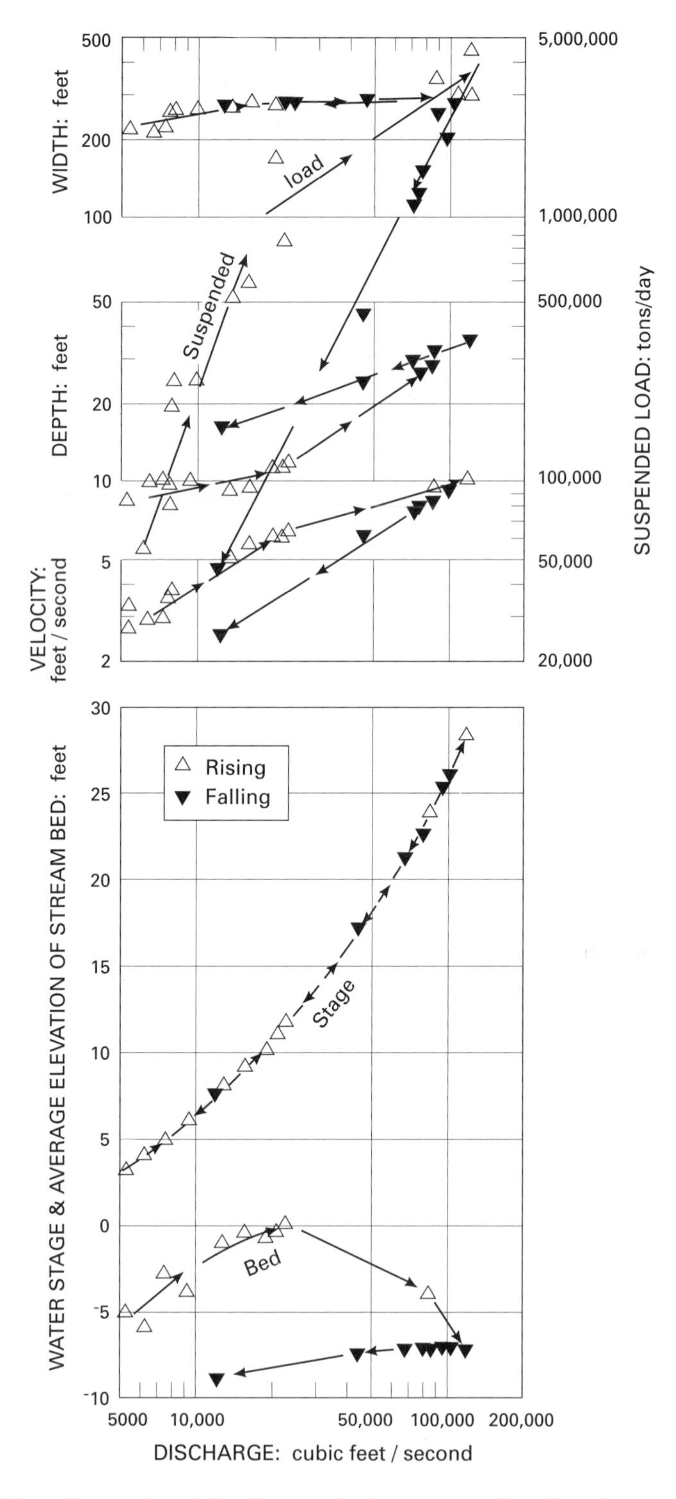


*Figure 13.10* Changes in width, depth, velocity, water-surface elevation, and suspended load with discharge during the spring snowmelt flood season of 1941. Colorado River at Grand Canyon, Arizona.

River. Before Glen Canyon Dam was constructed upstream, the spring runoff in the Colorado typically began in early April, peaked in late May, and receded to normal flow by the end of July. During such a spring rise in 1941 (Figure 13.10), the discharge increased from 5,000 cfs to 120,000 cfs. The hydrograph rise carried a much larger suspended load than did the recession or falling side of the hydrograph that had low load, low velocity, and large depth. At a discharge of 20,000 cfs on the rising stage, the suspended load was measured as 1 million tons per day, whereas on the falling stage at the same discharge the load was 120,000 tons per day.

The bed elevation rose by channel bed filling or deposition during the first part of the flood rise, then lowered or scoured as the discharge continued to increase toward peak flow. It continued to lower during the falling stage. At 20,000 cfs the bed elevation was 8.5 feet lower on the falling stage than on the rising stage. The gage height or stage was the same in both rising and falling stages at each discharge. In stream gaging terms, the control did not shift but remained stable; that is, the bar of gravel downstream of the cable or measuring section did not change, although fill was followed by scour at the measuring section. The reality of this change can be observed in the successive cross sections of Figure 13.11. The river bed was lower on January 12 (5,210 cfs) than on March 6 (23,400 cfs). Scour followed the filling when the discharge rose above 20,000 cfs.

It is evident that at the same discharge, with width and slope the same, large depth and low velocity involve a larger value of roughness coefficient than small depth and high velocity. This change in resistance at the same discharge on rising and falling implies a change in bed configuration, presumably dunes with a high resistance factor and plane bed with a low resistance factor. A summary of observed and computed data for a single value of discharge is shown in Table 13.2.

It should be noted that the scour of the bed did not result merely from high velocity. This can best be illustrated by looking at the changes with time of discharge, streambed elevation, and velocity, an example of which is shown in Figure 13.12. As at the Grand Canyon measuring station in 1941, the spring flood of the Colorado River at Lees Ferry in 1948 had higher suspended-load concentrations on the rising stage than on the falling stage at the same discharge. As shown in Figure 13.11, the high discharge in the fourth week of May associated with the principal deepening of the river bed was coincident with a decrease in mean water velocity. At the end of August 1948, when the discharge was the same as

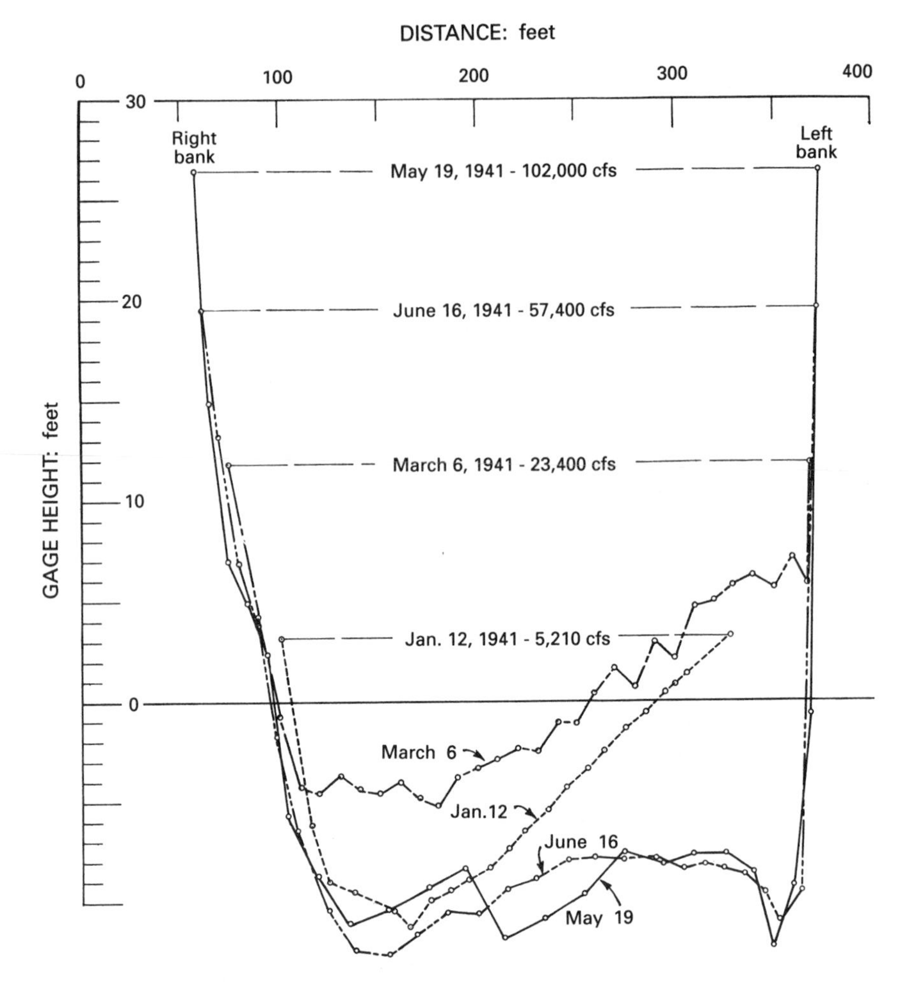


*Figure 13.11* Channel cross sections during the 1941 spring flood of the Colorado River at Grand Canyon. The hydraulic relations during the same flood are shown in Figure 13.10.

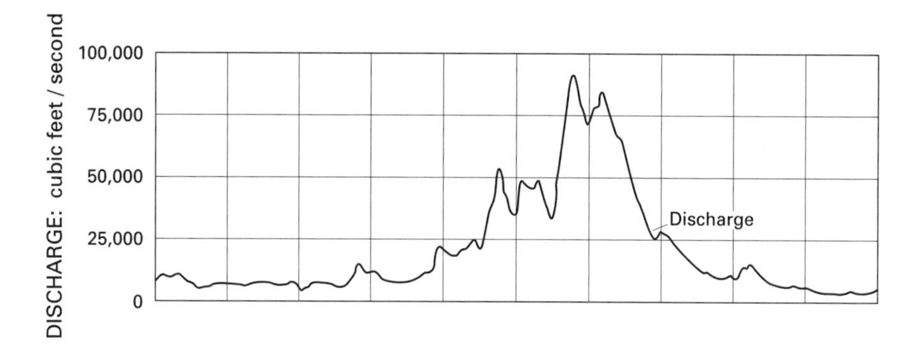


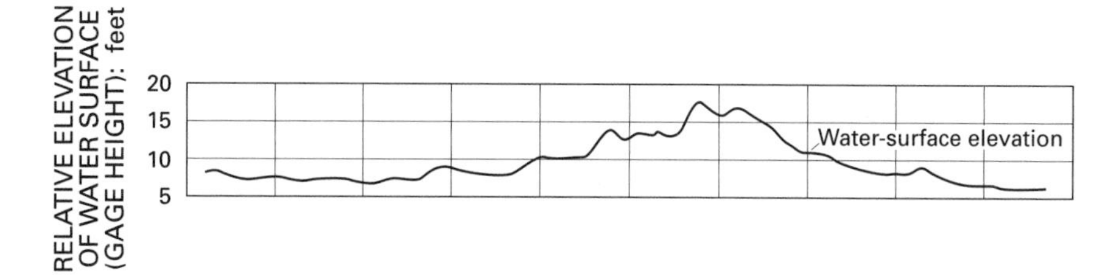


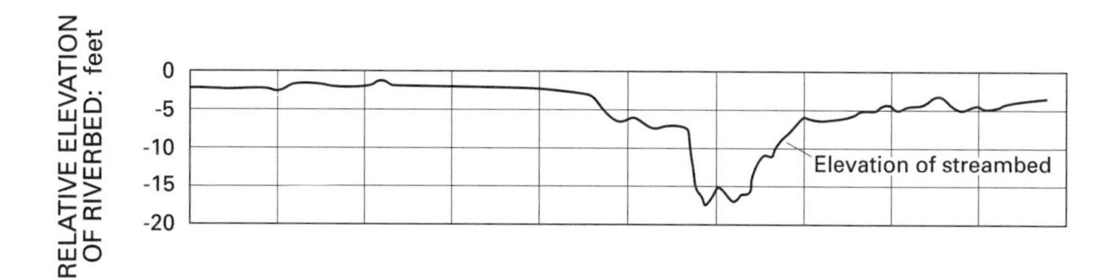


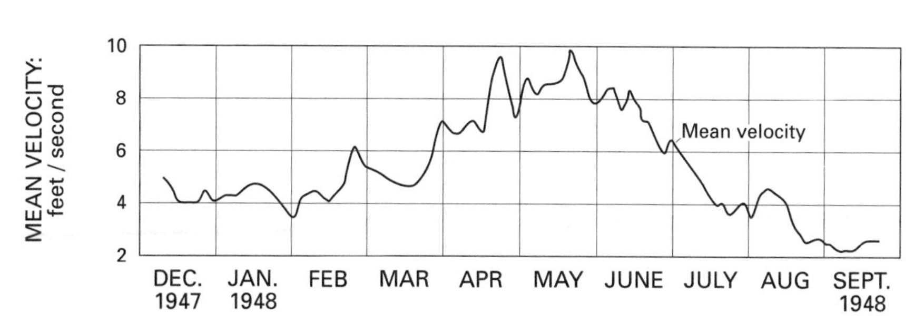


*Figure 13.12* Changes in discharge, water-surface elevation, bed elevation, and mean velocity during a 10-month period including the spring snowmelt season of 1948. Measurements were made on the Colorado River at Lees Ferry. Note that the streambed scoured 16 feet between mid-April and late May as the discharge increased.

*Table 13.2* Values of hydraulic parameters at the same discharge on rising and falling stages, Colorado River at Grand Canyon, spring 1941

| Parameter | Unit | Rising stage | Falling stage |
|---|---|---|---|
| Discharge | Cfs | 50,000 | 50,000 |
| Width | Feet | 300 | 300 |
| Depth | Feet | 20 | 26.5 |
| Velocity | Feet per second | 8.3 | 6.2 |
| Suspended load | Tons per day | 2,050,000 | 700,000 |
| Elevation of streambed above arbitrary datum | Feet | –2.4 | –7.5 |
| Slope[a] | — | 0.0006 | — |
| Darcy-Wiesbach | f | 0.0373 | 0.0907 |
| Manning's *n* | $Feet^{1/6}$ | .029 | .047 |

a. Taken from detailed river profile and applied mile 86.7 to mile 87.5, the reach immediately above the gaging station.

the preflood values of the preceding January, the mean velocity was lower than in January.

To summarize, the changes measured on the Colorado River at Grand Canyon and Lees Ferry illustrate the large variation of sediment load carried by this river during a flood passage. The elevation of its streambed and the cross-sectional area of its flow changed simultaneously, along with its mean water velocity and depth. These hydraulic adjustments were associated with an altered bed roughness in response to changes in sediment transport.

CHAPTER FOURTEEN

# River Morphology: The Most Probable State

## The Mutual Adjustment

The nature of a river channel is not closely constrained within a narrow range of characteristics by physical laws that must be fulfilled. The river responds to physics, but there remains much latitude in the morphology that a channel may assume. Many combinations of parameters occur, but tendencies lead toward a statistically most probable state.

A singular characteristic of rivers is that discharge varies rapidly and through a large range with the passage of a flood wave. As discharge changes, there is a corresponding change in depth, velocity, roughness, and width. To accommodate an increased discharge, a channel may maintain a constant width and a constant velocity, and absorb the increase in discharge merely though an increase in depth. Or it may maintain its cross section and change only its velocity and flow resistance. But channels do not accommodate a discharge increase by change of one factor alone. Rather, all dependent factors change—and at very specific rates dependent on the discharge.

Channel characteristics mutually adjust to changes in the independent discharge and the load of debris. Interestingly, the mode or allocation of adjustment among the dependent variables is highly consistent from one location to another in a given river system, and from one river to another. All that is known of rivers, wherever they are located, indicates similarity among them. A working hypothesis to explain such consistency should be based on a physical principle of great generality, because there exists such a variety of rivers, large and small, on various kinds of rocks, and in a variety of climatic settings. The characteristics of rivers, though consistent, have considerable variability around a mean. The hypothesis should allow for central tendencies without requiring a unique or single solution. The adjustments observed are all in factors related to the dissipation or conservation of energy, and to the distribution of energy expenditure.

## The Concept of Entropy

The hypothesis I propose involves the concept of entropy. Entropy, as used in the second law of thermodynamics, is a measure of the energy distribution through a system. As energy becomes more dispersed or more evenly distributed in a system, the possibility of that energy's being used for mechanical work is decreased, and entropy increases.

In an open system in steady state, there is no net increase in entropy. With the utilization of energy and its dissipation into heat, an increase in entropy must be equaled by a subtraction of entropy somewhere in the system. This is accomplished by the introduction of energy available for mechanical work. As the steady state condition in an open system is approached, the most probable distribution of energy expenditure would be characterized by equal expenditure in all parts of the system and minimum total work in the system. But these two conditions cannot be simultaneously fulfilled.

Consider a reach of river. The water flowing downhill is constantly converting its potential energy of elevation into kinetic energy and, by turbulence and friction, thence into heat. This continual change of potential energy into heat involves an increase in entropy in the reach. The introduction of water from upstream is a subtraction of entropy in the system. As the steady state condition is approached, the increase in entropy is minimum within the constraints, that is, total work in the system is mimimum.

The rate of energy expenditure is power, which is

$$\text{force} \times \text{distance} \div \text{time} = \text{power}$$

The force, a product of mass times acceleration, is in pounds and represents the weight of a given volume of water. The volume of water in cubic feet dropping through a vertical distance is work. If the weight of a volume of water is $\gamma(\text{vol})$, the vertical distance it drops per foot of horizontal path is $s$, and the time required to move this vertical distance is $t$ seconds, then the power expenditure per unit of horizontal distance is specific weight times cubic feet per second times slope:

$$\frac{\gamma(\text{vol})}{t} = \gamma Qs$$

Then $Qs$ is proportional to the rate of power expenditure per unit of distance and also is a measure of the rate of transfer of potential energy

*Table 14.1* Comparison of four channel systems having different profiles ($z$ is exponent in $s \propto Q^z$)

| | | | Case 1 $z = 0$ | | | Case 2 $z = -0.5$ | | | Case 3 $z = -0.75$ | | | Case 4 $z = -1.0$ | | |
|---|---|---|---|---|---|---|---|---|---|---|---|---|---|---|
| Reach | $Q$ | w | $s$ | Qs | $\frac{Qs}{w}$ | s | $Qs$ | $\frac{Qs}{w}$ | s | $Qs$ | $\frac{Qs}{w}$ | s | $Qs$ | $\frac{Qs}{w}$ |
| 1 | 100 | 20 | 0.01 | 1.00 | 0.050 | 0.018 | 1.8 | 0.090 | 0.022 | 2.2 | 0.110 | 0.026 | 2.6 | 0.130 |
| 2 | 200 | 40 | .01 | 2.00 | .050 | .012 | 2.4 | .060 | .013 | 2.6 | .065 | .013 | 2.6 | .065 |
| 3 | 400 | 80 | .01 | 4.00 | .050 | .009 | 3.6 | .045 | .008 | 3.2 | .040 | .006 | 2.4 | .030 |
| 4 | 800 | 160 | .01 | 8.00 | .050 | .007 | 5.6 | .035 | .005 | 4.0 | .050 | .003 | 2.4 | .015 |
| 5 | 1,600 | 320 | .01 | 16.00 | .050 | .005 | 8.0 | .025 | .003 | 4.8 | .015 | .0018 | 2.9 | .009 |
| Totals | | | .050 | 31.00 | .250 | .051 | 21.4 | .255 | .051 | 16.8 | .280 | .050 | 12.9 | .250 |

into heat. Accordingly, power expenditure is proportional to the rate of increase of entropy. As a steady state is approached, the sum of *Qs* in various parts of the system tends to be minimized and the rate of increase of entropy tends to be minimized.

With regard to equal energy expenditure in all parts of the system, the evolution toward a steady state in an open system tends toward uniform distribution of work, or equal rate of production of entropy per unit area, within the conditions imposed on the system.

The power or energy expenditure (entropy production) per unit area of the channel bed is $\gamma Qs/w$. This tends to be minimized.

The two requirements of of minimum total work and uniform distribution of work oppose each other and cannot be fully met in the river system. The tendency to minimize total work done in the system and the tendency to equalize the power per unit bed area are incompatible goals. The result is a compromise in which they are jointly minimized. An example will elucidate the concept.

## Alternative Longitudinal Profiles of a River

Consider a river system in which we compare the rate of energy utilization for different longitudinal profiles, that is, for different changes of channel slope downstream. Assume a particular downstream rate of increase of discharge due to tributaries entering. Let there be five numbered reaches as in the first column of Table 14.1. Assume that the discharge increases reach to reach as specified in the second column. Let width increase at the same rate as discharge, thus as shown in column 3.

Now consider four alternative distributions of slope *s* along the river, as plotted in Figure 14.1. Four hypothetical cases are postulated. In each case all reaches are of the same length. Thus, the sum of slopes *s* is the same in the four cases, as in Table 14.1. The total fall is then equal in the four cases.

The relation of slope to discharge is expressed by the exponent *z* in the equation $s \propto Q^z$. While the total fall in elevation is equal in the four cases, the slope changes downstream in different relations to discharge. In Case 1, the slope is constant at 1 foot per 100, or 0.01, expressed as $z = 0$. In Case 2, the slope varies as a power function of discharge, $s \propto Q^{-0.5}$; in Case 3, $s \propto Q^{-0.75}$; and in Case 4, $s \propto Q^{-1.0}$. In the last case, slope is inversely related to discharge.

The tabulation shows the comparative values. The constraints necessary to make the quantities dimensional have not been included. The

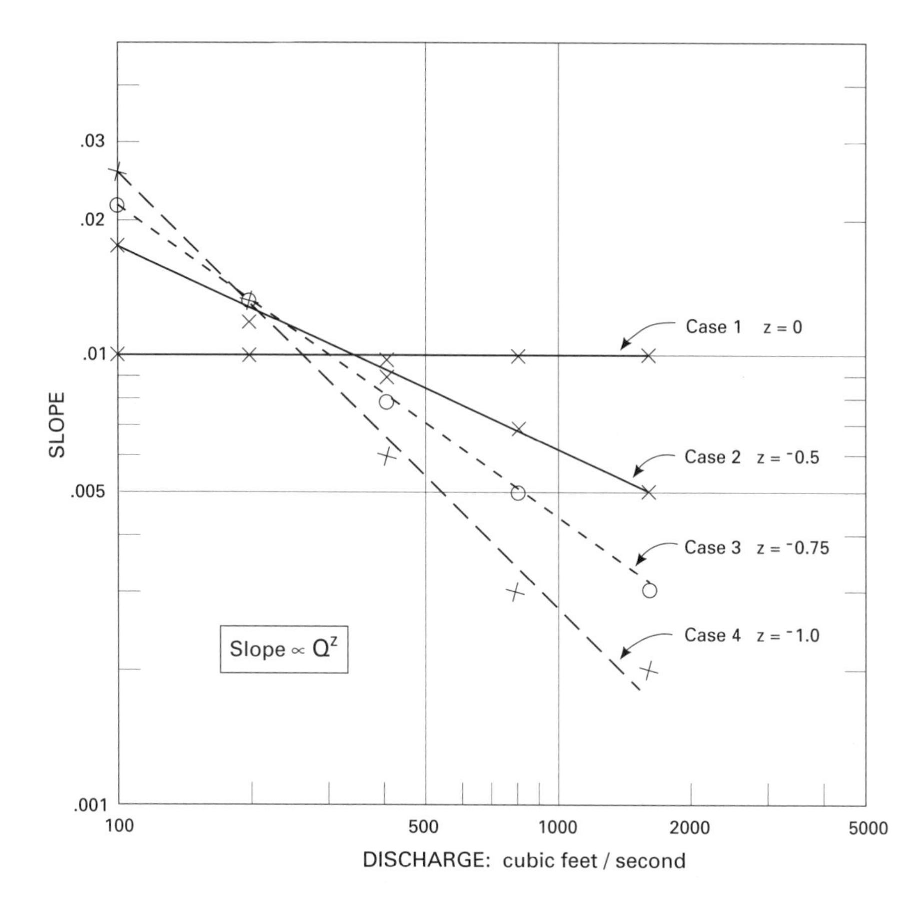


*Figure 14.1* The assumed variation of slope as discharge increases downstream in the four cases of Table 14.1

quantity $Qs$ is proportional to the rate of gain in entropy in a unit length along the river and the sum ($\Sigma\ Qs$) is the total rate of gain of entropy or total work done in the entire length. Among the cases considered, the total is smallest when slope is an inverse function of discharge, that is, when $s \propto 1/Q$, Case 4. This is also the case where the rate of entropy gain $Qs$ is virtually equal in each unit length, as seen in Case 4.

The most uniform rate of power expenditure is Case 1, in which slope is constant ($z = 0$) and all values of $Qs/w$ are equal.

Figure 14.2 is a plot of the profiles of the four cases described. Case 1 has a straight profile in which there is a uniform rate of power expenditure. Case 4 is a concave profile in which most of the elevation drop occurs upstream where the discharge is small, and the total work done

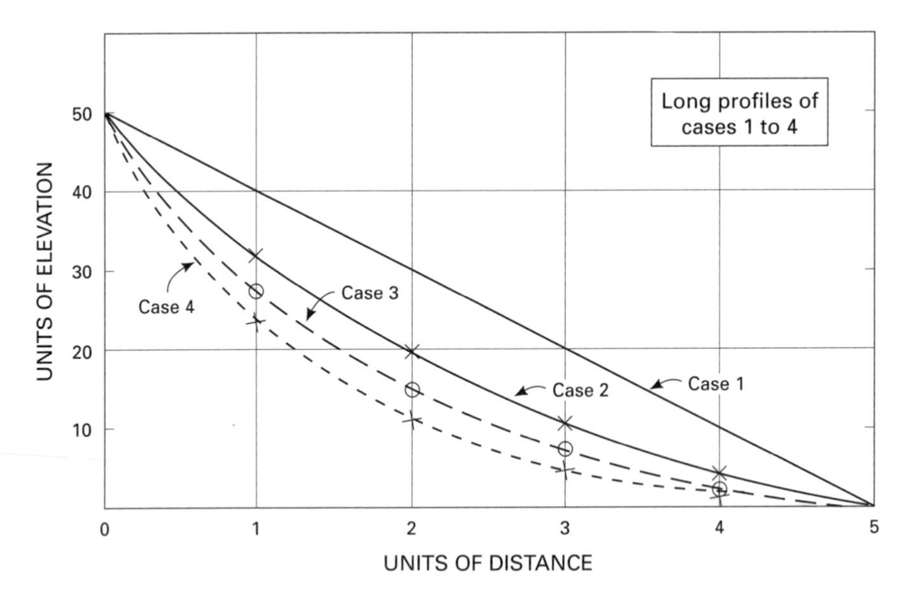


*Figure 14.2* Long profiles of the four cases of Table 14.1. Case 1 has a straight profile; the profile of Case 4 is the most concave.

is minimum. The minimum total work, the sum of *Qs*, is smallest when the profile is very concave, the slope changing radically from steep in the upstream part to a low gradient downstream ($z = -1.0$), Case 4.

The tendencies for minimum work and most uniform distribution of work cannot be attained simultaneously. In nature, rivers exhibit a profile between those extremes. The usual value of the slope exponent $z$ is $-0.70$ to $-0.75$, so real rivers look like Case 3. The compromise between uniformity of energy loss and minimum work there can be seen in several aspects of channel morphology.

## Minumum Variance

The hypothesis examined here states that the river system will tend to adjust its dependent, or adjustable, variables to meet the requirement that the imposed discharge and load will be accommodated, and that the adjustment will seek to minimize work done in accordance with the definition of an open system in steady state. But the system will also tend toward uniformity in the rate of expenditure of energy. These requirements will cause the adjustment to seek a compromise that will not completely meet either the smallest value of work done or the greatest uniformity in rate of doing work.

Let us attempt to conceptualize how the adjustment will move toward a compromise. Consider first two parameters that are correlated, but less than perfectly. Simultaneous values can be plotted in the form of a regression. The graph shown in Figure 5.4 is typical. The line of best fit is the line from which the sum of the squares of the deviations, the variance, is minimum.

Next add a third parameter, where two dependent variables are functions of the independent one. The line of best fit is the line from which the sum of the squares of the deviations is minimum, or the joint variance is minimum. This is called the least squares line and is the most probable relation between the independent and dependent variables. The variance in statistics is the weighted average of the squared deviation of values from the population mean.

When two variables are involved, the standard deviation of their sum is the square root of the sums of the squares of their respective standard deviations or, if $F = f(x,y)$, then

$$\sigma_F = \sqrt{(\sigma_x^2 + \sigma_y^2)}$$

If the properties $x$ and $y$ are independent, the most probable combination will be the one in which the sum of the variances is minimum, or

$$\sigma_x^2 + \sigma_y^2 = \text{minimum}$$

When several variables are involved, the most probable conbination will be when the sum of their respective variances is minimum.

The next question is how to measure the variance of the factors pertinent to rivers. It is measured by the square of the exponent in the hydraulic geometry. The ratio of the standard deviation of the logarithms of a dependent variable to the standard deviation of the logarithms of an independent variable, when squared, is the variance. That the ratio of the standard deviation is the exponent follows from the typical relations in the hydraulic geometry:

$$u \propto Q^m$$

$$\log u = \text{constant} + m \log Q$$

$$m \propto \frac{\log u}{\log Q}$$

from which it follows that the exponent $m$ is also equal to the ratio of the standard deviations of the logs of the variables:

$$\frac{\sigma^{\log u}}{\sigma^{\log Q}}$$

Because the variance is the square of the standard deviation, the square of the exponent is the variance. When several dependent variables mutually adjust, the most probable relation among them is the relation that provides the minimum sum of the squares of the variances.

Using this basic idea that the most probable state is the one in which the sum of variances is minimized, Langbein computed the values of the exponents in the hydraulic geometry. His computed values agree reasonably well with the values obtained by analysis of actual river data.

This effort reinforces the hypothesis developed in preceding chapters that the river channel adjusts toward a condition in which minimum work is balanced by the tendency for uniform distribution of work. In the condition of steady state of an open system, these are tendencies; the two conditions cannot simultaneously be achieved.

## A Summary of the Evidence

River curves and meanders incorporate and exhibit the results of the tendencies toward minimum work and uniform distribution of energy utilization. The straight channel has a longitudinal profile at high discharge consisting of alternate steep and flatter water-surface segments, in concert with the pool and riffle sequence. Introduction of curves increases the slope over the pool and results in a straight and uniform water-surface profile at high flow (Figure 4.15). The achievement of uniform rate of energy expenditure is bought at the expense of a steeper profile or a retreat from minimum work.

The shape of a meander curve is different from some other shapes, such as a half-circle, in that the sum of all the angular deviations from the mean downvalley direction is smaller than the sum of deviations of other shapes. The meander shape also appears in the shapes of a bent steel spring or the convulsion of a wrecked railroad track. The sine-generated curve that closely fits meander curves is a minimum variance curve. As in other examples in statistics, minimum variance is also a most probable condition.

That curves in river channels are the most probable condition is also indicated by the fact that meandering is the pattern most prevalent in nature—far more common than straight, sinuous, and braided patterns. Meandering represents the condition that channels approach as erosion and deposition proceed.

The relation of channel width and radius of curvature to meander wavelength is linear through five orders of magnitude. The radius of curvature has a value of 2.3 channel widths for meanders of all sizes, from small brooks to the Mississippi. Known in hydraulics as the ratio of least frictional loss, this value reveals a tendency toward energy conservation or minimum work.

Analysis of the hydraulic factors of slope, depth, velocity, friction factor, and bed shear has compared meandering with straight reaches of channel. The hydraulic parameters in curved reaches are adjusted so that there is greater uniformity (less variability) among them. This trait has been demonstrated for river channels and for channels of meltwater streams on the surface of a glacier (Dozier, 1976).

The compromise between opposing tendencies is also seen in the profiles of rivers. The longitudinal profile of a river is concave to the sky. Slope decreases in the downstream direction. The tendency for uniform rate of energy expenditure along the river would be met by a straight rather than a concave profile. The tendency for minimum total work would be met by a very concave profile. Actual river profiles are intermediate between those extremes.

The drainage network exhibits characteristics also seen in the morphology of its channels. The branching pattern is a structure that tends to minimize the total length of stems within certain constraints. It is also a pattern that is random in structure and can be duplicated by a variety of random procedures. Trees and other plants have branching patterns nearly identical to river networks. Examples show that the branching pattern of trees provides at least one form of efficiency in that it tends to minimize the total length of stems needed to serve the photosynthetic leaves. There are other constraints that limit the extent of this minimization.

Occurrences in nature exhibit tendencies that may be described by the most probable case, the case in which the variances about the mean are minimum. In river morphology, the exponents in the hydraulic geometry cluster around certain values. Most striking is the tendency for a common rate of increase of width downstream as discharge increases. Width increases as the square root of discharge, or has an exponent of 0.5. These exponents have been derived from theoretical considerations by finding exponent values that jointly have minimum variance. Thus the widely recognized hydraulic geometry of channels is an approach to the most probable state.

## Progression toward the Most Probable Form

In nearly every aspect of the scientific endeavor as well as in the practical experience of everyday living, users of information are told—whether or not they take it into account—the probability that a given statement is trustworthy, whether the daily weather forecast, political polls, or scientific papers. A measure of reliability is the scatter or deviation from a norm. In a regression of two variables, we speak of the line of best fit, or the least squares line, or the line of minimum variance. This tool is the basis of the procedures here discussed for computing, from basic principles, the most probable relation among dependent and independent variables in the hydraulic geometry.

Estimation of the most probable condition is identical to other procedures in statistics, computing the condition of minimum variance that results not in a forecast, but in a weighing of the inherent variability. The recognition of variability makes more vivid the results of Park's analysis of worldwide values of the exponents of the hydraulic geometry as summarized in his triaxial diagrams reproduced here in Figure 10.6. Scatter among various rivers is expected and the average or most probable results are both shown in that figure.

A conundrum of fluvial science is how tendencies such as those described here—toward minimum work and toward equal distribution of power expenditure—are expressed in physical operations and changes in a river channel. Successive local acts of erosion or deposition are dictated by the available forces exerted by the flowing water and countered by resisting forces of bed, bank, and transported load.

At any moment of time and at each location in the channel, if the available stress is greater than the resisting force, erosion will occur; if the stress is less than the resisting force, sediment in motion will be deposited. As these local events occur, the stress structure of the channel is altered until, as suggested by Gilbert, there is an equality of action along the channel. The steady state is an average condition: the hydraulic parameters are constantly adjusting, rapidly and materially, as the water discharge and the sediment it carries vary through time. Low flow is followed by flood followed by low flow, each of different duration depending on the nature and location of the rainfall or snowmelt. To accommodate these various changes the interdependent hydraulic variables will change in any of several combinations of values.

There is not just one way these factors will change. The immutable physical laws of conservation of energy and conservation of mass can be satisfied by many combinations—in fact, the particular values that will exist at any moment of time and place are indeterminate. Moreover, adjustment to the initial perturbation takes time and may not be completed before another chance event disrupts the condition, causing readjustment to begin anew. Indeterminacy is a principle long recognized in physics, but applicable also to fluvial science.

Just as turbulence in a fluid has a large random component, local variations in erosion and deposition also exhibit random variations having physical effects that result in balances in energy utilization. In this manner, local velocity distribution and its effect on shear force persistently cause erosion or deposition that cumulatively forms and maintains the stream channel in all its complexities. The river, then, is the carpenter of its own edifice.

# References


Ackers, P., and W. R. White. 1973. Sediment transport: new approach and analysis. *Jour. Hydraul. Div. Amer. Soc. Civil Engrs.* 99: 2041–60.

Andrews, E. D. 1980. Effective and bankfull discharges of streams in the Yampa River Basin, Colorado and Wyoming. *Jour. Hydrol.* 46: 311–330.

Bagnold, R. A. 1973. The nature of saltation and of bedload transport in rivers. *Proc. Roy. Soc. Lond.* A332: 473–504.

——— 1980. An empirical correlation of bedload transport rates in flumes and natural rivers. *Proc. Roy. Soc. Lond.* A372: 453–473.

——— 1986. Transport of solids by natural water flow: evidence for a worldwide correlation. *Proc. Roy. Soc. Lond.* A405: 369–374.

Bakhmeteff, B. A., and W. Allan. 1945. The mechanics of energy loss in fluid friction. Amer. Soc. Civil Engrs. Trans. Paper No. 2288.

Balling, R. C., and S. G. Wells. 1990. Historical rainfall patterns and arroyo activity within the Zuni River drainage basin, New Mexico. *Annals Assoc. Amer. Geog.* 80: 603–617.

Barnes, H. H. 1967. Roughness characteristics of natural channels. U.S. Geol. Survey Water Supply Paper 1849.

Belt, C. B. 1975. The 1973 flood and man's constriction of the Mississippi River. *Science* 189: 681–684.

Bray, D. I. 1979. Estimating average velocity in gravel-bed rivers. *Jour. Hydraul. Div. Amer. Soc. Civil Engrs.* 105: 1103–22.

——— 1980. Flow resistance in gravel-bed rivers. Proc. Internatl. Workshop on Engineering Problems in the Management of Gravel-bed Rivers, Newton, Dowys, United Kingdom, June 23–28.

Broscoe, A. J. 1959. Quantitative analysis of longitudinal stream profiles of small watersheds. Office of Naval Research, Project NR-389-49, Tech. Rept. 18.

Burrows, R. L., W. W. Emmett, and B. Parks. 1981. Sediment transport in the Tanana River near Fairbanks, Alaska, 1977–79. U.S. Geol. Survey Water Resources Investigations 81-20. 56 pp.

Clarke, F. E. 1970. Tunisian flood a major disaster. U.S. Dept. Interior, press release dated Jan. 27, 1970.

Dietrich, W. E., and Thomas Dunne. 1993. The channel head. In *Channel network hydrology,* ed. K. Beven and M. J. Kirkby, pp. 175–219. New York: John Wiley & Sons.

Dozier, J. 1976. An examination of the variance minimization tendencies of a supraglacial stream. *Jour. Hydrol.* 31: 359–380.

Dunne, T. 1980. Formation and controls of channel networks. *Progr. in Phys. Geog.* 4: 211–239.

Dunne, T., and L. B. Leopold. 1978. *Water in environmental planning.* San Francisco: W. H. Freeman Co. 818 pp.

Emmett, W. W. 1975. The channels and waters of the upper Salmon River area, Idaho. U.S. Geol. Survey Prof. Paper 870A.

——— 1980. A field calibration of the sediment trapping characteristics of the Helley-Smith bedload sampler. U.S. Geol. Survey Prof. Paper 1139.

——— 1990. Little Granite Creek, Wyoming, analysis of April 1989. U.S. Geol. Survey memo, Denver. Mimeo.

Ferguson, R. J. 1973. Sinuosity of supraglacial streams. *Geol. Soc. Amer. Bull.* 84: 251–256.

Fields, F. L. 1975. Estimating streamflow characteristics for streams in Utah using selected channel geometry parameters. U.S. Geol. Survey Water Resources Investigations 34-74. 19 pp.

Fuglister, F. C. 1955. Alternative analysis of current surveys. *Deep-Sea Research* 2: 213–229.

Hack, J. T. 1957. Studies of longitudinal stream profiles in Virginia and Maryland. U.S. Geol. Survey Prof. Paper 294B.

Haible, W. W. 1980. Holocene profile changes along a California coastal stream. *Earth Surface Processes* 5: 249–264.

Harenberg, W. R. 1980. Using channel geometry to estimate flood flows at ungaged sites in Idaho. U.S. Geol. Survey Water Resources Investigations 80-32. 39 pp.

Hedman, E. R. 1970. Mean annual runoff as related to channel geometry of selected streams in California. U.S. Geol. Survey Water Supply Paper 1999E. 17 pp.

Hedman, E. R., and W. M. Kostner. 1972. Mean annual runoff as related to channel geometry of selected streams in Kansas. Kansas Water Resource Board Tech. Report 9. 25 pp.

——— 1977. Streamflow characteristics related to channel geometry of the Missouri River Basin. *U.S. Geol. Survey Research* 5 (3): 285–300.

Hedman, E. R., and W. R. Osterkamp. 1982. Streamflow characteristics related to channel geometry of streams in western United States. U.S. Geol. Survey Water Supply Paper 2193. 17 pp.

Hedman, E. R., W. M. Kostner, and H. R. Hejl. 1974. Selected streamflow characteristics as related to active channel geometry of streams in Kansas. Kansas Water Resource Board Tech. Report 10. 32 pp.

Hedman, E. R., D. O. Moore, and R. L. Livingston. 1972. Selected streamflow characteristics as related to channel geometry of perennial streams in Colorado. U.S. Geol. Survey Open File Report. 14 pp.

Henderson, F. N. 1978. *Open channel flow.* New York: Macmillan Co. 522 pp.

Hicks, D. M., and P. D. Mason. 1991. Roughness characteristics of New Zealand rivers. New Zealand Dept. Sci. Indust. Research, Marine and Freshwater, Natural Resources Survey, Wellington.

Horton, R. E. 1945. Erosional development of streams and their drainage basins: hydrophysical approach to quantitative morphology. *Geol. Soc. Amer. Bull.* 56: 275–370.

Hoyt, W. G., and W. B. Langbein. 1955. *Floods.* Princeton: Princeton University Press. 461 pp.

Hydraulics Research Station (Wallingford). 1977. *Resistance in gravel bed rivers, United Kingdom.* Annual Report of Director. 41 pp.

Kirchner, J. W. 1986. Strahler stream order and the structure of channel networks: a preliminary investigation. Dept. Geol., University of California, Berkeley. Mimeo.

——— 1993. Statistical inevitability of Horton's Laws and the apparent randomness of stream channel networks. *Geology* 21: 591–594.

Knighton, A. D. 1974. Variation in width-discharge relation and some implications for hydraulic geometry. *Geol. Soc. Amer. Bull.* 85: 1069–76.

——— 1982. Longitudinal changes in the size and shape of stream bed material: evidence of variable transport conditions. *Catena* 9 (1/2): 25–34.

Komar, P. D. 1987. Selective grain entrainment by a current from a bed of mixed sizes. *J. Sedim. Petrol.* 57: 203–211.

Lamb, Hubert. 1991. *Historic storms of the North Sea, British Isles and Northwest Europe.* New York: Cambridge University Press.

Langbein, W. B. 1940. Channel storage and unit hydrograph studies. *Amer. Geophys. Union Trans.* 21: 620–627.

——— 1949. Annual floods and the partial duration flood series. *Amer. Geophys. Union Trans.* 30: 879–881.

——— 1964. The geometry of river channels. Amer. Soc. Civil Engr. Hydr. Div. Paper 3846, pp. 297–313.

Langbein, W. B., and W. G. Hoyt. 1959. *Water facts and the nation's future.* New York: Ronald Press. 288 pp.

Langbein, W. B., and L. B. Leopold. 1964. Quasi-equilibrium states in channel morphology. *Amer. Jour. Sci.* 262: 782–794.

——— 1966. River meanders: theory of minimum variance. U.S. Geol. Survey. Prof. Paper 422H.

Leopold, L. B. 1942. Areal extent of intense rainfall, New Mexico and Arizona. *Amer. Geophys. Union Trans.*, pt. 2, 558–563.

——— 1951. Rainfall frequency: an aspect of climatic variation. *Amer. Geophys. Union Trans.* 32: 347–358.

——— 1959. Probability analysis applied to a water-supply problem. U.S. Geol. Survey Circular 410.

——— 1962. Rivers. *Amer. Sci.* 50: 511–537.

——— 1968. Hydrology for urban land planning. U.S. Geol. Survey Circular 554.

——— 1969. The rapids and the pools—Grand Canyon. U.S. Geol. Survey Prof. Paper 669, pp. 131–145.

——— 1971. Trees and streams: the efficiency of branching patterns. *Jour. Theoret. Biol.* 31, pt. 2, 335–354.

——— 1973. River channel change with time—an example. *Bull Geol. Soc. Amer.* 84: 1845–60.

——— 1991. Lag times for small drainage basins. *Catena* 18: 157–171.

——— 1992. Sediment size that determines channel morphology. In *Dynamics of gravel-bed rivers,* ed. P. Billi et al., pp. 287–311. New York: John Wiley & Sons.

Leopold, L. B., and W. W. Emmett. 1976. Bedload measurements, East Fork River, Wyoming. *Proc. Natl. Acad. Sci.* 73 (4): 1000–4.

Leopold, L. B., and W. B. Langbein. 1962. The concept of entropy in landscape evolution. U.S. Geol. Survey Prof. Paper 500A.

——— 1963. Association and indeterminacy in geomorphology. In *Fabric of geology,* ed. C. A. Albritton, pp. 184–192. San Francisco: W. H. Freeman Co.

Leopold, L. B., and T. Maddock. 1953. The hydraulic geometry of stream channels and some physiographic implications: U.S. Geol. Survey Prof. Paper 252.

——— 1954. *The flood control controversy.* New York: Ronald Press. 278 pp.

Leopold, L. B., and J. P. Miller. 1954. A postglacial chronology for some alluvial valleys in Wyoming. U.S. Geol. Survey Water Supply Paper 1261.

——— 1956. Ephemeral streams—hydraulic factors and their relation to the drainage net. U.S. Geol. Survey. Prof. Paper 282A.

Leopold, L. B., and D. L. Rosgen. 1991. Movement of bed material clasts in gravel streams. Paper presented at Federal Interagency Sedimentation Conference, Las Vegas.

Leopold, L. B., and M. G. Wolman. 1957. River channel patterns: braided, meandering, and straight. U.S. Geol. Survey Prof. Paper 282B.

——— 1960. River meanders. *Bull. Geol. Soc. Amer.* 71: 769–794.

Leopold, L. B., W. W. Emmett, and R. M. Myrick. 1966. Channel and hillslope processes in a semiarid area, New Mexico. U.S. Geol. Survey Prof. Paper 352G.

Leopold, L. B., M. G. Wolman, and J. P. Miller. 1964. *Fluvial processes in geomorphology.* San Francisco: W. H. Freeman Co. 511 pp.

Leopold, L. B., R. A. Bagnold, M. G. Wolman, and L. M. Brush. 1960. Flow resistance in sinuous or irregular channels. U.S. Geol. Survey Prof. Paper 282D.

Limerinos, J. T. 1970. Determination of the Manning coefficient from measured bed roughness in natural channels. U.S. Geol. Survey Water Supply Paper 1898B.

Lisle, T. E., and M. A. Madej. 1992. Spatial variation in armoring in a channel with high sediment supply. In *Dynamics of gravel-bed rivers,* ed. P. Billi et al., pp. 277–293. New York: John Wiley & Sons.

Lowham, H. W. 1976. Techniques for estimating flow characteristics of Wyoming streams: U.S. Geol. Survey Water Resources Investigations, 76-112. 83 pp.

Maybeck, M. 1976. Total mineral dissolved transport by world major rivers. *Hydrol. Sci. Bull.* 21: 265–284.

Miller, J. P. 1958. High mountain streams: effects of geology on channel characteristics and bed material. New Mexico Bur. Mines and Mineral Resources Memoir 4. 53 pp.

Miller, T. K., and L. J. Onesti. 1977. Multivariate empirical test of the Leopold and Miller stream order—hydraulic geometry hypothesis. *Geol. Soc. Amer. Bull.* 88: 85–88.

Moore, D. O. 1974. Estimating flood discharges in Nevada using channel geometry measurements. Nevada Highway Dept. Hydrol., Report No. 1. 43 pp.

Nace, R. L. 1970. World hydrology: status and prospects. Internat. Assoc. Sci. Hydrol. Publ. 92, Symposium of Reading (England), pp. 1–10.

Nixon, M. 1959. A study of the bankfull discharges of rivers in England and Wales. Inst. Civil Engrs. Proc. Paper No. 6322, pp. 157–174.
Park, C. C. 1977. World-wide variations in hydraulic geometry exponents of stream channels: an analysis and some observations. *Jour. Hydrol.* 33: 133–146.
Prigogine, I. 1955. *Introduction to the thermodynamics of irreversible processes.* Springfield, Illinois: C C Thomas. 115 pp.
Robinson, A. R., S. M. Glenn, et al. 1989. Forecasting Gulf Stream meanders and rings. *EOS* Nov. 7: 1464–65.
Rosgen, D. L. In press. A classification of natural rivers. *Catena.*
Schick, A. P. 1970. Desert floods. Internat. Assoc. Sci. Hydrol. UNESCO Symposium of Wellington, N.Z. Publ. No. 96, pp. 478–493.
Scott, A. G., and J. L. Kunkler. 1976. Flood discharge of streams in New Mexico as related to channel geometry. U.S. Geol. Survey Open File Report 76-414. 29 pp.
Searcy, J. K. 1959. Flow duration curves. U.S. Geol. Survey Water Supply Paper 1542A. 33 pp.
Shreve, R. L. 1967. Infinite topologically random channel networks. *Jour. Geol.* 75: 176–186.
Smith, T. R. 1974. Derivation of the hydraulic geometry of steady state channels from conservation principles and sediment transport laws. *Jour. Geol.* 82: 98–104.
Stommel, H. 1954. Circulation in the North Atlantic Ocean. *Nature* 173: 886–893.
Strahler, A. N. 1957. Quantitative analysis of watershed geomorphology. *Amer. Geophys. Union Trans.* 38: 913–920.
Stuart, T. A. 1962. The leaping behavior of salmon and trout at falls and obstructions. Freshwater and Salmon Fisheries Research, Dept. Agric. and Fisheries, Scotland, Publ. No. 28.
U.S. Geological Survey. 1970. *National Atlas of the United States.* Washington, D.C.: Government Printing Office.
Von Arx, W. S. 1952. Notes on the surface velocity profile and horizontal shear across the width of the Gulf Stream. *Tellus* 4: 211–214.
Webb, B. W., and D. E. Walling. 1982. The magnitude and frequency characteristics of fluvial transport in a Devon drainage basin and some geomorphological implications. *Catena* 9 (1/2): 9–24.
White, W. R., H. Miller, and A. D. Crabbe. 1975. Sediment transport theories: a review. *Proc. Instn. Civil Engrs.* 59, pt. 2: 265–292.
Woldenberg, M. J. 1966. Horton's laws justified in terms of allometric growth and steady state in open systems. *Geol. Soc. Amer. Bull.* 77: 431–434.
——— 1969. Spatial order in fluvial systems: Horton's laws derived from mixed hexagonal hierarchies of drainage basin areas. *Geol. Soc. Amer. Bull.* 80: 92–112.
Wolman, M. G., and J. P. Miller. 1960. Magnitude and frequency of forces in geomorphic processes. *Jour. Geol.* 68: 54–74.

# Symbols

$a$ a coefficient
$b$ exponent of width
$Q_{bkf}$ bankfull discharge
$d$ depth
$f$ exponent of depth
$g$ gravity
$i$ sediment transport rate per unit of width
$j$ exponent of transport rate
$k$ a coefficient
$m$ rank order; exponent of velocity
$n$ number of cases; Manning roughness coefficient
$p$ probability; a coefficient
$r$ radius; a coefficient
$s$ slope
$t$ time
$u$ velocity
$u_*$ shear velocity
$w$ width
$x$ deviation from mean
$y$ exponent of roughness
$z$ exponent of slope
$BP$ years before present
$C$ concentration of sediment
$D$ diameter of grain
$D_A$ drainage area
$I$ inflow
$K$ roughness height
$L$ suspended sediment transport rate; length of channel

$M$ total path length
$N$ normal component of force
$O$ outflow; stream order
$Q$ discharge
$R$ hydraulic radius
$RI$ recurrence interval
$R_g$ grain Reynolds number
$S$ storage volume; distance along path
$T$ tangential component of force
$W$ weight
$\mathbf{d}$ pipe diameter
$\mathbf{F}$ Froude number
$f$ Darcy Weisbach friction factor
$\alpha$ angle of friction
$\propto$ proportional to
$\gamma$ specific weight
$\rho$ density
$\sigma$ grain density; standard deviation
$\tau$ shear stress
$\tau_0$ shear stress at initial motion
$\upsilon$ kinematic viscosity
$\omega$ power per unit width; maximum deviation angle
$\omega_0$ power per unit width needed for initial motion
$\theta$ dimensionless shear stress; angle of deviation
$\Sigma$ summation of
$\Omega$ total stream power

# Author Index

# General Index